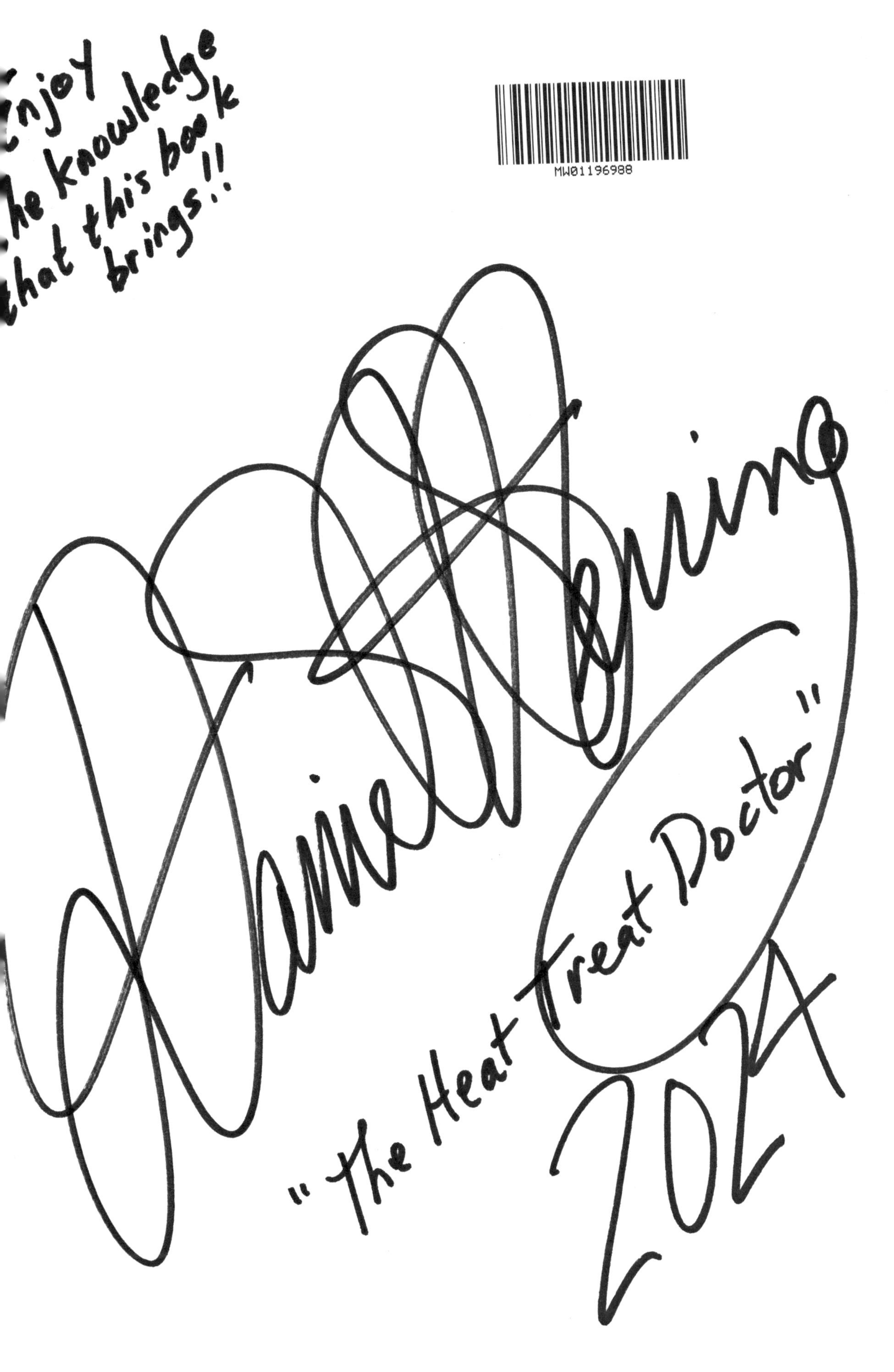
Enjoy
he knowledge
hat this book
brings!!
MW01196988
" The Heat Treat Doctor "
2024

VACUUM HEAT TREATMENT

PRINCIPLES | PRACTICES | APPLICATIONS

VACUUM HEAT TREATMENT

PRINCIPLES | PRACTICES | APPLICATIONS

DANIEL H. HERRING

BNP Media II, LLC | CUSTOM MEDIA GROUP
Troy, Michigan

Printed in the United States of America

Published in 2012 by BNP Media II, LLC, Troy, MI 48084

Vacuum Heat Treatment: Principles | Practices | Applications
Daniel H. Herring
Includes indexes
ISBN 978-0-9767565-0-7

Project management by Melanie Kuchma
Art direction and book jacket design by Breanna Fong
Book designed by Clare Johnson and Chris Pirrone
Cover draft by Greg Gonzalez, Mainly Sunny Marketing Communications

Photos reprinted with permission

DEDICATION

To the past…

- Mom (Margaret) and dad (Daniel),
  who inspire me to write and invent

To the present…

- Daughters (Jennifer and Beth) and sons (Daniel, Timothy and Howard),
  who inspire me to dream
- Wife (Jeanne),
  who inspires me to achieve my dreams

To the future…

- Grandchildren (Michael, Katrina, Samuel, Rose Anne, Daniel, Tobias, Nikolai, Sebastian and Margaret),
  who inspire me to hope

To the reader…

Sage advice from days of yore:

"A little learning is a dangerous thing
Drink deep, or taste not the Pierian spring:
There shallow draughts intoxicate the brain,
And drinking largely sobers us again.
Fired at first sight with what the muse imparts,
In fearless youth we tempt the heights of arts
While from the bounded level of our mind
Short views we take nor see the lengths behind
But more advanced behold with strange surprise,
New distant scenes of endless science rise!"

– Alexander Pope, 1711

CONTENTS

FOREWORD

Having read Dan's book and having offered a number of technical suggestions, I highly recommend *Vacuum Heat Treatment* to industry readers as an excellent, authoritative reference tool and an absolute must-read for anyone involved in the heat treatment of metals and other materials. Prior works have been disjointed, involving multiple authors and failing to address the total subject. This book explores a broad range of topics, from furnace particulars to specific process issues, from the perspective of an acknowledged expert.

Dan has worked in this field for 40 years and has spent the last five years gathering data and writing this book. The subject matter is well-researched, well-documented and logically presented. His strongest chapters are process-related, like vacuum brazing and vacuum carburizing and general heat treating of materials. They are addressed from his metallurgical training and experience, particularly from his role as "The Heat Treat Doctor,"® wherein he has consulted around the globe for numerous companies and in many industries on a wide variety of metallurgical and furnace-related problems.

Vacuum-furnace technology encompasses many fields of discipline, including metallurgy, vacuum technology, mechanical engineering, electrical engineering, electronics, physics, chemistry, hydraulics and more. Those who design and build vacuum furnaces will encounter all of these fields. Users must also gain a grasp of these issues in order to apply the vacuum furnace to their processes successfully and avoid possible catastrophic maintenance problems. Dan's book covers these points well and in depth.

Because *Vacuum Heat Treatment* is so broad, up-to-date and forward-looking, I expect it to be on company and personal bookshelves for the next 50 years and referenced around the world.

William R. Jones, FASM, CEO
Solar Atmospheres Inc.
Souderton, Pa.

PUBLISHER'S PREFACE

Of all the technical books dealing with vacuum technology, *Vacuum Heat Treatment,* authored by Dan Herring and published by *Industrial Heating,* will be the seminal work for decades to come. Here's why.

1. This book is practical. It is not too technical, nor is it devoid of pertinent technical content. It is intended to help those who live day in and day out with vacuum-furnaces operate them more effectively and profitably. The goal is not just to fill minds with theoretical information, but to provide all the content necessary to turn users of vacuum-furnace equipment with moderate knowledge into proficient and profitable users of the equipment. Some textbooks of this type are so heavy on technical theory that they miss the mark of helping the reader improve day-to-day business. This work by Dan Herring is not an average textbook. It is more than theory; it is theory put to work.
2. This book is authored by the leading expert in the thermal-processing market. Dan Herring is an engineer. He's worked on both the equipment-manufacturing side and the equipment-using side of the desk. He's also been an invaluable resource for numerous vacuum-furnace builders and users for over a decade as an independent consultant, "The Heat Treat Doctor."® As the publisher of the leading magazine and website in the industry, I can personally attest that Dan has helped all the major vacuum-furnace manufacturers in some fashion. All of them trust Dan and rely on his expertise. Dan is also a talented teacher. Some brilliant people can't teach. This is not true in Dan's case. He knows his material, and he is able to teach it in a way that is clear and interesting.
3. This book is perfectly timed. The publishing of this book comes at a "sweet spot" in the history of vacuum-furnace technology. The convergence of new sensors and controls, together with a push for more environmentally friendly thermal processes and the development of new materials needing to be thermally processed in vacuum, make this a moment of critical importance to vacuum technology. Dan's book will push the understanding and acceptability of vacuum thermal processing rapidly ahead as newer and newer processes are found that require vacuum treatments.

This book also differs from other technical works in that it is a commercial venture. As a for-profit business, *Industrial Heating* has never published a book before. This is our first. Since we believe in minimizing risk, we asked many of the companies that have benefited from Dan's expertise to help us underwrite a portion of the publishing costs. Throughout this book, you'll see advertisements from these sponsors. Please do what you can to support these companies – they are leaders in the vacuum thermal-processing industry, and they understand that a well-informed industry makes better purchasing decisions.

Finally, I'd like to extend my sincere appreciation to the following individuals who contributed significantly to this book.

- Reed Miller, editor and associate publisher of *Industrial Heating* and the technical brain in the organization. Reed spent many hours working with this manuscript. His positive attitude and perseverance are much appreciated.
- Bill Mayer, *Industrial Heating*'s associate editor. Bill's unflagging dedication and oversight of the entire proofreading process was challenging at times, but he stuck with it.
- Melanie Kuchma and Chris Wilson from BNP Media's Custom Media Division for their excellent and professional management of this project. My thanks also goes to the art directors and others contracted by Melanie and Chris for their work on this book.
- Kathy Pisano, Larry Pullman, Steve Roth and Patrick Connolly – outstanding sales people for *Industrial Heating* – for maintaining such good relationships with leading companies in this industry and for arranging many of the sponsorships in this book.
- Bill Jones, founder and CEO of Solar Atmospheres and Solar Manufacturing, for reviewing this work and offering seasoned counsel and technical insights.
- The many sponsors of this work (see a complete list of sponsors on page xi). Without your faith in both Dan Herring and *Industrial Heating*, this work would have remained unpublished.
- Dan Herring, "The Heat Treat Doctor,"® for the idea and the execution of this book. Dan has been an invaluable resource and contributor to *Industrial Heating* for well over a decade and a genuinely good man. My thanks also goes to Jeanne Herring, Dan's better half, for her unflagging support of this project and her sometimes-absent husband.
- And last, but certainly not least, God, from Whom all blessings flow, including technical knowledge like that found in this book.

Enjoy, and benefit from this work.

Doug Glenn
Publisher, *Industrial Heating*

SPONSOR ACKNOWLEDGEMENTS

SPONSOR	BAR	PAGE
ALD-Holcroft	20-Bar	11, 403, 450
Solar Manufacturing	20-Bar	6, 76, 421
G-M Enterprises	15-Bar	64, 362
SECO/WARWICK	15-Bar	243, 327
ECM USA, Inc.	15-Bar	101, 404
Surface Combustion	10-Bar	155
Agilent Technologies	10-Bar	33
Yokogawa Corp. of America	10-Bar	216
Dry Coolers	10-Bar	178
Wirco	10-Bar	287
North American Cronite	10-Bar	286
Praxair	10-Bar	304
Schunk Graphite Technology	2-Bar	288
AmeriKen Die Supply Inc.	2-Bar	263
VAC AERO International	2-Bar	228
Oerlikon Leybold Vacuum Products	2-Bar	21
Gasbarre Products Inc.	2-Bar	263
Met-Tek Inc.	2-Bar	131
AVS	2-Bar	422
National Element	2-Bar	83
Televac, a Fredericks Company	2-Bar	56

PREFACE

What began as a simple attempt to answer the question "What is Vacuum?" has evolved into a full-fledged book on the subject of vacuum heat treatment. The aim of this book is to present need-to-know information on vacuum methods, processes and equipment with an emphasis on understanding how vacuum technology is being employed throughout the thermal-processing industry.

This book is written by a heat treater who also happens to be a metallurgist. The author is someone who has operated, damaged, repaired, designed, built and maintained vacuum furnaces – on all three shifts. As such, this book addresses the subject from both a scientific/engineering standpoint as well as from a practical hands-on approach. The engineer in all of us wants to know how a particular vacuum furnace does or should work, but the heat treater must also know how to best run their equipment and establish their recipes so as to have reasonable certainty of the outcome. It is this latter need that differentiates this book from many others that have come before it.

Everyone involved in manufacturing hopes that their efforts result in useful engineering products. As such, we need to strive for absolute control of the heat-treating process so that we can predict with a high level of confidence the result of this vital manufacturing step and how it will impact post-heat-treatment operations. Only vacuum technology assures us that we can accomplish these goals in a controlled, predictable and repeatable manner.

While each chapter is self-contained and presented in a logical order, they are also arranged by common theme. We begin by considering the fundamentals of vacuum followed closely by a look at the basic components of a vacuum system. Next, we discuss equipment features and maintenance, and then we consider the unique characteristics of the various heat-treatment processes run in vacuum. Finally, we look ahead at the exciting future the technology offers and its unique place in manufacturing.

Vacuum Heat Treatment is all about giving the reader a resource that will endure the test of time and provide valuable insights into, for example, the types of systems in use and their features along with the process application knowledge needed to do their job – whether he or she is the CEO, owner or president; a manufacturing, process or quality engineer; a department manager; a heat-treat supervisor; or a furnace operator curious to learn more about the equipment they operate and the processes they run. The reader will gain a great deal of knowledge about the process and equipment variables that must be controlled, monitored and recorded.

Another unique aspect of this book is the many pictures and illustrations contained herein. The author is incredibly grateful to all of the many companies that provided them. There were literally thousands of images from dozens and dozens of companies to select from, and the choices are my way of providing additional insight to the reader. Study them carefully, and ask yourself questions about what you see. In many cases, these images are as important as the written words.

You will find units of measure expressed in both the metric and English system throughout the book. It should be noted that the temperature conversions are approximate, by design, as the heat-treat industry does not often use the same temperature-conversion precision as the scientific community. If exact conversions are necessary, the English units should be the starting point.

Any errors or omissions in the book are those of the author alone. Your feedback and comments are highly encouraged, so please do not hesitate to offer opinions, comments or criticisms. Send them directly to the author's principle e-mail address: dherring@heat-treat-doctor.com.

You may be asking yourself how valuable this technical reference can be given the fact we have sponsors and promotional pieces strategically placed within the book. This is a unique approach – one the publisher and I feel provides the reader with yet more resources, which is why we have chosen this path. Original equipment manufacturers and suppliers of components and ancillary products for the heat-treat industry are valuable sources for information and help. Rely on them to supplement what is presented in these pages.

Like any labor of love, this work has taken on a life of its own, and no work is the effort of just one person alone. Literally hundreds of people have contributed in large and small ways, and it is impossible in this short space to list them all. You know who you are, and know that you have my eternal gratitude. There are a few people, however, who must be mentioned: Doug Glenn, Bill Jones and Alan Charky stand out for their respective vision, technical brilliance and inspiration.

To my son Tim, a special "thank you" for helping in the book's preparation, and to my lovely wife and partner, Jeanne, for her support and keen mind. She has been an invaluable asset, asking intriguing questions, offering excellent suggestions and making improvements on my poor attempt at expressing the English language in clear and concise terms. Love you!

Daniel H. Herring
"The Heat Treat Doctor"®
Chicago, Ill.
2012

VACUUM HEAT TREATMENT

PRINCIPLES | PRACTICES | APPLICATIONS

CHAPTER 1

ALL ABOUT VACUUM

A vacuum system (Fig. 1.1) provides a space in which the pressure can be maintained below atmospheric pressure for a portion of or for the entire time. The primary advantage of vacuum heat treatment is its versatility. In almost all cases it provides a "safe" environment with respect to the surface of the components being treated, is self-contained and uses cycles/recipes that can be reproduced consistently. When not in use, like an electric light, it is simply turned off. When turned back on, minimal conditioning time is required.

A principal difference between vacuum heat treating and all other forms of heat treatment is the absence of, or the precise

FIGURE 1.1 | Typical vacuum-furnace system

control of, surface reactions. In addition, vacuum processing can remove contaminants and, under certain circumstances, degas or convert oxides found on the surface of a material.

The word vacuum comes from the Latin "vacuus" meaning empty or "vacare" meaning to be empty. When we think of an empty space, what comes to mind is something entirely devoid of matter. Such a space does not exist, nor can it be produced. In practical terms, a vacuum must be considered a space with a highly reduced gas density. In heat treating, gas molecules and contaminants are removed from a vacuum vessel using a pump. Air (Table 1.1) is the most important of all gases to be eliminated since it is present in every system.

TABLE 1.1 [1] | Composition of air [a]

Chemical constituent	Volume (%)	Volume (ppm)	Pressure (Pa)
Nitrogen (N_2)	78.084 ± 0.004		79,117
Oxygen (O_2)	20.946 ± 0.002		21,223
Argon (Ar)	0.934 ± 0.001		946.357
Carbon Dioxide (CO_2)	0.033 ± 0.001		33.437
Neon (Ne)		18.18 ± 0.04	1.842
Helium (He)		5.24 ± 0.004	0.510
Methane (CH_4)		2.0	0.203
Krypton (Kr)		1.14 ± 0.01	0.116
Hydrogen (H_2)		0.5	0.051
Nitrogen Dioxide (N_2O)		0.5 ± 0.1	0.051
Xenon (Xe)		0.087 ± 0.001	0.009

Note:

[a] The chemical composition of air includes water vapor, the percentage of which is dependent on the relative humidity.

Common Vacuum Units

In heat-treat applications, pressure is commonly measured in torr (U.S.) or millibar (Europe and Asia). A torr is 1/760th of atmospheric pressure. In other words, atmospheric pressure – the pressure all around us – at standard temperature (0°C) and pressure (sea level) is 760 torr (1013 mbar).

One of the mystifying things about vacuum and vacuum furnaces, especially in the U.S. heat-treating industry, is the confusing way in which vacuum units are used. Devices installed on furnaces often measure in different units, which force us to speak in terms of microns, torr, millitorr, millimeters of mercury, millibar,

bar, Pascal, inches of water column and inches of mercury! This is extremely confusing, especially to those who are not familiar with vacuum terminology. If at all possible, try to stay with one unit of measure, converting everything to that common base. Conversions between common vacuum units (Table 1.2) are available from a number of sources.

TABLE 1.2 [1] | Conversions between common vacuum units

Millibar (mbar)	Torr (mm Hg)	Microns	Hg Absolute	Atmo-spheres	psia	Millibar (mbar)	Torr (mm Hg)	Microns	Pascal	Microbars
1013.25	760.00	760,000	29.9	1	14.7	0.27	0.20	200	26.66	266.65
982.05	736.60	736,600	29	0.97	14.2	0.13	0.10	100	13.33	133.32
948.19	711.20	711,200	28	0.94	13.8	0.12	0.09	90	12.00	119.91
914.32	685.80	685,800	27	0.90	13.3	0.11	0.08	80	10.67	106.66
880.46	660.40	660,400	26	0.87	12.8	0.09	0.07	70	9.33	93.32
846.60	635.00	635,000	25	0.84	12.3	0.08	0.06	60	8.00	79.99
812.73	609.60	609,600	24	0.80	11.8	0.07	0.05	50	6.67	66.66
778.87	584.20	584,200	23	0.77	11.3	0.05	0.04	40	5.33	53.33
745.00	558.80	558,800	22	0.74	10.8	0.04	0.03	30	4.00	40.00
711.14	533.40	533,400	21	0.70	10.3	0.03	0.02	20	2.67	26.66
677.28	508.00	508,000	20	0.67	9.8	0.013	0.01	10	1.33	13.33
643.41	482.60	482,600	19	0.64	9.3	0.012	0.009	9	1.20.	12.0
609.55	457.20	457,200	18	0.60	8.8	0.011	0.008	8	1.07	10.67
575.69	431.80	431,800	17	0.57	8.3	0.009	0.007	7	0.93	9.33
541.82	406.40	406,400	16	0.53	7.8	0.008	0.006	6	0.80	7.99
507.96	381.00	381,000	15	0.50	7.4	0.007	0.005	5	0.67	6.66
474.09	355.60	355,600	14	0.47	6.9	0.005	0.004	4	0.533	5.33
440.23	330.20	330,200	13	0.43	6.4	0.004	0.003	3	0.400	4.00
406.37	304.80	304,800	12	0.40	5.9	0.003	0.0020	2	0.267	2.66
372.50	279.40	279,400	11	0.37	5.4	0.0013	0.0010	1	0.133	1.33
338.64	254.00	254,000	10	0.33	4.9	0.0012	0.0009	0.9	0.012	1.20
304.78	228.60	228,600	9	0.30	4.4	0.0011	0.0008	0.8	0.010	1.06
270.91	203.20	203,200	8	0.27	3.9	0.0009	0.0007	0.7	0.093	0.93
237.04	177.80	177,800	7	0.23	3.4	0.0008	0.0006	0.6	0.080	0.80
203.18	152.40	152,400	6	0.20	2.9	0.0007	0.0005	0.5	0.067	0.67
169.32	127.00	127,000	5	0.17	2.4	0.0005	0.0004	0.4	0.053	0.53
135.46	101.60	101,600	4	0.13	2.0	0.0004	0.0003	0.3	0.040	0.40
101.59	76.20	76,200	3	0.10	1.5	0.0003	0.0002	0.2		
67.73	50.80	50,800	2	0.066	0.98	0.00013	0.00010	0.1		
33.87	25.40	25,400	1	0.033	0.49	0.00012	0.00009	0.09		
1.33	1.00	1,000	0.039	0.0013	0.019	0.00011	0.00008	0.08		
1.20	0.90	900	0.035	0.0011	0.017	0.00009	0.00007	0.07		
1.07	0.80	800	0.031	0.0010	0.015	0.0000800	0.00006	0.06		
0.93	0.70	700	0.028	0.0009	0.013	0.0000667	0.00005	0.05		
0.80	0.60	600	0.024	0.0008	0.012	0.0000533	0.00004	0.04		
0.67	0.50	500	0.020	0.0006	0.010	0.0000400	0.00003	0.03		
0.53	0.40	400	0.016	0.0005	0.008	0.0000267	0.00002	0.02		
0.40	0.30	300	0.012	0.0004	0.006	0.0000133	0.00001	0.01		

A Little Vacuum Theory

A gas is a collection of molecules in constant motion. The higher the temperature, the faster these molecules move. As one might expect, the motion of gas molecules dramatically slows down or stops near or at absolute zero. As molecules speed up with an increase in temperature, there is an increase in their kinetic energy (or energy of motion). Molecular collisions occur between molecules. If contained, these molecular collisions against the walls of their container result in a pressure rise (which always occurs in a closed container when a gas is heated). In other words, pressure is simply the force per unit area that a gas exerts on the walls of its container.

In 1811, Avogadro determined that a mole of *any* gas occupies a volume of 22.4 liters and contains 6.02 x 10^{23} molecules at standard temperature and pressure. To create a vacuum in any closed vessel, therefore, some of these gas molecules must be removed.

At atmospheric pressure, 1 cubic centimeter (1 cc) of air contains 2.69 x 10^{19} molecules all moving around in a random motion. This results in a fantastic number of collisions. The mean free path between molecules (or the average distance a molecule can travel before colliding with another molecule) is 6.6 x 10^{-5} mm (2.5 x 10^{-6} inches). So, if we pump a 1 cubic centimeter volume down to 1 x 10^{-3} torr (1 micron), which is a vacuum level commonly used in heat treating, we still have a whopping 4 x 10^{13} molecules remaining!

How then can this be an acceptable condition for heat treating, especially when about 20% of the remaining molecules are oxygen? The answer is that the mean free path of the molecules increases tremendously, thus reducing the probability of molecular collision with each other, and since the number of molecules (density) has been reduced, fewer collisions occur with the surface of the workpiece. This results in even fewer collisions of oxygen molecules with the work surface and no visible oxidation forms.

For example, if we pumped down to 1 x 10^{-9} torr, or one millionth of a micron, the number of molecules per cubic centimeter is decreased to 4 x 10^{7} and the mean free path is 5,000,000 cm (over 30 miles!). The path of the molecules is now limited by the walls of the vessel and not by the collisions between molecules. Flow, as we know it, doesn't exist. This is the reason for the relatively large opening and piping used for diffusion pumping systems. If the opening to the diffusion pump weren't that large, any molecules that migrated into the pump suction stream would rebound and move to other parts of the chamber.

Vacuum Level

The more molecules that are removed from a vessel, the better the vacuum. The quality of a vacuum is described by the degree of reduction in gas density (i.e. gas pressure). In the vacuum field, we distinguish six different vacuum levels or quality of vacuum (Table 1.3). The heat treatment of steel is carried out in three of these: rough, soft and high. The majority of applications are processed in the soft vacuum range. By contrast, at 200 miles above Earth, the vacuum in space is 10^{-8} torr. At 400 miles it is 10^{-10} torr, and in deep (or outer) space it is 10^{-17} torr.

TABLE 1.3 [7] | Classification of vacuum

Classification	Vacuum level [a, b, c, d]	
	Pa	Torr
Low (rough) vacuum	133.3 to 1.33 x 10^{-1}	1 to 1 x 10^{-3}
Medium (soft) vacuum	<1.33 x 10^{-1} to 1.33 x 10^{-3}	< 1 x 10^{-3} to 10^{-5}
High (HV) vacuum	<1.33 x 10^{-3} to 1.33 x 10^{-6}	< 1 x 10^{-5} to 10^{-8}
Ultrahigh (UHV) vacuum	<1 x 10^{-7} to 1 x 10^{-8}	7.5 x 10^{-10} to 7.5 x 10^{-11}
Extreme ultrahigh vacuum	< 1 x 10^{-10}	< 7.5 x 10^{-13}
Interstellar space	10^{-17}	7.5 x 10^{-20}

Notes:

[a] The SI unit of pressure is the Pascal (1 Pa = 1 N m^{-2}).

[b] Normal atmospheric pressure of 1 atmosphere is 101,325 Pa or 1013 mbar (1 bar = 10^5 Pa).

[c] Normal atmospheric pressure of 1 atmosphere is 760 Torr (1 Torr = 133.3 Pa).

[d] Ultrahigh vacuum is defined as the pressure range of 10^{-6} Pa (Europe) and/or 10^{-7} Pa (U.S.) to 10^{-10} Pa.

REFERENCES

1. Herring, D.H., "What is Vacuum?" *Industrial Heating*, November 2006
2. *Webster's Ninth New Collegiate Dictionary*, Merriam-Webster Inc., Springfield, Mass., 1987
3. *CRC Handbook of Chemistry and Physics*, 90th Edition, David R. Lide (Ed.), 2009
4. Kimball, William H., *Vacuum...is it really nothing?*, C.I. Hayes Inc., 1977
5. Brunner Jr., William F. and Thomas H. Batzer, *Practical Vacuum Techniques*, Robert E. Krieger Publishing Company, 1974
6. *Steel Heat Treatment Handbook*, edited by Totten, George E. and Maurice A. Howes, "Chapter 7: Vacuum Heat Treating," Marcel-Dekker, 1997
7. Herring, D.H., "High and Ultra-High Pressure," white paper, 2012
8. Dushman, Saul, *Scientific Foundations of Vacuum Technique*, John Wiley & Sons Inc., 1949
9. *"Vacuum Heat Treating Processes,"* ASM International, Practical Heat Treating Course Nos. 27361C and 273642812

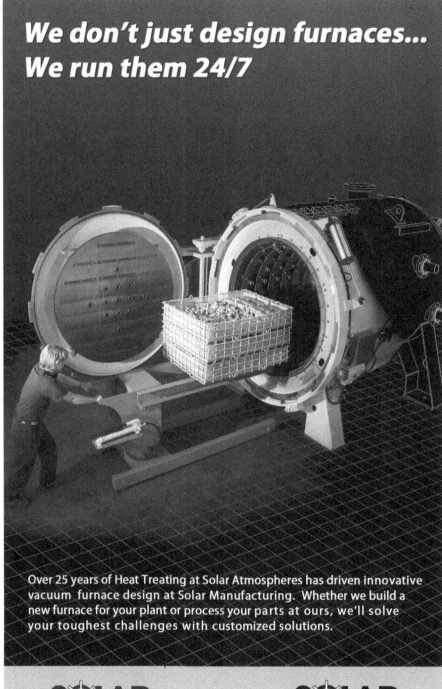

CHAPTER 2

THE THEORY OF GASES

In vacuum heat treating, we are always dealing with the movement of gases. So, everyone needs to understand something about the nature (theory) of gases and how they behave, especially in vacuum. The main difficulty, however, is that too much theory tends to become a distraction. Our mission is to learn how these principles help us understand what goes on inside a vacuum vessel.

FIGURE 2.1 | Molecules on the move [1]

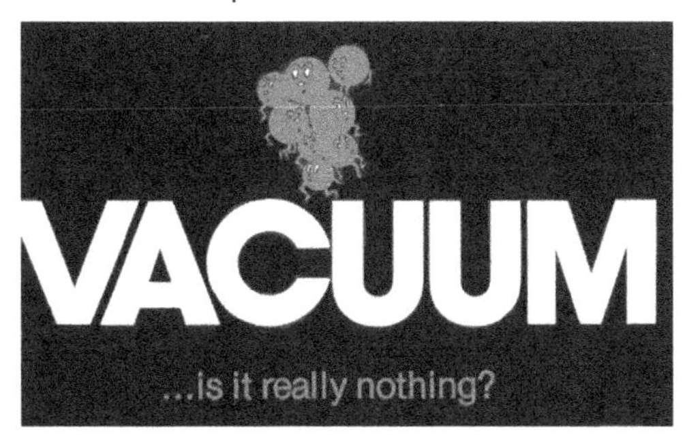

We begin with the realization that any gas can be completely described in terms of the following quantities:

- Pressure (P)
- Volume (V)
- Temperature (T)
- Number of molecules (n)

The Gas Laws, therefore, are formulas that simply allow one to find the value of one of these quantities if you know the others. Let's review them.

Boyle's Law: $P_1V_1 = P_2V_2$

This formula tells us that if the temperature is held constant, the increase in pressure is exactly proportional to the decrease in volume.

Charles' Law: $P_1/T_1 = P_2/T_2$

This formula tells us that an increase in pressure is directly proportional to the increase in the absolute temperature if the volume is held constant.

Avogadro's Law: $P_1/n_1 = P_2/n_2$

This formula tells us that equal volumes of any gas at the same temperature and pressure contain the same number of molecules. A formula that relates all four of the quantities needed to completely describe the state of a gas is:

Ideal Gas Law: $PV = nRT$

Here R is a constant known as the universal gas constant and has a value of 62.4 torr-liter/mole-°K. Another useful gas law is the law of partial pressure, which is known as:

Dalton's Law: $P_{TOTAL} = P_1 + P_2 + \ldots + P_n$

This formula tells us that in a mixture of gases where the gases do not react chemically, each gas exerts its own pressure independently as if no other gases were present.

Other characteristics that help describe the behavior of gases in a vacuum environment include:

- The Kinetic Theory of Gases
- Mean free path of molecules
- Phase change
- Evaporation
- Condensation
- (Dynamic) equilibrium
- Vapor (partial) pressure
- Gas flow
- Resistance to gas flow
- Gas conductance

Let's briefly consider several of these characteristics.

Characteristics

THE KINETIC THEORY OF GASES

This theory is used to explain the behavior of gases in terms of the behavior of the individual gas molecules. Gases are in constant motion. Hence, it makes sense to discuss their kinetic nature. Remember also that gases are free to wander throughout the space available to them. So, the temperature of a gas is simply a measure of this kinetic nature (i.e. kinetic energy) of the particles. The higher the temperature, the faster and faster the molecules move and the greater their kinetic energy.

By comparison, pressure is related to the number of collisions of molecules against the walls of their container. An increase in temperature causes molecules to hit harder and more often, creating even higher pressure (in accordance with Charles' Law). Similarly, a decrease in temperature causes molecules to hit less hard and less often, resulting in lower pressure.

If you were to remove some of the gas from the vacuum vessel, fewer molecules are left to make contact with the walls and the pressure is lower (in accordance with Avogadro's Law and the Ideal Gas Law). Decreasing the volume of gas (at constant temperature) results in a reduced area where the original number of molecules strikes and causes the pressure to increase (in accordance with Boyle's Law).

MEAN FREE PATH

The mean free path, or distance between molecules, can be calculated from the kinetic theory in Equation 2.1.

$$2.1)\quad \lambda = \frac{1}{\sqrt{2}\pi D^2 n}$$

where λ is the mean free path in centimeters, D is the diameter of the gas molecule in centimeters and n is the number of gas molecules per cubic centimeter.

What is important is that this formula tells us that pressure, which is proportional to the number of molecules per unit volume (in accordance with the Ideal Gas Law), is inversely proportional to the mean free path. As the pressure decreases, the mean free path between molecules increases.

PHASE CHANGE

If we change the state of matter by changing the temperature or pressure (or both), we change the phase in which it exists. In vacuum, as we pump down, we reduce the pressure and temperature and convert some of the moisture in the air to ice crystals.

DYNAMIC EQUILIBRIUM AND VAPOR PRESSURE

When the number of molecules leaving a part surface is equal to the number of molecules returning to it (when the rate of evaporation equals the rate of condensation), the system is said to be in dynamic equilibrium. The partial pressure of the vapor at which it occurs is the vapor pressure of the material.

GAS FLOW

The rate at which gas flows through the vacuum vessel into the pump is important in vacuum systems. This determines the time required to reach operating pressure and may ultimately determine the system's tolerance to leaks and outgassing.

What happens when we pump down a vessel?

IN THE RANGE OF 760 TO 1 TORR

This is the initial pump-down stage from atmospheric pressure. The air in the vessel, with its associated high relative humidity, begins to be removed. As the pressure decreases, water vapor condenses due to the cooling effect of the sudden drop in pressure. A "fog" develops – that is, a cloud swirls around the interior with a turbulence that is characteristic of gas flow at high pressure and high flow rate. The net result is a loss of pumping efficiency. This is why we should not keep a vacuum furnace open longer than absolutely necessary and that the size of the mechanical pump and blower is important in mitigating this phenomenon.

Over time, a (slow) change in the composition of the gas remaining in the vessel takes place. Initially, air is the major component of the gases. Certain oils, grease and water exist, usually on the sides of the vessel. Eventually, almost all the air is pumped out – the grease and water will continue to evaporate and their partial pressures will constitute a much larger proportion of the total pressure. The main concern during this phase is workload contamination due to impurities.

IN THE RANGE OF 1 TO 1 x 10^{-4} TORR

The ability of the gases remaining in the vessel to conduct heat begins to decrease rapidly. There is also a change in the electrical characteristics of the gas, and the voltage necessary to start a discharge decreases.

IN THE RANGE OF 1 x 10^{-4} TO 1 x 10^{-6} TORR

As we lower the vacuum level, we see a decrease in molecular density and an increase in sliding friction. In this vacuum range, molecules collide with the sides of the vessel as often as with each other.

REFERENCES

1. Kimball, William H., *Vacuum...is it really nothing?*, C.I. Hayes Inc., 1977
2. Brunner Jr., William F. and Thomas H. Batzer, *Practical Vacuum Techniques*, Robert E. Krieger Publishing Company, 1974

CHAPTER 3

VACUUM PUMPING SYSTEMS (PART ONE)

In order to create a vacuum within a closed container, or vessel, we need to remove the molecules of air and other gases that reside inside by means of a pump. The vacuum vessel and pumps (mechanical, booster, diffusion, holding) together with the associated piping manifolds, valves (mechanical pump, high-vacuum isolation, vacuum release, backing), vacuum measurement equipment (molecule counters) and traps comprise a typical vacuum system (Fig. 3.1).

FIGURE 3.1 | Typical vacuum system [1]

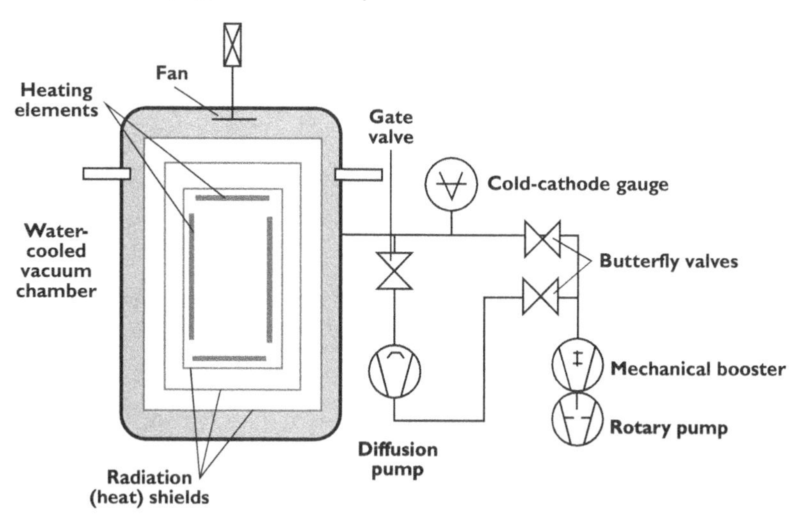

The Importance of Properly Sized Vacuum Pumps

The Ideal Gas Law (Equation 3.1) allows us to calculate the multiplying factors that we must deal with as the pressure drops in our vacuum system (Table 3.1). One can immediately see the large throughput necessary in the pumping equipment in order to carry off these huge gas volumes as a consequence of lowering the overall pressure of the system.

3.1) $V = nRT/P$

TABLE 3.1 | Effect of pressure on a fixed quantity of gas

Pressure		Volume	
mbar	torr	Cubic meters (m^3)	Cubic feet (ft^3)
1013	760	0.028	1
1.33	1	21.52	760
0.13	0.1	2,152.08	7,600
0.013	0.01	21,520.80	76,000
0.0013	0.001	21,520.80	760,000
0.00013	0.0001	215,208.03	7,600,000
0.000013	0.00001	2,152,080.25	76,000,000

Industrial vacuum systems in the heat-treatment industry are typically classified by their pressure ranges as follows:

- Rough (>1 x 10^{-1} torr)
- Low (1 x 10^{-1} torr to 1 x 10^{-3} torr)
- High (1 x 10^{-3} torr to 1 x 10^{-4} torr)
- Very high (1 x 10^{-4} torr to 1 x 10^{-6} torr)
- Ultrahigh (<1 x 10^{-6} torr)

The selection of the proper vacuum pumping system is highly application-dependent and is complicated by the wide variety of operational, process and equipment variables possible. However, each system must meet a specific set of requirements that are imposed by the process and production requirements. These requirements determine the size and type of pumps needed for the successful operation of the system.

Among the process variables, which should be known before attempting to size or select a pumping system, are the internal volume of the vacuum vessel, the time and pressure required and the size of the gas load that must be pumped off. In an ideal world, we know the answers to all of these questions. In the real world, we often compromise by applying safety factors to allow for the uncertainty of the information available. Here are several important formulas to help.

Equations for a Vacuum Pumping System

3.2) Pumping speed: $1/S_t = 1/S_p + 1/C_t$

where:

S_p = pump with speed measured at the pump inlet

S_t = resultant pumping speed

C_t = total conductance pumping onto a vessel through a passage

3.3) Pump-down time: $t = 2.3 \times K \times [V/S_p] \times \log [P_1/P_2]$

where:

t = time to pump a vessel of volume, V, in liters from a pressure, P_1, to a pressure, P_2, with a pump whose speed is S_p in liters/second. For most practical applications P_1 and P_2 can be chosen as the upper and lower limits of the entire pressure range that the pump must cover. If extreme accuracy is required and the pump speed varies considerably throughout the operating range, the formula can be applied in successive increments of the overall range and the values obtained added to determine the total cycle time.

and where:

K is given in Table 3.2.

TABLE 3.2 | K factor for pump-down time calculation

Pressure (torr)	Factor (K)
1,000-100	1.00
100-10	1.25
10-1	1.50
1-0.1	2.00
0.1-0.01	4.00

3.3) Gas flow: $[75 A_p/(A_p - A)] \times A$

The molecular conductance, C_t, of an orifice of area A (in^2) in a passage of cross-sectional area A_p (in^2) is given in this equation.

Mechanical Pumps

To reach the various vacuum levels, different vacuum pumping systems are required. The foundation of any of these systems is the positive-displacement mechanical or roughing pump (Fig. 3.2). The roughing pump, so-called because it is used to produce a "rough" vacuum, is used in the initial pump-down from atmospheric pressure to around 2×10^{-2} torr, depending on the type of pump. Mechanical pumps operate under the principle that they take in a large volume of air at the beginning of the cycle, compress it to a small volume and then exhaust it to the outside atmosphere. A thin layer or film of oil creates the actual seal between the moving parts in this so-called "wet" pump. Gas is exhausted under pressure against a valve disk at the outlet.

FIGURE 3.2 | Typical mechanical-pump cross-sectional view

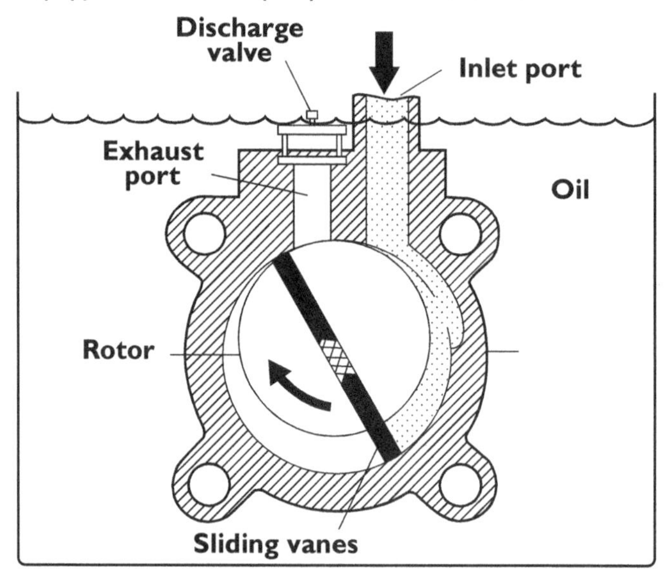

The internal components of the mechanical pump (Fig. 3.3) help us understand its operation. Basically, it is an eccentric cylinder driven about an axis by an electric motor. During operation, the rotor turns with the shaft, which causes the piston to sweep the volume between it and the stator. The piston does not turn in this case, but the vane-like extension on the piston (called the slide or slide valve) moves up and down in an oscillating seal (called the slide pin or slide-valve pin).

FIGURE 3.3 | Typical mechanical-pump operation [1]

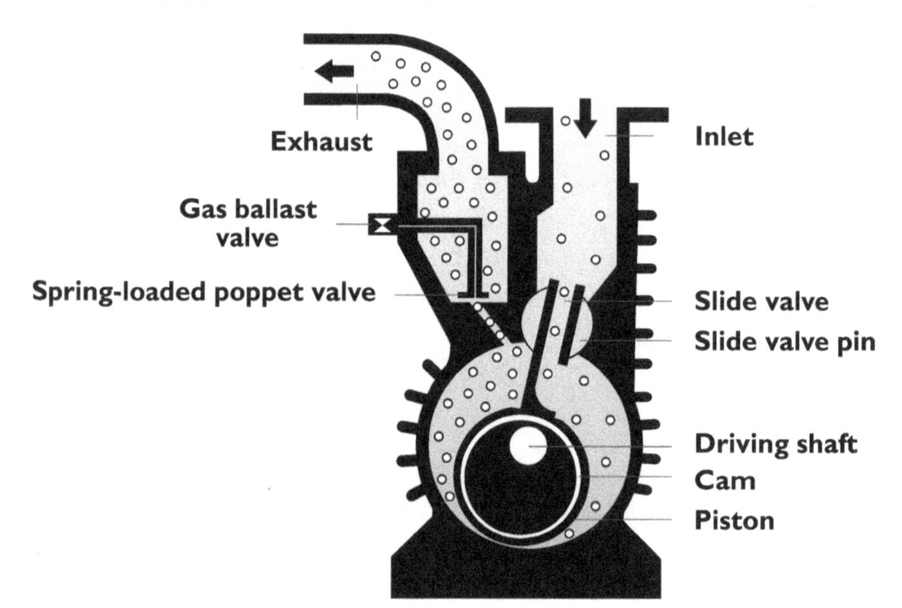

At the start of a rotation, the ported slide valve is open. As the rotation occurs, the slide valve closes, trapping a given volume of gas. This volume is compressed as the revolution continues. Near the end of the revolution, the pressure is above atmospheric and the gas discharges through a spring-loaded poppet valve. On the completion of the revolution, the slide valve opens and another increment of gas is admitted.

A vacuum pump will remove a number of molecules with each rotation. How many molecules are actually removed will depend largely on the actual pump displacement, rotational speed and vacuum-system pressure. Each time molecules are removed, the remaining molecules spread out in the vacuum chamber to occupy the available volume. This repeats, the pump removes molecules, the pressure reduces and there are less and less molecules to expand into the pump inlet with each rotation.

Mechanical pumps can be single or dual stage. A single-stage design will achieve a pressure of about 2×10^{-2} torr, while a dual-stage pump is capable of reaching pressures around 5×10^{-3} torr. A two-stage, or compound, pump has two pumping chambers connected in series. The exhaust of the first stage is coupled to the inlet of the second stage.

Lower pressure, fewer molecules and increased movement in the same volume results in less pumping efficiency. Why mechanical pumps start off with high efficiency and fall off at these pressure ranges can be explained as follows. Consider 1 cubic foot of volume at atmospheric pressure (760 torr). If we were to put this volume of gas in a container that was twice as large, the pressure would be exactly half, or 380 torr. That is, if we double the volume, we halve the pressure. Doubling the volume again to 4 cubic feet, therefore, results in a pressure of 190 torr. So, to evacuate a chamber to 1×10^{-3} torr theoretically requires that we remove a volume of 760,000 cubic feet.

In everyday operation, a mechanical roughing pump will have great difficulty achieving this ultimate pressure (lowest attainable pressure) since its efficiency begins to fall off at 1×10^{-1} to 8×10^{-2} torr. In other words, mechanical pumps will reach "blank-off" pressure – where they are no longer capable of pumping gas. In real-world operation, mechanical pumps will become inefficient at pressures considerably higher than blank-off pressures.

An alternative to "wet" mechanical pumps (those that use mechanical pump oil) are the so-called "dry" mechanical pumps. These pumps are used in applications where pumping efficiency and process contamination concerns are important. They have positive environmental impact (due to reduced oil consumption and minimal disposal issues) and operate with less noise and vibration.

Dry pumps operate on the compressor principle (Fig. 3.4). As the two rotors rotate, gas is drawn in through an inlet slot aligned with the cavity in one of the rotors. Further rotation closes the inlet, while the lobes (or claws) compress the trapped volume of gas until the cavity in the second rotor exposes the outlet or exhaust slot. A small volume of gas remains trapped and is carried over into the next pumping cycle. These designs produce high compression ratios and operate at high efficiency.

FIGURE 3.4 | Typical dry-pump operation [5]

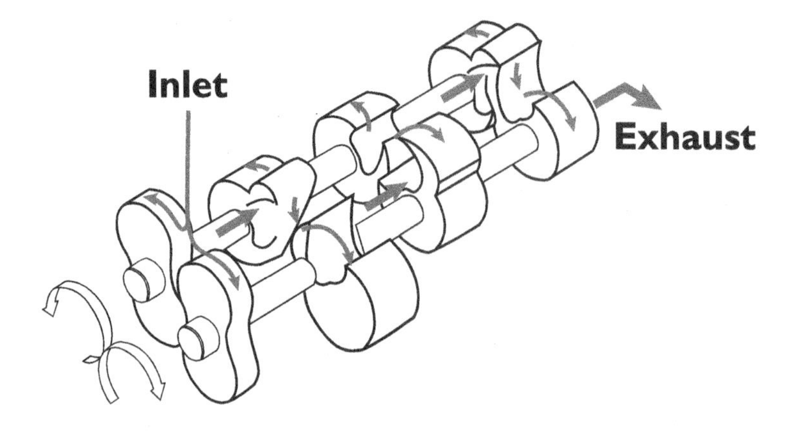

Blowers (Booster Pumps)

The booster pump (Fig. 3.5), or blower, is a different type of mechanical pump that is placed in series with the roughing pump and designed to "cut in," or start, at less than 100 torr. It is designed to provide higher speeds in the pressure range under 100 torr to 1 x 10^{-3} torr. In this intermediate pressure range, the roughing pump is losing efficiency while the diffusion (vapor) pump has yet to start to reach full efficiency.

FIGURE 3.5 | Typical Roots blower [2]

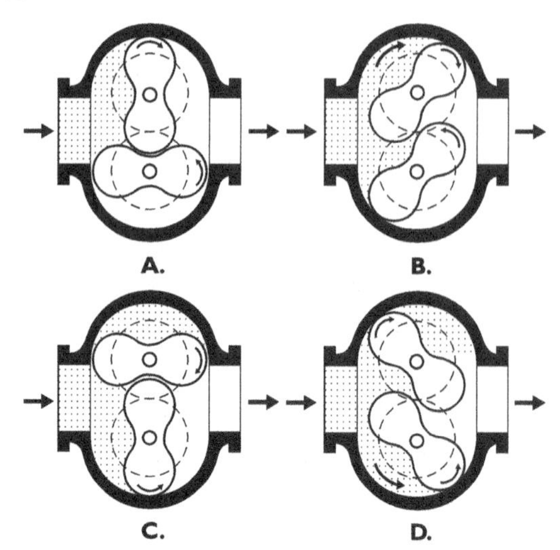

The operation of the booster (Fig. 3.6) is as follows. Two impellers are mounted on parallel shafts and rotate in opposite directions. They are geared together so that the correct relative position of each impeller to the other can be maintained. The impellers do not touch each other, and no sealing fluid is used. The back leakage is small compared to the total speed of the pump in its useful range.

FIGURE 3.6 | Booster-pump operation [1]

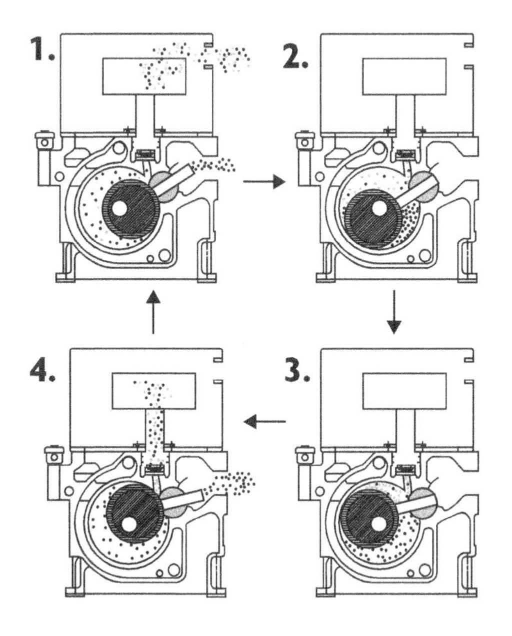

1. Air ingested starts
Volume trapped

2. Air ingested increases
Gas compressed

3. Air ingested increases
Gas compressed

4. Volume trapped
Gas expelled

During operation, gas from the inlet side is trapped between the impeller and housing. No compression takes place as this gas is moved from the inlet to the discharge port. When the leading lobe of the impeller passes the discharge port, gas from the discharge area (which is at higher pressure) enters but is swept away by the trailing lobe.

Mechanical booster pumps have a useful compression ratio of 10:1, so they must be backed by a mechanical roughing pump in order to reach their maximum efficiency. The mechanical booster pump is highly efficient in reducing the time required to evacuate a large or "gassy" system to the operating pressure at which the diffusion pump is efficient.

Ways to Help Pumping Efficiency

Individuals familiar with vacuum furnaces know the importance of having absolutely leak-tight vacuum chambers, doors, feed-throughs and penetrations. For critical applications, such as processing of superalloys or reactive metals, a leak rate of less than 5 microns/hour is mandatory. For normal vacuum applications, the leak rate should not exceed 20 microns/hour. Even a slight air or water leak will overwhelm the vacuum pumps and cause the vacuum level to rise several decades in pressure. Proper attention, therefore, must be given to the entire vacuum system. Leak detection of all joints, welds, seals, valves and pumps as well as the vessel itself is critical to success (c.f. Chapter 12).

An important operational consideration is to limit the amount of time a furnace chamber is exposed to room air, either during loading/unloading or when the unit is not in production. The effects of humidity (water vapor) are often devastating to the pumping system, decreasing its efficiency and creating an oil/water mixture in the pump that requires it to be ballasted. When not in production, vacuum furnaces should be pumped down to several hundred microns and then turned off.

Finally, moisture trapped in the hot zone or heat-exchanger tube bundles (if internal) is extremely difficult to overcome by pumping alone. Oftentimes backfilling with nitrogen or, if available, argon will help minimize this effect.

REFERENCES

1. Herring, D.H., "How to Move Molecules in Vacuum Systems (Part One: Mechanical and Booster Pumps)," *Industrial Heating*, November 2007
2. Dave Morris, BOC Edwards - Stokes Vacuum, private correspondence
3. Brunner, William F. and Thomas H. Batzer, *Practical Vacuum Techniques*, Robert E. Krieger Publishing Company, 1974
4. Jones, William R., "Pumping and the Vacuum Furnace," *Heat Treating*, July 1986
5. "Understanding the Effects of Air Exposure," *R&D Magazine*, April 2005
6. Brad Weber, Tuthill Corporation, Kinney Vacuum Division, private correspondence
7. *Steel Heat Treatment Handbook*, edited by Totten, George E. and Maurice A. Howes, "Chapter 7: Vacuum Heat Treating," Marcel-Dekker, 1997
8. Kimball, William H., *Vacuum...is it really nothing?*, C.I. Hayes Inc., 1977
9. Jim Wolford, Dresser Corporation, Roots Blower, private correspondence
10. Hoffman, Dorothy, Bawa Singh and John H. Thomas III, *Handbook of Vacuum Science and Technology*, Academic Press, 1997

CHAPTER 4

VACUUM PUMPING SYSTEMS (PART TWO)

Vacuum pumps are the heart of any vacuum system. While mechanical pumps have the ability to work against atmospheric back pressure and booster pumps improve the speed and level to which we pump down, these pumps have the disadvantage of losing efficiency as the system pressure continues to lower. In order to reach extremely low vacuum levels, the use of diffusion pumps is required (Fig. 4.1).

In simplest terms, diffusion pumps use a high-speed jet of vapor to direct gas molecules that enter the pump throat down into the body of the pump and out the exhaust. Diffusion pumps are also known as gas-jet pumps and are a type of momentum transfer pump.

FIGURE 4.1 | Typical diffusion pump (courtesy of Agilent Technologies)

TABLE 4.1 | Typical vacuum heat-treating and joining processes

Material	Process	Operating temperature, °C (°F)	Approximate pressure range, torr [a, b, c]
Aluminum alloys	Brazing	595-650 (1100-1205)	10^{-5} to 10^{-6}
Beryllium	Annealing	730-900 (1350-1650)	10^{-4}
	Sintering	1040-1065 (1900-1950)	$< 10^{-4}$
Cast iron	Annealing	870-925 (1600-1700)	10^{-3}
Copper alloys	Brazing	1095-1120 (2000-2050)	10^{-2} to 10^{-3}
Iron	Sintering	1120-1150 (2050-2100)	10^{-1} to 10^{-2}
Nonferrous alloys	Brazing (gold)	1040-1065 (1900-1950)	10^{-3} to 10^{-4}
	Brazing (silver)	615-980 (1140-1800)	10^{-3} to 10^{-4}
Stainless steels			
Ferritic	Annealing	630-870 (1160-1600)	10^{-2} to 10^{-3}
Martensitic	Annealing	830-900 (1515-1650)	10^{-2} to 10^{-3}
	Brazing (copper)	1090-1230 (2000-2250)	2 to 10^{-1}
	Brazing (nickel)	1090-1260 (2000-2300)	10^{-4} to 10^{-5}
	Brazing (silver)	675-980 (1250-1800)	10^{-1} to 10^{-2}
	Hardening	775-1175 (1425-2150)	$< 10^{-3}$
	Sintering	1205-1315 (2200-2400)	$< 10^{-2}$
Superalloys	Annealing	1260-1315 (2300-2400)	10^{-3}
	Brazing	1260-1315 (2300-2400)	10^{-3}
	Hardening	1260-1315 (2300-2400)	10^{-3}
Steel, carbon	Annealing	760-815 (1400-1500)	10^{-3}
Steel, high-speed			2 to 10^{-1}
Cobalt	Hardening	1275-1315 (2325-2400)	2 to 10^{-1}
Molybdenum	Hardening	1175-1230 (2150-2250)	2 to 10^{-1}
Tungsten	Hardening	1230-1290 (2250-2350)	2 to 10^{-1}
Superalloys	Aging	620-845 (1150-1550)	10^{-4} to 10^{-5}
	Brazing	1040-1250 (1900-2275)	$< 10^{-4}$
	Solution treating	1050-1250 (1925-2275)	10^{-4} to 10^{-5}
Tool steels			
A, D, H	Hardening	1275-1315 (2325-2400)	2 to 10^{-2}
Tantalum alloys	Annealing	1050-1290 (1920-2350)	10^{-4} to 10^{-5}
	Degassing	1050-1290 (1920-2350)	10^{-4} to 10^{-5}
Titanium alloys	Annealing	900-1010 (1650-1850)	$< 10^{-3}$
	Degassing	790-955 (1450-1750)	$< 10^{-3}$
Tungsten	Stress relief	595-1010 (1100-1850)	$< 10^{-3}$

Notes:

[a] Partial pressure may be required during a portion of the cycle to prevent de-alloying from occurring.

[b] The partial-pressure gas is typically nitrogen, although argon is used if there is a concern about nitriding a particular alloy at temperatures above 980°C (1800°F).

[c] Partial pressure of hydrogen (dry or wet) is also used by industry, typically in the range of 2 to 5 torr for applications such as magnetic annealing of low-carbon steel to produce decarburization, thus enhancing magnetic properties.

Diffusion Pumps

The diffusion pump (Fig. 4.2) consists of a boilerplate system in which a high-grade fluid is heated and then subsequently vaporized during boiling. This is typically accomplished by means of an external heating element and a stack assembly, or chimney (commonly referred to in the industry as a "Christmas tree assembly"), through which the vapors pass. These vapors exit the chimney through one or more levels of annular converging/diverging nozzles directed radially outward and downward at an angle of approximately 45 degrees and at speeds in excess of 400 km/hour (250 miles/hour). The hot vapors are accelerated by the action of the compression stacks within the diffusion pump, which serves as a venturi, creating supersonic velocities. As they travel outward and downward, they collide with molecules of the gases being drawn into the pump inlet by the pressure differential created during the boiling of the oils. This gives them an effective downward velocity toward the exit (foreline) from which they are removed efficiently by the mechanical pumping system.

Diffusion pumps are, therefore, a type of vapor pump (without moving parts) and are used to help achieve lower system pressures than can be achieved by a mechanical-pump/blower combination alone. The diffusion pump is capable of pumping gas loads with full efficiency with/at inlet pressures not exceeding 8×10^{-2} torr and discharge (or foreline) pressures not exceeding 3×10^{-1} torr. The diffusion pump cannot operate independently. It requires a separate pump to reduce the chamber pressure to or below the maximum intake pressure of the pump before it will operate. Also, while operating, a separate or holding pump is required to maintain the discharge pressure below the maximum tolerable pressure.

A separate holding pump operates continuously with its holding valve in the open position pumping on the foreline to the diffusion pump. Even when the diffusion pump is in a standby mode, the holding valve is open and the holding pump maintains the discharge pressure at or below the maximum tolerable operational pressure of the diffusion pump.

The operation of the diffusion pump (Figs. 4.3, 4.4) is as follows. The diffusion line is pumped down and the pump heaters are turned on, thus heating a fluid in the boiler portion of the pump. The inlet of the pump, which is attached directly to the vessel via a right-angle main poppet valve assembly, and a mechanical pump attached to outlet allows the pressure of the entire system to be reduced to about 5×10^{-1} torr. The rise in pressure forces the vapors up the chimney of the pump, where it is directed out the compression jet ring assemblies into the surrounding area of lower pressure. The nozzles deflect the vapor as a jet downward and outward to the walls, where the vapor condenses on the water-cooled inner walls of the pump body.

Gas molecules from the vessel enter the pump throat and diffuse through the less-dense fringe at the edge of the vapor stream. When a gas molecule has penetrated into the high-density core of the stream, the probability of it being knocked backward toward the inlet is less likely than the probability of it being carried (entrained) along the vapor stream toward the outlet. Thus, the predominant direction of molecular travel is away from the inlet and toward the outlet. In a multi-stage pump, the gas molecules are directed toward the next compression stage, where the action is repeated. Several succeeding stages will compress the low-pressure gas at the inlet to a higher pressure at the outlet, where it is removed to atmosphere by the mechanical pumping system.

FIGURE 4.2 | Diffusion-pump operation: schematic view (left); cross-sectional view (right) (courtesy of Agilent Technologies)

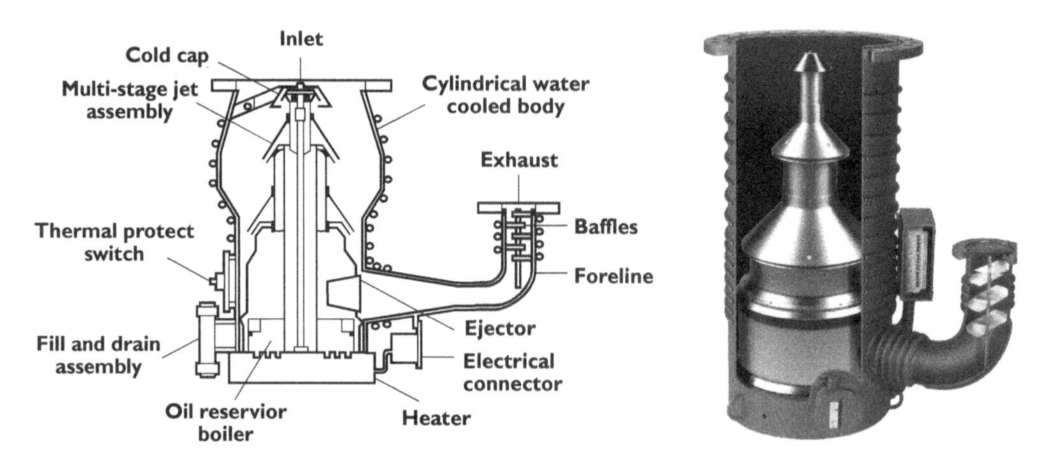

FIGURE 4.3 | Basics of how a diffusion pump works

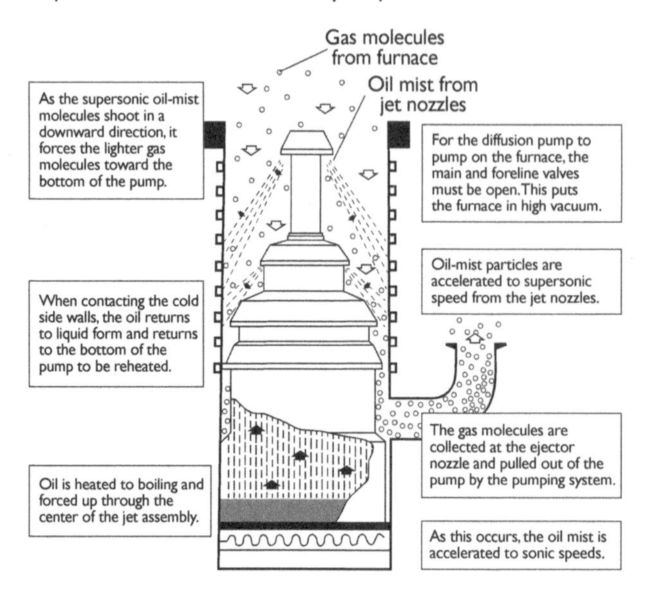

The movement of molecules from an area of low pressure to an area of higher pressure will only continue as long as the region of higher pressure (or fore-pressure) does not exceed a critical limit. Consequently, it is necessary for a diffusion pump to be "backed" by a mechanical pump. In practice, the backing pump has two or three times the minimum capacity required.

Today, several types of oil, based on silicones, hydrocarbons, esters, perfluorals and polyphenyl ethers, can be used as diffusion-pump fluids being vaporized in the range of 190-280°C (375-535°F). Each fluid has specific properties (Table 4.2). Mercury is no longer used in vacuum pumping systems due in large part to its toxicity. The choice of the pump fluid depends on the required application (vacuum level) of the pumping system.

TABLE 4.2 | Typical properties of certain fluids used in diffusion pumps (ultrahigh-vacuum applications) [3]

Fluid property	Polyphenyl ether	Silicone	Hydrocarbon
Vapor pressure, torr (25°C)	4×10^{-10}	2×10^{-8}	5×10^{-6}
Molecular weight	446	484	420
Density (25°C)	1.20	1.07	0.87
Flashpoint (°C)	288	221	243
Boiling point (1.3 mbar, °C)	29	223	220
Viscosity, cSt (25°C)	1,000	40	135
Viscosity, cSt (100°C)	12.0	4.3	7.0
Surface tension (dynes/cm)	49.9	30.5	30.5
Refractive index (25°C, 589 nm)	1.67	1.56	1.48
Thermal stability	Excellent	Good	Fair
Oxidation resistance	Excellent	Good	Fair
Chemical resistance	Excellent	Good	Fair
Radiation resistance	Excellent	Good	Fair

In order to produce an ultrahigh vacuum, one must be concerned with the characteristics of the pump (i.e. cleanliness and oil selection) and the vacuum vessel. Contamination from the chamber can be reduced by an intelligent choice of materials (ones that are less susceptible to outgassing), fabrication techniques and operating procedures. The gas load originating in the pump may be reduced by trapping methods or processing procedures.

Although diffusion pumps have been replaced in some applications by more advanced designs (such as cryogenic or turbomolecular pumps), they are still widely used due to their reliability, simple design and operation without noise or vibration, and they are relatively inexpensive to operate and maintain. Industrial and laboratory vacuum levels (Table 4.3) are achievable using these types of pumps.

FIGURE 4.4 | Vacuum levels by type of pump [8]

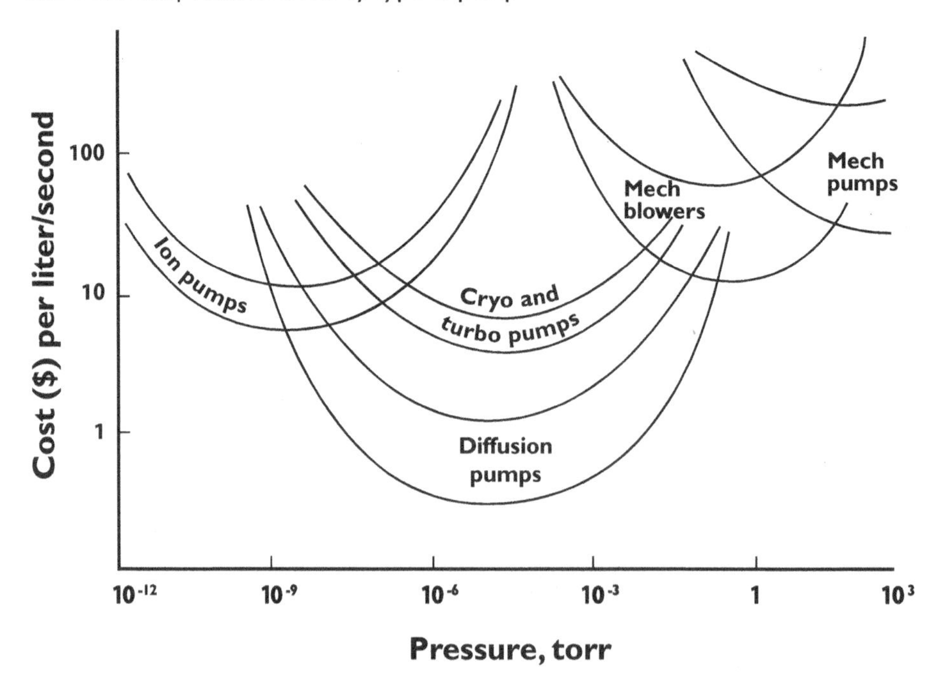

TABLE 4.3 | Characteristics of vacuum [5]

Characteristic	Atmospheric pressure	Beginning of high-vacuum range	Very high vacuum range	Ultrahigh vacuum range	Ultra-ultrahigh vacuum range
Pressure, torr	760	10^{-3}	10^{-6}	10^{-8}	10^{-9} to 10^{-12}
Number of molecules per second bombarding each square centimeter of vessel walls	3×10^{23}	4×10^{17}	4×10^{14}	4×10^{12}	4×10^{11} to 4×10^{8}
Mean free path of one molecule between collisions with another molecule (cm)	6.5×10^{-6}	5×10^{0}	5×10^{2}	5×10^{5}	5×10^{6} to 5×10^{9}
Number of molecules per cubic centimeter	2×10^{19}	3×10^{13}	3×10^{10}	3×10^{8}	3×10^{7} to 3×10^{4}
Pump type	Mechanical + blower	Mechanical + blower	Diffusion pump or ion pump	Diffusion pump or ion pump	Ion pump or cryogenic pump

Evacuation Effects

In general, the effects of evacuating a vessel can be summarized as follows. [4]

A. The effects of evacuating a vessel from 760 torr (atmospheric pressure) to 1 torr are:

1. Removing (high relative-humidity) air
   a. Water-vapor condensation (due to cooling effect associated with a sudden drop in pressure)
   b. "Fog" develops (a cloud "swirls" around with a turbulence that is characteristic of a gas flow at high pressure and high flow rate)

2. (Slow) change in the composition of the gas remaining
   a. Initially, air is the major component of the gas (certain other contaminants such as oils, grease and water exist on cold surfaces such as vessel walls).
   b. Eventually, almost all of the air is pumped out. The grease and water will continue to evaporate and their partial pressure will constitute a much larger portion of the total pressure. This is called outgassing.

B. The effects of evacuating a vessel from 1 torr to 1 x 10^{-4} torr are:

1. The ability of the gases remaining in the vessel to conduct heat begins to decrease rapidly.

2. A change in the electrical characteristics of the gas begins (voltage to start a discharge decreases).

C. The effect of evacuation from 1 x 10^{-4} torr to 1 x 10^{-6} torr is:

1. Decreasing molecular density
   a. Molecules collide with the sides of the vessel as often as they will with each other.
   b. There is an increase in sliding friction.

Pump Problems – Troubleshooting Guide

Common problems with mechanical pumps require routine maintenance and inspection. They include:

- ❖ Oil contamination
- ❖ Sludge buildup
- ❖ Loose or slipping belts
- ❖ Improper oil level (too low or too high)
- ❖ Stuck discharge valve

- ❖ Clogged oil lines or valves
- ❖ Damaged discharge valve
- ❖ Ingestion of foreign contaminants (metal fines, metal chips, etc.)
- ❖ Excessive vibration (pipe connection or floor mounting)
- ❖ Exhaust filters (more than 12 months in age)
- ❖ Oil temperature not being regulated between 60-70°C (140-160°F)

Of the various mechanical-pump problems that can arise, contamination of the oil is the most common. Vapors present in the gas being pumped may condense and mix with the oil. Moisture (water vapor) is especially problematic and if not removed will flash to vapor and tie up a large portion of the pump's gas load capacity, thus creating a significant loss in pumping efficiency (resulting in either extremely long pump-down times, failure to achieve a low vacuum level or both).

In order to rid the oil of water and other liquid condensates, a gas ballast is used. A gas ballast may be used in conjunction with correctly regulating the operating temperature of the oil with a water-control valve assembly. A ballast valve on the pump can be opened (manually or automatically) to admit air, nitrogen or argon into the pump, disrupting its operating efficiency and resulting in a reduction in the compression necessary to exhaust the gases and, correspondingly, a decrease in the amount of vapor that condenses. The use of a gas ballast increases the amount of oil carried out in the exhaust. The gas-ballast valve is very effective in removing water vapor, but it is actually very ineffective in cleaning dirty oil or fixing oil that has cracked (fractionated) due to mixing with other downstream by-products.

In addition, the oil may break down chemically, forming a sludge that causes numerous (short- and long-term) problems with pump operation, especially as it relates to severe wear on internal components, often to the point where rebuilding is not possible. Disassembly and cleaning of the pump is the only solution to this problem.

Mechanical pump oil must be changed on a routine basis (typically every 300 hours). In replacing pump oil, be careful to use the type of oil recommended for the pump and be equally careful to apply precisely the right amount of oil. Either too much or too little oil in the pump reservoir will lead to serious difficulties. Checking the amount of fluid in the pump reservoir during normal operation is strongly recommended. It is possible, due to improper operation, to have the pump oil backstream into the vessel in considerable quantities.

Common problems with diffusion pumps include:

- ❖ Power failures
- ❖ Excessively high foreline pressures

- Backstreaming
- Process by-products clogging oil returns in boilerplate
- Defective heaters and/or broken wiring on the boiler
- Water inlet temperature above 45°C (115°F)
- Water exit temperature above 65°C (150°F)
- Mixtures of hydrocarbon-based oils with silicone-family oils
- High leak rates on the system when being pumped on
- Water-cooled copper lines full of mineral (calcium) deposits negating proper heat transfer

Of the various diffusion-pump problems, exposure of the hot pump oil to the atmosphere or interruption/loss of the coolant flow is of most concern. Accidentally introducing air when the diffusion pump is at too high a temperature almost inevitably leads to a pump malfunction or failure and often requires expensive and lengthy repairs (most often at the manufacturer). Severe cracking (breakdown) of the oil and oxidation will occur depending on the type of oil. This leads to excessive back pressure, and the products of the oil breakdown will deposit on the jet structure blocking the openings or deposit in the area of the oil heater, burning it out. Overheating due to inadequate coolant flow also decomposes the oil and can cause excessive backstreaming into the vacuum-furnace chamber. Depending on the actual amount of air in the hot pump, coupled with previously deposited materials in the base of the pump, the oil may expand excessively in vapor form with a significant pressure buildup.

Backstreaming

As the vacuum pressure continues to be lowered, some pump-fluid gas molecules attempt to reverse course and move up and back toward the vacuum chamber. This phenomenon is called backstreaming, and it increases in frequency as the inlet pressure exceeds 0.001 mbar (1 micron). Examples of situations that can lead to backstreaming include diffusion pumps run at too low a pressure (e.g., 1 torr) or poppet valves that are not fully closed before backfilling the furnace.

The effects of backstreaming can be negated to a degree by use of a cold trap located in the line connecting the inlet of the diffusion pump and the vessel being pumped. Cold traps are cooled by water, refrigerants or liquid nitrogen and serve to limit backstreaming and stop the flow of contaminants coming from the chamber, which would otherwise be carried into the pumping system proper and contaminate the pump fluids. Other methods – such as baffles and cold caps located above the top jet of the diffusion pump – have also been used. Each family of cold traps offers both favorable and unfavorable impacts on specific furnace operations, thus making selection of a trap methodology that best suits your specific operations of paramount importance.

REFERENCES

1. Steve Palmer, Agilent Technologies, private correspondence
2. James Grann, Ipsen Inc., private correspondence
3. Mike Moyer and Don Jordan, Solar Atmospheres Inc., private correspondence
4. McCarthy, Dave, *Diffusion Pumps for Vacuum Furnace Applications,* Varian Vacuum Technologies Inc.
5. Joaquim, M.E. and W. Foley, "Inside a Vacuum Diffusion Pump, Application Note," March 2003
6. Brunner Jr., William F. and Thomas H. Batzer, *Practical Vacuum Techniques,* Robert E. Krieger Publishing Company, 1974
7. Steinherz, H.A. and P.A. Redhead, "Ultrahigh Vacuum," *Scientific American,* Vol. 206, March 1962
8. Hablanian, Marsbed H., *High Vacuum Technology, A Practical Guide, 2nd Edition,* Marcel Dekker Inc., 1997

CHAPTER 5

VACUUM MEASUREMENT SYSTEMS (PART ONE)

One, two, three…

Counting molecules is a job for vacuum gauges. Depending on the type of vacuum systems and the required operating vacuum level, different vacuum gauges are required – often in combination with one another – to accurately determine and/or control the vacuum level of the chamber at any given moment in time. The criteria for selecting a vacuum gauge are dependent on various conditions, such as:

- The vacuum range to be detected
- The gas composition (inert, reactive, corrosive)
- Required accuracy and repeatability
- Environmental conditions

Vacuum gauges are divided into three basic categories based on their working pressure (Fig. 5.1). These include:

- Absolute-pressure gauges
- Medium-vacuum gauges for use down to approximately 0.001 mbar (1 micron)
- High-vacuum gauges for use below 0.001 mbar (1 micron)

FIGURE 5.1 | Working pressure range for vacuum gauges (courtesy of Oerlikon Leybold Vacuum) [1]

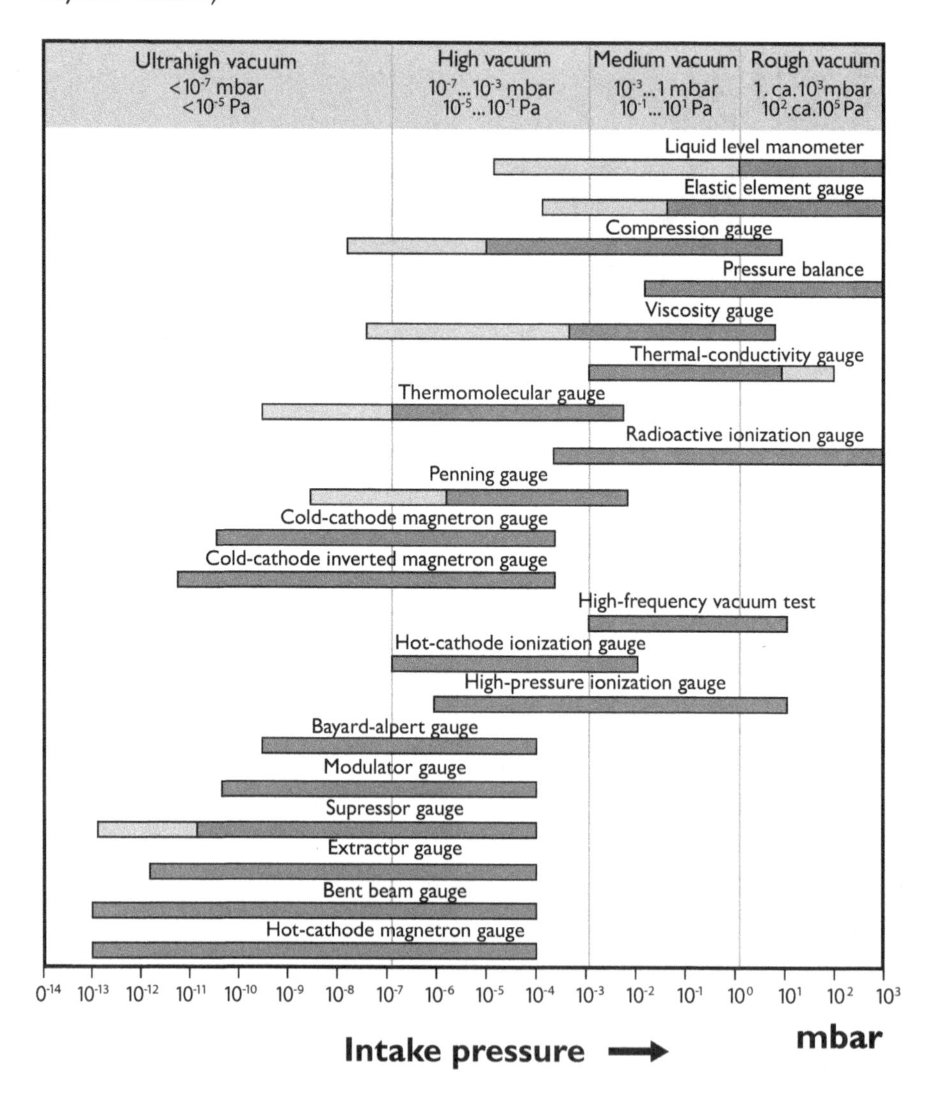

What is a vacuum gauge?

A vacuum gauge is an instrument for measuring pressures below atmospheric pressure. There are many types, each designed for a specific function. Some of the more common types of vacuum gauges are shown below, listed in order of descending pressure range.

MANOMETER

A manometer is a relatively simple device that usually consists of a tube or column filled with a liquid. The pressure is found by measuring the column height or the difference in heights of several columns.

The U-tube manometer is the most common type of manometer today because the difference in height between the two columns is always a true indication of the pressure regardless of variations in the internal diameter of the tube. With both ends of the tube open, the liquid is at the same height in each leg (Fig. 5.2a). When positive pressure is applied to one leg (Fig. 5.2b), the liquid is forced down in that leg and up in the other. The difference in height, "h," which is the sum of the readings above and below zero, indicates the pressure. When a vacuum is applied to one leg (Fig. 5.2c), the liquid rises in that leg and falls in the other. The difference in height, "h," which is the sum of the readings above and below zero, indicates the amount (or degree) of vacuum.

FIGURE 5.2 | Principles of the manometer [2]

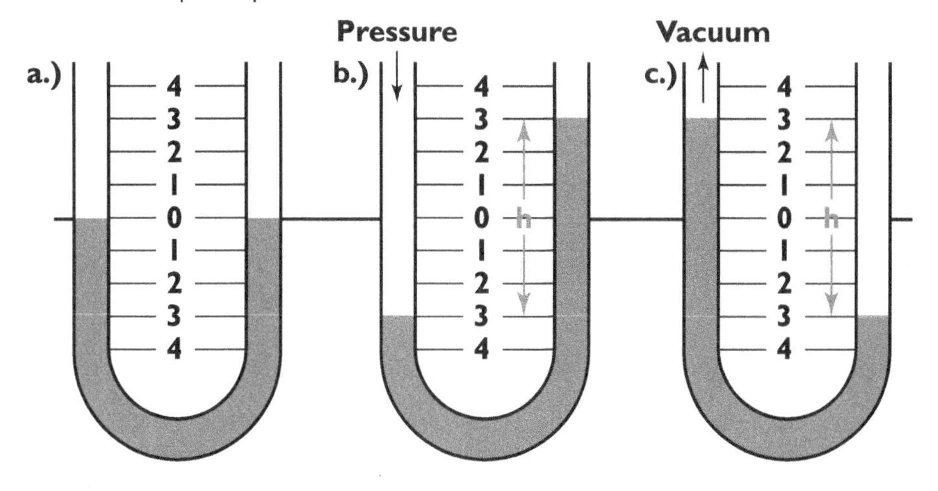

THERMAL-CONDUCTIVITY GAUGE

These devices operate on the principle that heat transported by gas molecules can be related to gas pressure. A heat source causes changes in surface temperature (or in the heating power required to maintain constant temperature), and this can be related to the pressure of the system.

Various types of thermal-conductivity gauges are distinguished according to the method of indicating the surface temperature (i.e. the way in which their wire temperature is measured). The most common types, with a typical operational range of 1 to 10^{-3} torr, include thermocouple gauges and Pirani gauges.

For their operation, thermocouple gauges depend on the fact that the conduction of heat away from a heated filament is dependent upon the pressure, with the cooling being greater at high pressures and less at low pressures. The system typically consists of a heated filament with thermocouples attached to an external circuit in such a manner that the temperature of the filament can be read in terms of the pressure of the gases surrounding it. In the readout devices, a special circuit is provided for standardization of the current passing through the filament.

Thermocouple gauges are simple, inexpensive and rugged. They can be contaminated by material from the vessel getting into the tube, but upstream filters are available, and they can be cleaned in some cases. Most operating systems make use of several of these devices in the roughing line, in the foreline of a diffusion pump and on the vessel itself.

Pirani gauges (Fig. 5.3), like thermocouple gauges, depend on the thermal conductivity of the gas surrounding the heated filament. However, the actual change in resistance of the heated platinum or tungsten filament wire is measured and used as a calibration means. These gauges have an extra element sealed away from the vacuum, employed in a bridge circuit, in order to compensate for ambient temperature (otherwise, large errors would ensue).

Pirani gauges do not suffer as much from inadvertent exposure to air with the filament heated and have a simple gauge readout circuit. The principal limitations include contamination from vapors from the vessel being pumped or the deposition of pump oil on the filament (which changes its emissivity and, hence, its temperature and pressure) and the fact that these gauges are nearly impossible to clean. Thermal conductivity is a function of pressure in the range of 1 mbar (750 microns) to atmospheric pressure in the laminar-flow range.

FIGURE 5.3 | Principles of the Pirani gauge [8]

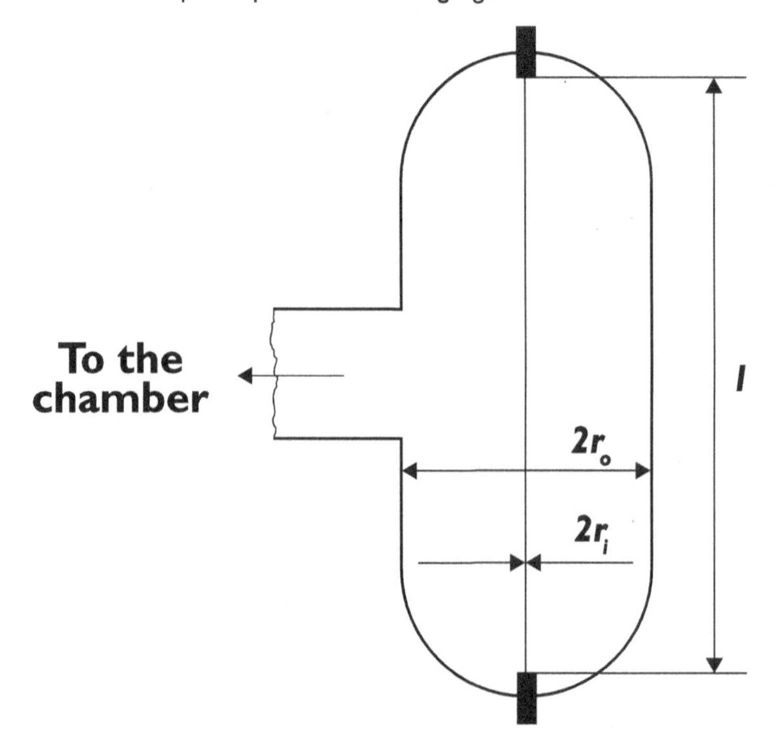

In the simplest case, a Pirani gauge consists of a thin wire (diameter $2r_i$) that is mounted in the axis of a cylindrical tube (diameter $2r_o$). The thin wire is heated by a constant electrical power source with heat transported to the walls of the tube.

KNUDSEN GAUGE

Knudsen (radiometer) gauges (Fig. 5.4) are typically accurate to 10^{-6} torr. These devices measure pressure in terms of the net rate of transfer of momentum by molecules between two surfaces maintained at different temperatures and separated by a distance smaller than the mean free path of the gas molecule.

FIGURE 5.4 | Principles of the Knudsen gauge [8]

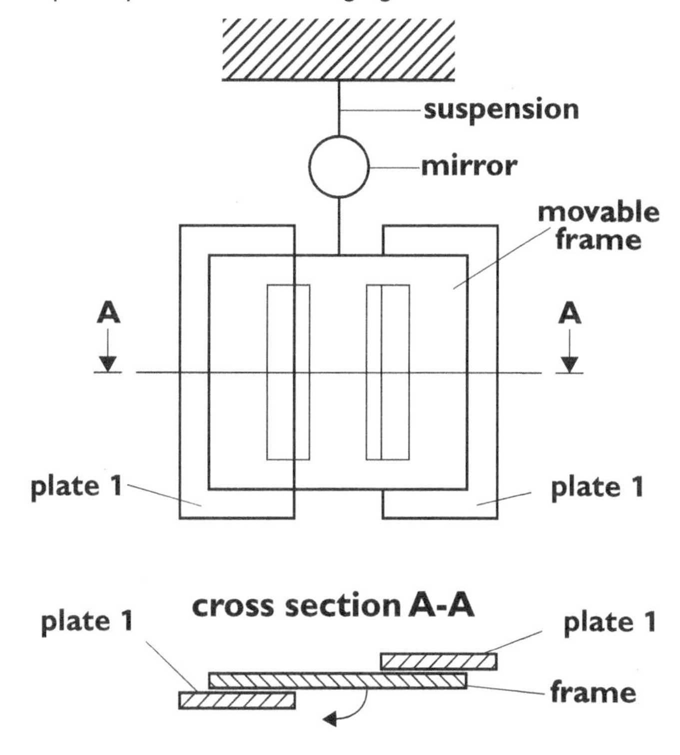

Two parallel plates of the same area are held at different temperatures. Gas molecules traveling from the higher- to the lower-temperature plate have higher impact energy than gas particles coming from the surroundings to the other side of the plate. Thus, different pressures are measured.

McLEOD GAUGE

The McLeod gauge (Fig. 5.5) is typically accurate to 10^{-6} torr. The principle of operation involves measuring the pressure of a gas by measuring its volume twice – once at the unknown low pressure and again at a higher reference pressure.

FIGURE 5.5 | Principles of the McLeod gauge [8]

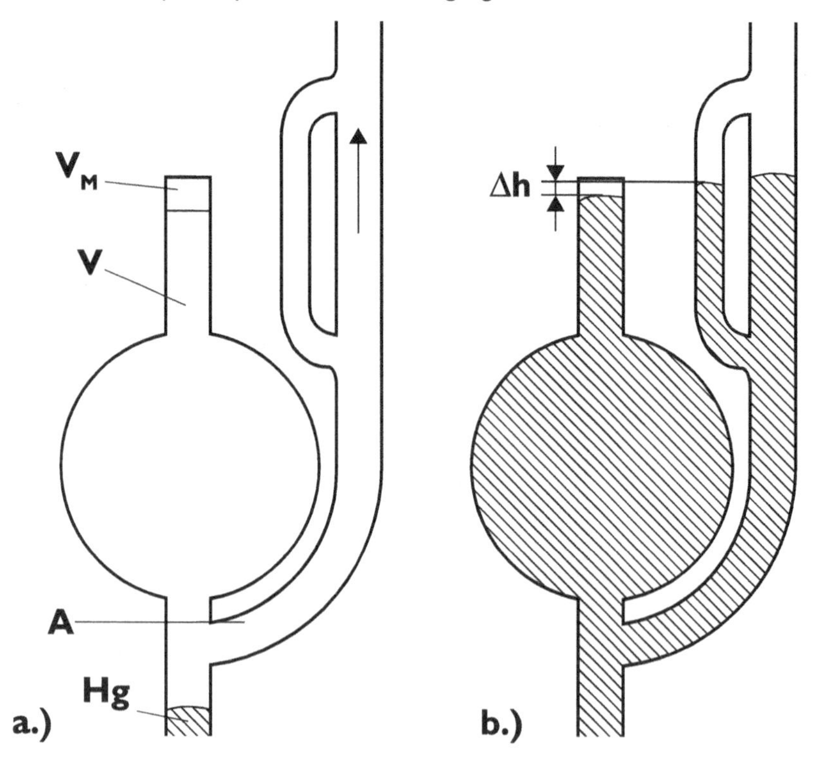

A liquid (mercury) piston is created with which a gas of a well-defined volume, V, is compressed into a known smaller volume, V_M. The pressure in the closed volume, V_M, is then increased from the unknown pressure to the (higher) measurable pressure that can be calculated (Δh).

The use of a McLeod gauge enables accurate readings to be obtained that are independent of any external calibration device. In other words, they provide absolute measurements. All other types of gauges must be externally calibrated and, therefore, are subject to drift in use. The principle disadvantages are that the device is not a continuous readout, is fragile and is not automatic. The gauge is also easily contaminated.

IONIZATION GAUGE

An ionization gauge is typically accurate to 10^{-9} torr (and beyond). These devices have a means of ionizing the gas molecules and a means of correlating the number and type of ions produced with the pressure of the gas. Various types of ionization gauges are distinguished according to the method of producing the ionization. The common types are:

- Hot-cathode gauge (typical range: 1×10^{-2} to 1×10^{-10} torr)
- Cold-cathode gauge (typical range: 1×10^{-2} to 1×10^{-11} torr)

FIGURE 5.6 | Principles of the cold-cathode gauge [5]

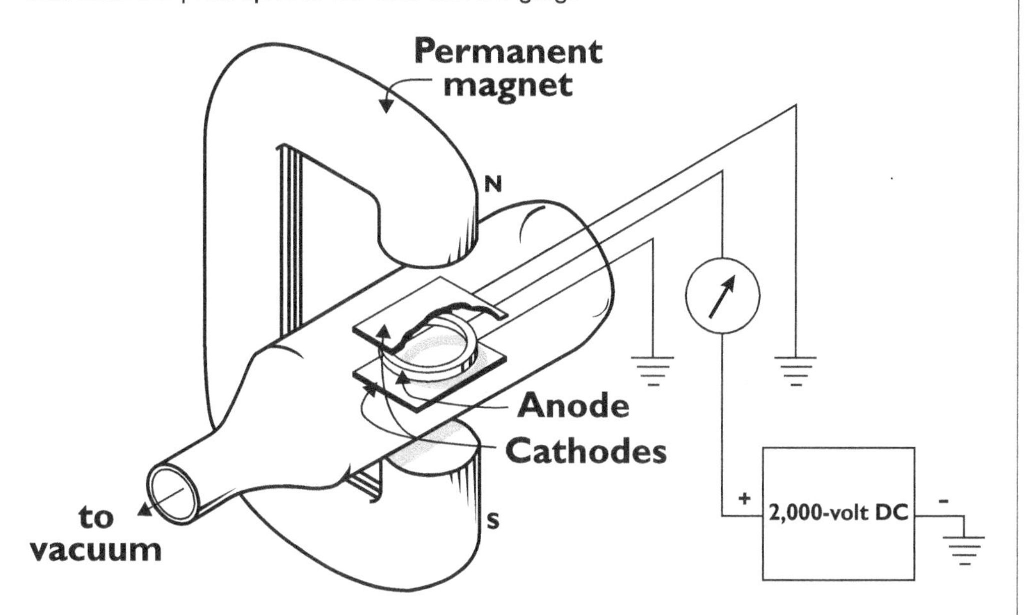

A cold cathode (Fig. 5.6) produces an ion current by utilizing a high-voltage discharge. The electrons emanating from the cathode are caused to spiral by a magnetic field as they move across the tube to the anode. This spiraling of electrons increases the probability of collision with any gas molecules present. Such collisions set free positively charged ions, which are attracted to a negatively charged plate.

The greatest advantage of this type of gauge is that it is not harmed by inadvertent exposure to atmospheric pressure. Issues involve hazards and insulating methods of high voltage (>2,000 volts) in the vicinity of the gauge. These gauges are also readily contaminated by either hydrocarbons or metallic ions but can be easily disassembled and cleaned.

Hot-cathode gauges can be employed in all types of applications. In these gauges, electrons are emitted from a heated filament and accelerated toward a positively charged cylindrical grid. Some of these electrons collide with gas molecules, causing the production of positively charged ions, which are collected at a negatively charged plate (as in the case of cold-cathode gauges). Since no magnetic field is employed, the ion currents produced are considerably lower than cold-cathode gauges. It is necessary to employ an amplifying stage in the readout device in order to increase the current sufficiently to provide a readout.

The advantages of these gauges are their availability over a wide range of pressures, relatively inexpensive sensing tubes, the availability of an outgassing cycle (to rid the tube of contaminants) and a linear output down to relatively low pressures. The principle limitations are exposure to atmospheric pressure and measuring errors arising from pumping effects. Indeed, exposure to any pressure much above 0.001 mbar (1 micron) while the filament is heated will burn it out.

A Little History

Engineers first became interested in vacuum measurements in the 1600s when they noted the inability of pumps to raise water more than about 9 meters (30 feet). The Duke of Tuscany in Italy commissioned Galileo to investigate the problem. Galileo, among others, devised a number of experiments to investigate the properties of air. After Galileo's death in 1642, Evangelista Torricelli continued the work, which included vacuum-related investigations and the invention of the mercury barometer. He discovered that the atmosphere exerts a force of 101.3 kPa (14.7 psi) and that, inside a fully evacuated tube, the pressure was enough to raise a column of mercury to a height of 760 mm (29.9 inches). The height of a column of mercury is therefore a direct measure of the atmospheric pressure. The value of 1/760th of an atmosphere is called a torr in honor of Torricelli.

Pressure is simply defined as a force per unit area, and the most accurate way to measure air pressure is to balance a column of liquid of known weight against it and measure the height of the liquid column so balanced. The units of measure commonly used in the U.S. are inches of mercury (Hg), using mercury as the fluid, and inches of water column (W.C.), using water or oil as the fluid.

Pressure Ranges

The correct choice of gauge depends on knowledge of the working principles of the gauge, the range of pressures it can measure and its accuracy over the required range. These factors have been determined by experience, and there is a vacuum gauge for every pressure range.

- For low-vacuum ranges (higher-pressures) between atmospheric and 10 torr, Bourdon tubes, bellows, active strain gauges and capacitance sensors are all suitable measurement devices.
- For mid-range vacuum requirements (those in the 10^1 to 10^{-3} torr range), there are several choices, including the capacitance manometer – a good choice for more accurate measurements – or the thermocouple or Pirani-type gauges.
- For high vacuum (10^{-3} to 10^{-6} torr), either cold-cathode or Bayard-Alpert hot-cathode gauges are used. There is some concern over accuracy and/ or stability, and both require frequent calibration.

Location

Finally, a number of factors must be considered when installing the devices discussed above. In particular, selection of the location for installation needs to:

- Avoid vacuum gradients
- Negate pumping effects
- Avoid strong magnetic and electrical fields
- Avoid contamination from product evaporation or oil vapors from diffusion or mechanical pumps

REFERENCES

1. American Vacuum Society
2. Copyright 1998 Putnam Publishing Company and OMEGA Press LLC, reproduced courtesy of Omega Engineering Inc.
3. Dwyer Instruments Inc., Michigan City, Ind.
4. Kimball, William H., *Vacuum...is it really nothing?*, C.I. Hayes Inc., 1977
5. Brunner Jr., William F. and Thomas H. Batzer, *Practical Vacuum Techniques*, Robert E. Krieger Publishing Company, 1974
6. *Webster's Ninth New Collegiate Dictionary*, Merriam-Webster Inc., Springfield, Mass., 1987
7. Considine, Douglas M. and Glenn D. Considine, *Van Nostrand's Scientific Encyclopedia*, Van Nostrand, 1997
8. Edelmann, C., "Pressure Measurement – Total Pressures," Otto von Guericke University
9. *The Vacuum Technology Book*, Pfeiffer Vacuum, 2009
10. *Much Ado About Nothing, or So You Want to Measure Vacuum?*, Televac, 1976

CHAPTER 6

VACUUM MEASUREMENT SYSTEMS (PART TWO)

Four, five, six...

Counting molecules is a job for vacuum gauges, and it's now time to understand the differences between these devices and when to use them.

Recall first that the vacuum level in a vessel is determined by the pressure differential between the evacuated volume and the surrounding atmosphere (Table 6.1). The two basic reference points in all these measurements are standard atmospheric pressure (760 torr) and perfect vacuum (0 torr), so calculating changes in volume in vacuum systems requires conversions to negative pressure (psig) or absolute pressure (psia).

TABLE 6.1 | Comparison of vacuum and pressure levels

% of absolute vacuum	mm of Hg (inches of Hg)	Pressure, mbar (psig)
10	-76.2 (-3.0)	-101.3 (-1.47)
15	-114.3 (-4.5)	-152.4 (-2.21)
20	-152.4 (-6.0)	-202.7 (-2.95)
25	-190.5 (-7.5)	-253.7 (-3.68)
30	-228.6 (-9.0)	-304.1 (-4.42)
35	-266.7 (-10.5)	-355.1 (-5.15)
40	-304.8 (-12.0)	-405.4 (-5.89)
45	-342.9 (-13.5)	-456.4 (-6.63)
50	-381.0 (-15.0)	-507.4 (-7.36)
55	-419.1 (-16.5)	-558.5 (-8.10)
60	-457.2 (-18.0)	-609.5 (-8.84)
65	-495.3 (-19.5)	-659.8 (-9.57)
70	-533.4 (-21.0)	-710.8 (-10.31)
75	-571.5 (-22.5)	-761.9 (-11.05)
80	-609.6 (-24.0)	-812.2 (-11.78)
85	-647.7 (-25.5)	-863.2 (-12.52)
90	-685.8 (-27.0)	-914.2 (-13.26)
95	-723.9 (-28.5)	-964.6 (-13.99)
100	-762.0 (-30.0)	-1,013.5 (-14.70)

Mechanical Gauge Designs

Mechanical gauges measure pressure or vacuum by making use of the mechanical deformation of tubes or diaphragms when exposed to a difference in pressure. For these reasons they are classified as differential-pressure gauges. Typically, one side of the element is exposed to a reference vacuum, and the instrument measures the mechanical deformation that occurs when an unknown vacuum pressure is exposed to the other side.

BOURDON GAUGES

These gauges work on the principle of the Bourdon tube (Fig. 6.1), which consists of a tube with an elliptical cross section formed in an arc. The tube is rigidly fixed at one end and closed at the other end. When the pressure in the tube increases, the radius of the arc increases (in other words, the tube tries to straighten itself out). When the pressure decreases, the radius decreases (thus, the free end of the tube moves in response to a change in pressure). A system of mechanical linkages attached to the free end moves a pointer over a calibrated scale (Fig. 6.2).

FIGURE 6.1 | Principles of the Bourdon tube: a.) principle of operation; b.) distribution of forces [1]

FIGURE 6.2 | Typical Bourdon vacuum gauge

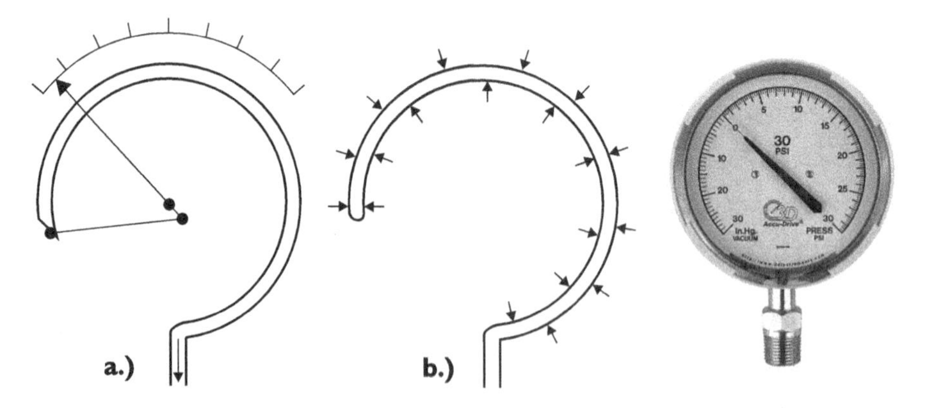

A quartz Bourdon tube uses a quartz helix element, and instead of moving linkages, the deformation rotates a mirror. When used for vacuum detection, two quartz Bourdon elements are formed into a helix. The reference side contains a sealed vacuum, and the measurement side is connected to the unknown process vacuum. The pressure difference between the two sides causes an angular deflection that is detected optically. The optical readout has a high resolution (about one part in 100,000). Advantages of this sensor are its precision and the corrosion resistance of quartz. Its main limitation is high price.

MANOMETER

The basic manometer consists of a reservoir filled with a liquid. When detecting vacuums, the top of the column is evacuated and sealed. A relatively small change in vacuum pressure will cause a relatively large movement of the liquid. Manometers are simple, low cost and can detect vacuums down to about 10^{-3} torr. The accuracy of the gauge is determined by how closely the difference in height of the two arms can be measured. Today, digital readout manometers (Fig. 6.3) have greatly improved accuracy. The sensitivity of the gauge depends on the density of the fluid used.

FIGURE 6.3 | Digital readout manometer

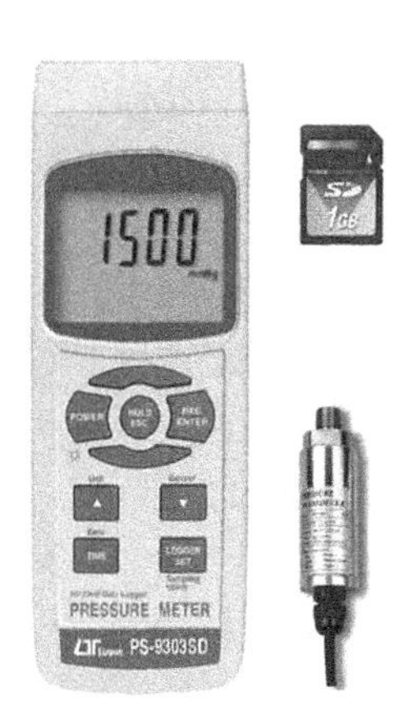

CAPACITANCE VACUUM MANOMETER

A capacitance sensor operates by measuring the change in electrical capacitance that results from the movement of a sensing diaphragm relative to some fixed capacitance electrodes (Fig. 6.4). The higher the process vacuum, the farther it will pull the measuring diaphragm away from the fixed capacitance plates. In some designs, the diaphragm is also allowed to move. In others, a variable DC voltage is applied to keep the sensor's Wheatstone bridge in a balanced condition. The amount of voltage required is directly related to the pressure.

FIGURE 6.4 | Principle of operation of a capacitance vacuum manometer [2]

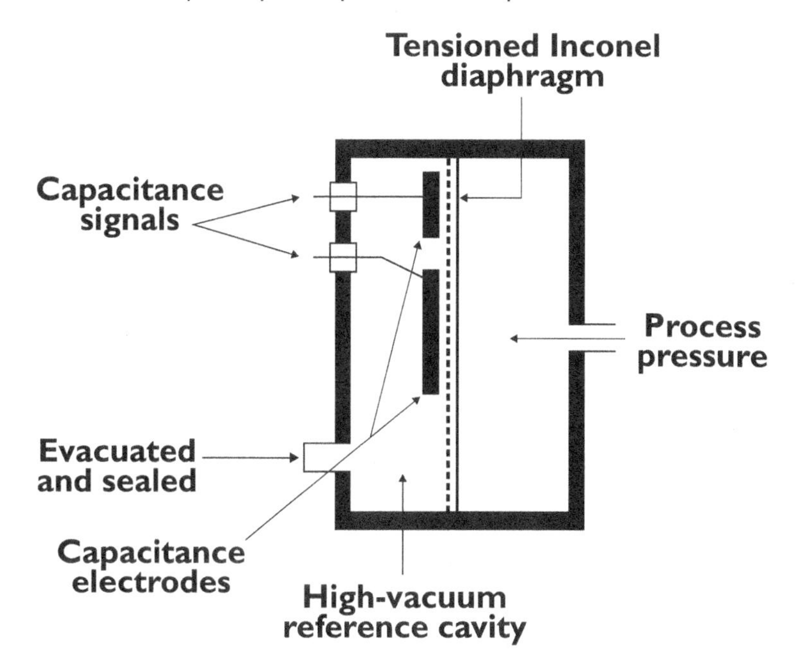

The great advantage of a capacitance gauge is its ability to detect extremely small diaphragm movements. Accuracy is typically 0.25-0.50% of the reading. Thin diaphragms can measure down to 10^{-5} torr, while thicker diaphragms can measure in the low vacuum to atmospheric range. To cover a wide vacuum range, one can connect two or more capacitance sensing heads to create a multi-range package.

McLEOD GAUGES

The McLeod gauge uses the principle of Boyle's law ($P_1V_1 = P_2V_2$; that is, if the temperature is held constant, the increase in pressure is exactly proportional to the decrease in volume) to amplify and measure pressures that are too small to be measured with a manometer. To do this, a sample of gas from the system is isolated and reduced in volume by a known amount. This reduction in volume causes a proportional increase in the pressure of the gas. Originally invented in 1878, the McLeod gauge is still used today, mainly for calibrating other gauges (although other techniques are making this obsolete). McLeod gauges can cover vacuum ranges between 1 and 10^{-6} torr.

MOLECULAR MOMENTUM GAUGES

This vacuum gauge is operated with a rotor that spins at a constant speed. Gas molecules in the process sample come in contact with the rotor and are propelled into the restrained cylinder. The force of impact drives the cylinder to a distance proportional to the energy transferred, which is a measure of the number of gas molecules in that space. The full scale of the instrument depends on the gas being measured. The detector has to be calibrated for each application.

VISCOUS FRICTION GAUGES

At high vacuum levels, viscosity and friction both depend on pressure. This instrument measures vacuum down to 10^{-7} torr by detecting the deceleration caused by molecular friction on a ball that is spinning in a magnetic field. Vacuum is determined by measuring the length of time it takes for the ball to drop from 425 to 405 revolutions per second after drive power is turned off. The higher the vacuum, the lower the friction and therefore the more time it will take. This design is accurate to 1.5% of indicated reading, resistant to corrosion and can operate at temperatures up to 4150°C (7500°F).

Thermal Gauge Designs

Below 1 torr, a change in the pressure of a gas will cause a change in its thermal conductivity (the ability of a gas to conduct heat). If an element heated by a constant power source is placed in a gas, the resulting surface temperature of the element will be a function of the surrounding vacuum. Because the sensor is an electrically heated wire, thermal vacuum sensors are often called hot-wire gauges. Because the characteristics of all gases are different, the response of a thermal-conductivity gauge will vary for each gas. To read accurately, the gauge must be calibrated for the gas being measured.

PIRANI GAUGES

The Pirani gauge utilizes the change in electrical resistance of a wire with temperature. A sensor wire is heated electrically, and the pressure of the gas is determined by measuring the current needed to keep the wire at a constant temperature by use of a Wheatstone bridge network (Fig. 6.5). The Pirani gauge is linear in the 10^{-2} to 10^{-4} torr range. Above these pressures, output is roughly logarithmic. Pirani gauges are inexpensive, convenient and reasonably accurate (within 2% at the calibration point and 10% over their operating range).

FIGURE 6.5 | Pirani gauge Wheatstone bridge network [3]

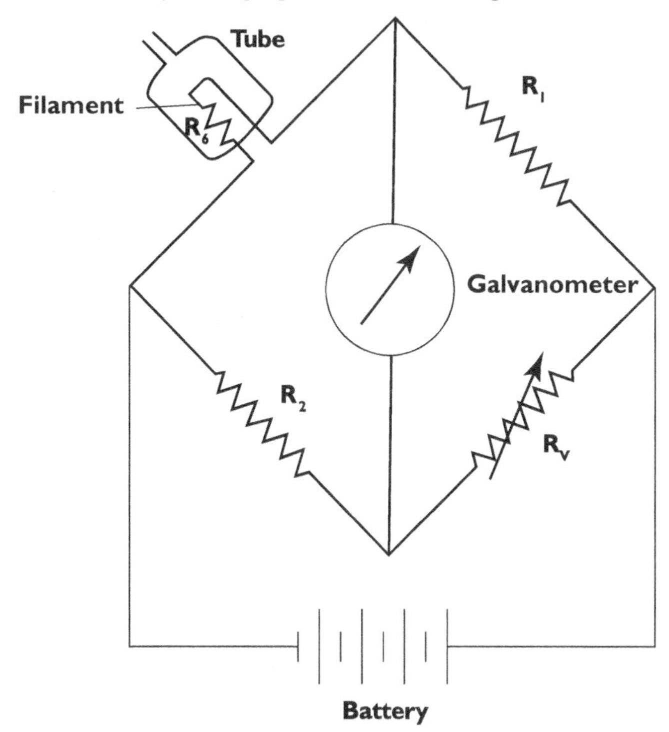

THERMOCOUPLE GAUGES

The thermocouple gauge relates the temperature of a filament in the process gas to its vacuum pressure (Figs. 6.6, 6.7). A constant current of 20-200 mA DC heats the filament, and the thermocouple generates an output of about 20 mV DC. The heater wire temperature increases as pressure is reduced.

Typical thermocouple gauges measure 10^{-3} to 2 torr. This range can be increased by use of a gauge controller with a digital/analog converter and digital processing. Using an industry-standard thermocouple sensor, such a gauge controller can extend the range of a thermocouple sensor to cover from 10^{-3} to 1,000 torr, thereby giving it the same range as a convection-type Pirani gauge but typically at a lower price.

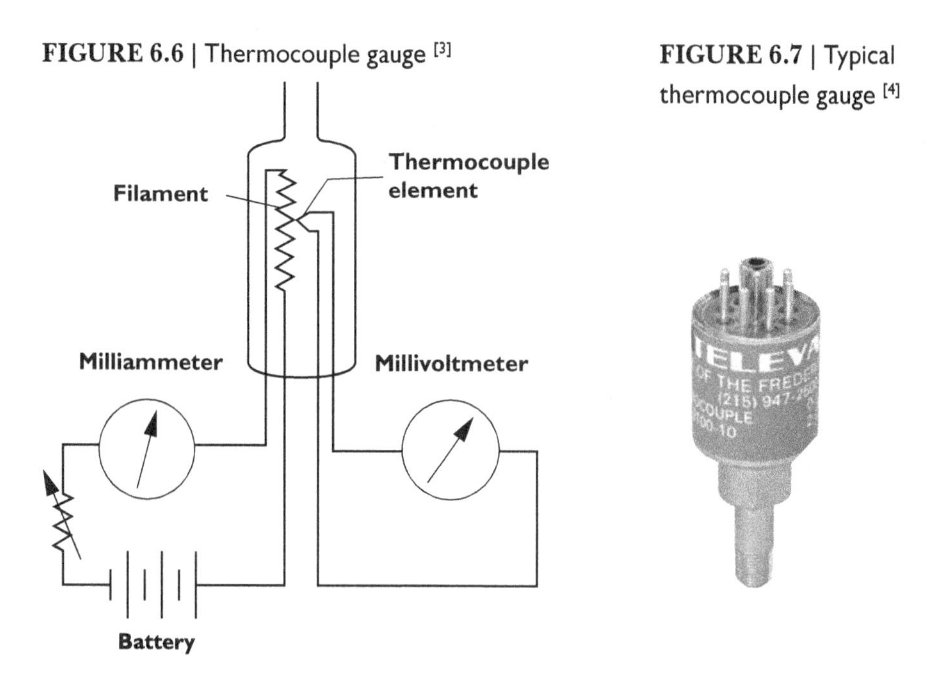

FIGURE 6.6 | Thermocouple gauge [3]

FIGURE 6.7 | Typical thermocouple gauge [4]

CONVECTION GAUGES

Similar to the Pirani gauge, this sensor uses a temperature-compensated, gold-plated tungsten wire to detect the cooling effects of both conduction and convection, thereby extending the sensing range (Fig. 6.8). At higher vacuum, response depends on the thermal conductivity of the gas, while it depends on convective cooling by the gas molecules at lower vacuums. The measurement range is from 10^{-3} to 1,000 torr. With the exception of its expanded range, features and limitations of this sensor are the same as those of Pirani and most thermocouple gauges.

The convection gauge measures absolute pressures by determining the heat loss from a fine wire filament maintained at a constant temperature. The response of the sensor depends on the gas type. A pair of thermocouples is mounted at a fixed distance from each other. One thermocouple is heated to a constant temperature by a variable-current power supply. Power is pulsed, and the temperature is measured between heating pulses. The second thermocouple measures convection effects and also compensates for ambient temperature. This sensor must be mounted vertically.

FIGURE 6.8 | Typical convection gauge [5]

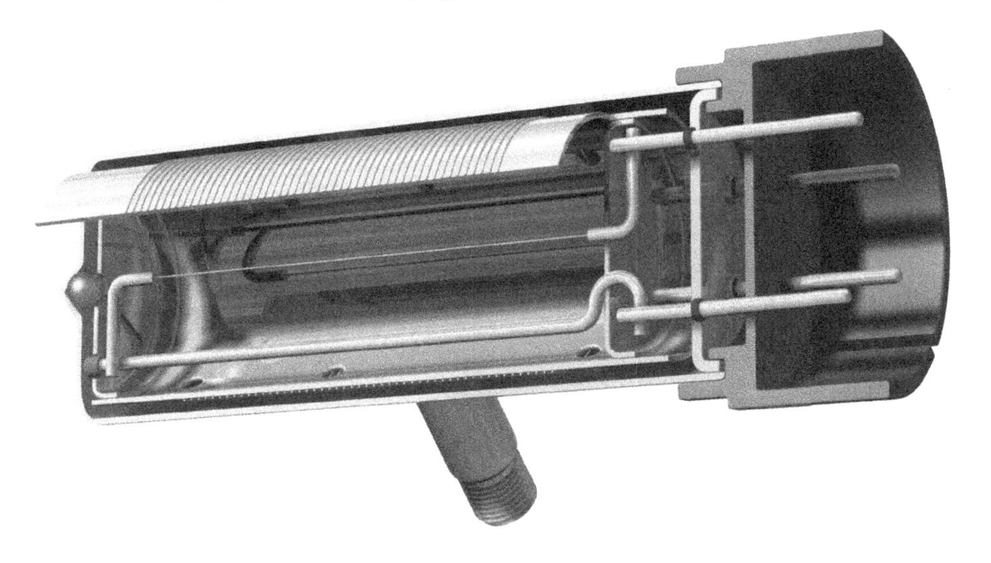

COMBINED GAUGES

To get around the range limitations of certain sensors, gauge manufacturers have devised means for electronically linking multiple sensor heads. For example, one manufacturer offers a wide-range vacuum gauge that incorporates two pressure sensors in one housing: a fast-response diaphragm manometer for measurements between 1,500 torr and 2 torr, and a Pirani gauge for measuring between 2 torr and 10^{-3} torr. The gauge controller automatically switches between the two sensors.

Ionization Gauge Designs

When fast-moving electrons pass through a gas, they can knock some of the outer electrons off of the gas molecules. The remaining part of the molecule (called an ion) then has a positive charge. The process is called ionization by bombardment. For a constant current of electrons at a given gas velocity, the rate at which these positive ions are formed is proportional to the concentration of the gas molecules (assuming the temperature remains constant). The ionization efficiency varies with the kind of gas, so these gauges must be calibrated for the gas for which they will be used. Two types are available: hot cathode and cold cathode.

Refined by Bayard-Alpert in 1950, the hot filament in the hot-cathode gauge emits electrons into the vacuum, where they collide with gas molecules to create ions (Figs. 6.9, 6.10). These positively charged ions are accelerated toward a collector, where they create a current in a conventional ion-gauge detector circuit. The amount of current formed is proportional to the gas density or pressure. Most hot-cathode sensors measure vacuum in the range of 10^{-3} to 10^{-10} torr.

FIGURE 6.9 | Hot-cathode gauge [5]

FIGURE 6.10 | Nude Bayard-Alpert ionization gauge [5]

Newer instruments extend this range significantly by using a modulated electron beam, synchronously detected to give two values for ion current. At pressures below 10^{-3} torr, there is little difference in the two values. At higher pressures, the ratio between the two readings increases monotonically, allowing the gauge to measure vacuums up to 1 torr.

Because most high-vacuum systems were made of glass in 1950, it made sense to enclose the electrode structure in glass. Today, however, a modern vacuum system may be made entirely of metal. One argument in favor of this is that glass decomposes during routine degassing, producing spurious sodium ions and other forms of contamination. Nevertheless, glass gauges remain the most popular choice for hot-cathode sensors.

COLD-CATHODE GAUGE

This gauge (also called a Philips gauge or Penning gauge) is based on the glow discharge that occurs in gas at low pressures in the presence of a magnetic field. Electrons that originate in one of the cathodes in the gauge do not go directly to the anode (because of the magnetic field). Instead they travel back and forth in helical paths between the cathodes multiple times before striking the anode. The increased path length provides a high probability of ionization (even at low gas pressures where a glow discharge normally does not occur). The total discharge current (negative and positive ions) is used to measure the pressure.

The major difference between hot- and cold-cathode sensors is in their methods of electron production. In a cold-cathode device, a high potential field draws electrons from the electrode surface. In the Philips design (Figs. 6.11-6.13), a magnetic field around the tube deflects the electrons, causing them to spiral as they move across the magnetic field to the anode. This spiraling increases the opportunity for them to encounter and ionize the molecules. Typical measuring range is from 10^{-10} to 10^{-2} torr. The main advantages of cold-cathode devices are that there are no filaments to burn out, they are unaffected by the inrush of air, and they are relatively insensitive to vibration.

FIGURE 6.11 | Cold-cathode gauge [2]

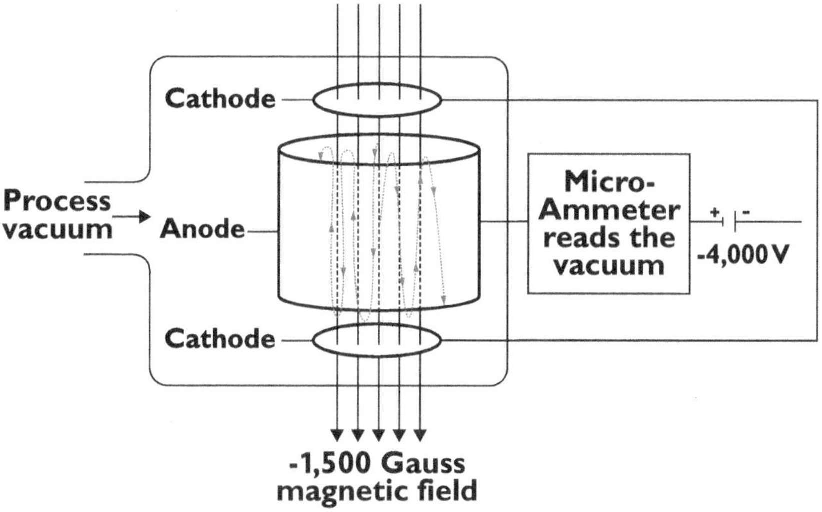

FIGURE 6.12 | Cold-cathode gauge [4]

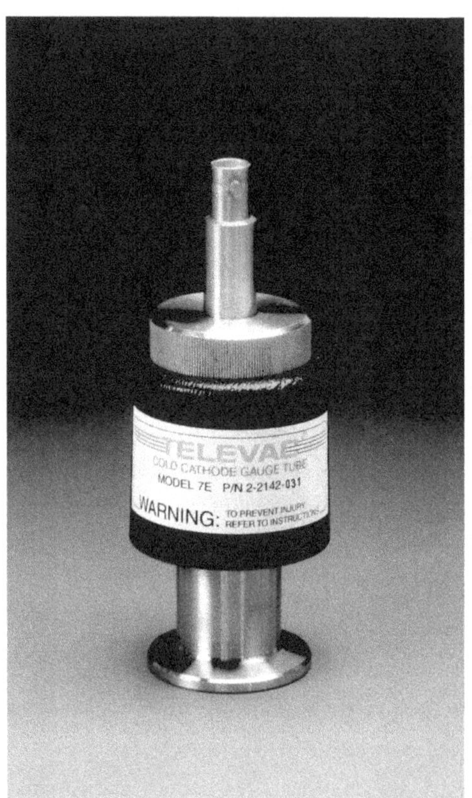

FIGURE 6.13 | Cross-sectional view of a cold-cathode gauge [4]

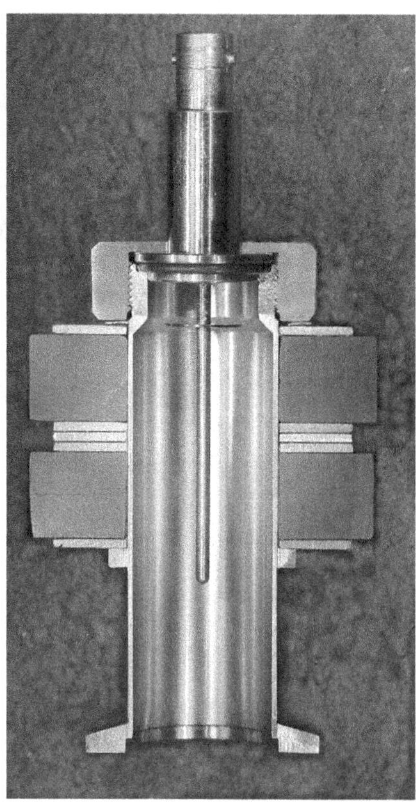

Limitation of Gauges

No discussion of vacuum gauges would be complete without mentioning the accuracies (and inaccuracies) of these devices. The most precise and only absolute gauge available is the McLeod gauge. It reads, and can only read, the pressure of the non-compressible gases at the inlet (which may or may not represent the true pressure inside the vessel). In the hands of a skilled operator, using dry gases, the McLeod gauge can be used as a calibrating device for other gauges. Thermocouple and Pirani gauges are dependent on proper calibration for their accuracy, and any contaminants introduced into these gauges will cause errors.

Ionization gauges, in addition to calibration issues, deal with linear and non-linear response ranges. In most cases, it is necessary to calibrate these gauges on a calibration manifold in the range of 10^{-4} torr and to assume that the decline in current will be essentially linear down to 10^{-8} torr. If any contamination is present, this assumption is invalid.

It is also worth mentioning that all gauges read the pressure in contact with their sensing elements. Again, this may not represent the true pressure in the vacuum vessel. For a newly calibrated thermocouple-type gauge, reading pressures between 0.001-0.013 mbar (1-10 microns) at an accuracy of 10% is typical. However, a similar gauge after just a few weeks of use may have deteriorated to a point where the accuracy is hardly a decade. Similarly, an ionization gauge can be expected to have an accuracy of about 25% of reading. Finally, at pressures below 10^{-7} torr, gauges begin to lose accuracy due to other effects, including connections, location and other considerations.

In Conclusion

There you have it – everything you wanted to know about vacuum gauges. The secret to success is having accuracy and repeatability so as to ensure that the process being run stays under control and calibration procedures in routine use.

REFERENCES

1. Edelmann, C., "Pressure Measurement – Total Pressures," Otto von Guericke University
2. Herring, D.H., "The Molecule Counters, Part Two: More About Vacuum Gauges," *Industrial Heating,* December 2008; original artwork reproduced courtesy of Omega Engineering Inc., www.omega.com
3. Brunner Jr., William F. and Thomas H. Batzer, *Practical Vacuum Techniques,* Robert E. Krieger Publishing Company, 1974
4. Televac, A Fredericks Company, Huntingdon Valley, Pa.
5. Brooks Automation, Granville-Phillips Product Center, Longmont, Colo.
6. Kimball, William H., *Vacuum...is it really nothing?*, C.I. Hayes Inc., 1977
7. *Webster's Ninth New Collegiate Dictionary*, Merriam-Webster Inc., Springfield, Mass., 1987
8. Avallone, Eugene A. and Theodore Baumeister, *Marks' Standard Handbook for Mechanical Engineers,* 10th Edition, McGraw-Hill, 1996
9. *McGraw-Hill Concise Encyclopedia of Science and Technology*, McGraw-Hill, 1998
10. Considine, Douglas M., *Process/Industrial Instruments and Controls Handbook*, 4th Edition, McGraw-Hill, 1993
11. Considine, Douglas M. and Glenn D. Considine, *Van Nostrand's Scientific Encyclopedia,* Van Nostrand, 1997
12. Herring, D.H., "The Molecule Counters, Part One: Vacuum Gauges," *Industrial Heating,* November 2008

CHAPTER 7

VAPOR PRESSURE

All solids and liquids have a tendency to evaporate into gaseous form, and all gases have a tendency to condense back into their liquid or solid form. In other words, all materials have a characteristic vapor pressure that varies with temperature. Formally, vapor pressure is the pressure of a vapor in (thermodynamic) equilibrium with its condensed phase(s) in a closed container or vessel.

Why is vapor pressure important in vacuum systems?

In a vacuum system, we must make sure that all component parts to be heat treated as well as fixtures and furnace materials of construction subjected to a vacuum environment will not experience significant evaporation or volatilization (boiling away) of their elemental constituents at the operating temperature and pressure of the process (or at the furnace bake-out temperature). Therefore, materials of construction for vacuum systems must be carefully chosen to avoid this condition, as any breakdown of these materials will have serious consequences to the products being run. Affected parts may find their surfaces, and in some cases their chemical composition, altered. This is of special importance in brazing, where the base or filler metal may be affected, changing the braze characteristics.

Materials of Construction

Material that we readily associate with having low boiling points (e.g., water, oils, greases) can be expected to give us trouble in a vacuum environment. Surprisingly, other materials (including most forms of rubber, plastics and certain insulating materials) can also readily break down. Tables 7.1 and 7.2 show the vapor-pressure characteristics of a number of fairly common materials used in and around vacuum systems.

TABLE 7.1 | Characteristics of selected solids in vacuum [1]

Chemical or trade name	Melting point, °C (°F)	Vapor pressure at 20°C (70°F), mbar	Remarks
Glyptal, red	[a]	$< 5 \times 10^{-6}$	Used for temporary thread and joint seals
Grease, Apiezon N	43 (109.4)	10^{-3} at 200°C (392°F)	Maximum temperature 30°C (86°F)
High-vacuum grease (Dow Corning)	[a]	$< 10^{-6}$	Common vacuum-sealing component used on O-rings and similar devices
Nylon (external)	[a]	$\approx 10^{-5}$	Insulators
Oil, Apiezon J	[a]	10^{-3} at 250°C (482°F)	Moderate viscosity
Rubber, natural	[a]	$\approx 10^{-5}$	Gasket material
Rubber, synthetic	[a]	$\approx 10^{-5}$	Gasket material
Teflon		$< 10^{-6}$	Insulators and gaskets
Vacuseal	50-60 (122-140)	10^{-5}	
Viton		10^{-8}	Gasket material
Wax, Apiezon W	85 (185)	10^{-3} at 180°C (356°F)	Permanent joints; maximum temperature 80°C (176°F)

Notes:

[a] Data not available.

TABLE 7.2 | Characteristics of selected liquids in vacuum [1]

Chemical or trade name	Boiling point, °C (°F)	Vapor pressure at 20°C (70°F), mbar (mm Hg)	Remarks
Acetone (CH_3COCH_3)	56.1 (133)	246.4 (184.8)	Industrial solvent or general lab use
Alcohol, Ethyl (C_2H_5OH)	78.3 (172.9)	58.5 (43.9)	Industrial solvent, liquid for leak detection, general lab uses
Alcohol, Methyl (CH_3OH)	64.5 (148.1)	128 (96)	Industrial solvent, liquid for leak detection, general lab uses
Benzene (C_6H_6)	80.1 (176.2)	99.5 (74.6)	Industrial solvent
Water (H_2O)	100 (212)	23.3 (17.5)	

Vapor Pressure – Theory

The vapor pressure of a material is the partial pressure present in the atmosphere that surrounds it. In other words, the vapor pressure tells us how much vapor a material will produce. A high vapor pressure means that the material will readily evaporate. Every material has a characteristic vapor pressure associated with it that varies with temperature. As the temperature increases, the vapor pressure increases.

All metals evaporate as a function of temperature (first-order effect) and vacuum level (second-order effect). Equation 7.1 allows us to determine the evaporation rate, Q, and shows us that the vapor pressure/temperature relationship is nearly logarithmic.

$$7.1)\quad Q_{max} = 0.058 P_v \sqrt{\frac{M}{T}}$$

where:

Q_{max} = evaporation rate (g/cm^3-sec)
P_v = vapor pressure (torr)
T = temperature (°K)
M = molecular weight

Relationships such as Equation 7.1 allow us to create vapor pressure versus temperature curves (Figs. 29.1, 29.2; Chapter 29) for a number of common elements (metals). For example, processing aluminum, cadmium, magnesium, manganese and zinc or their alloys at temperatures as low as 400°C (752°F) may be marginal or totally impractical. This is why processing brass (a mixture of copper and zinc) is normally not done in vacuum systems. If it is, it's done at partial or slightly positive pressures. As the temperature increases, fewer and fewer materials can be run without being affected. These curves tell us that materials such as carbon, niobium (columbium), molybdenum, tantalum and tungsten are preferred for interior hot zone components on high-temperature vacuum furnaces operating up to 3000°C (5432°F).

It is important to note that alloys do not behave precisely in accordance with the curves for pure metals. The tendency in these types of systems is for the lower vapor-pressure elements (those that are more volatile) to vaporize out and deposit on components at lower temperature (e.g., cold walls, element terminals, etc.).

How can we prevent vaporization from happening?

One way to overcome the problem of vaporization is to introduce a gas partial pressure in excess of the material's vapor pressure. Different gas choices, introduction methods and controls are possible. The natural questions are how and when they should be used.

Metals tend to volatize at temperatures *below* their melting points in vacuum furnaces. Table 7.3 shows this relationship for a number of common metals. The longer parts are held at the temperature and at the vacuum level shown, the greater the loss of the metallic element by evaporation. As noted above, where the element is part of a metal alloy system, the vapor-pressure relationship will change. The total vapor pressure of the alloy is said to be the sum of the vapor pressures of each constituent times the percentage in the alloy (although this relationship has been debated by those knowledgeable in the field).

TABLE 7.3 | Selected element vapor pressures [3]

Element	Temperature (°C) at which vapor pressure (torr) is:						
	0.001	0.01	0.1	1.0	10	100	760
Al	889	996	1123	1279	1487	1749	2327
Be	1029	1212	1367	1567	1787	2097	2507
B	1239	1355	1489	1648	3030	3460	2527
Cd	220	264	321	394	484	611	765
Ca	528	605	700	817	983	1207	1482
C	2471	2681	2926	3214	3946	4373	4552
Cr	1090	1205	1342	1504			2222
Co	1494	1649	1833	2056	2380	2720	3097
Cu	1141	1273	1432	1628	1879	2207	2595
Ga	965	1093	1248	1443	1541	1784	2427
Ge	1112	1251	1421	1635	1880	2210	2707
Au	1316	1465	1646	1867	2154	2521	2966
Fe	1310	1447	1602	1783	2039	2360	2727
Pb	625	718	832	975	1167	1417	1737
Mg	383	443	515	605	702	909	1126
Mn	878	980	1103	1251	1505	1792	2097
Hg	18	48	82	126	184	216	361
Mo	2295	2533	2880	3102	3535	4109	4804
Nd	1192	1342	1537	1775	2095	2530	3090
Ni	1371	1510	1679	1884	2007	2364	2837
Pd	1405	1566	1759	2000	2280	2780	3167
P	160	190	225	265	310	370	431
Pt	1904	2090	2313	2582	3146	3714	3827
K	161	207	265	338	443	581	779
Re	2790	3060	3400	3810			5630
Rh	1971	2149	2358	2607	2880	3392	3877
Se	200	235	280	350	430	550	685
Si	1223	1343	1585	1670	1888	2083	2477
Ag	936	1047	1184	1353	1575	1865	2212
Na	238	291	356	437	548	696	914
S	66	97	135	183	246	333	444
Ta	2820	3074	3370	3740			6027
Sn	1042	1189	1373	1609	1703	1968	2727
Ti	1384	1546	1742	1965	2180	2480	3127
U	1730	1898	2098	2338			3527
V	1725	1888	2079	2207	2570	2950	3527
Zn	292	343	405	487	593	736	907
Zr	1818	2001	2212	2459			3577

Chromium is an example of an element that will vaporize in an intermediate vacuum level during the processing of stainless and tool steels (or more exotic alloys). In vacuum, the chromium present in these materials evaporates noticeably at temperatures and pressures within normal heat-treatment ranges. Processed above 990°C (1815°F), chromium will vaporize if the vacuum level is less than 1 x 10^{-4} torr and parts are held for a prolonged time. Heat treaters often observe a greenish discoloration (chromium oxide) on the interior of their vacuum furnaces or on their parts, which is the result of chromium vapor reacting with residual air or water vapor in the hot zone. Otherwise, the evaporation deposit is bright and mirror-like. To avoid this, an operating partial pressure between 1 and 5 torr is recommended for most chromium-bearing parts.

TABLE 7.4 | Maximum vapor pressure for Type 304 stainless steel at 815°C (1500°F) [4]

Element	Weight percent	Maximum vapor pressure	
		mm Hg	microns
Carbon	0.07	6 x 10^{-32}	
Manganese	1.33	1.85 x 10^{-4}	18.5
Silicon	0.46	6.37 x 10^{-7}	0.06
Chromium	19.12	7.56 x 10^{-6}	0.76
Nickel	9.62	8 x 10^{-8}	0.008
Copper	0.25	2.30 x 10^{-8}	0.002
Iron	69.11	7 x 10^{-8}	0.007

For vacuum brazing (silver, copper, nickel), depletion of the filler-metal alloy can be avoided by raising the pressure in the furnace to a level above the vapor pressure of the alloy at brazing temperature. For example, copper having an equilibrium vapor pressure of 1 x 10^{-3} torr at 1120°C (2050°F) is usually run at a partial pressure between 1 and 10 torr. Nickel brazing is normally done in the 10^{-3} to 10^{-4} range. In the 10^{-5} to 10^{-6} torr range, however, you run the risk of losing some of the nickel, which has an equilibrium vapor pressure of 1 x 10^{-4} torr at 1190°C (2175°F).

When vacuum furnaces are used for brazing operations, even at intermediate temperatures, selective volatilization must be taken into account. Besides copper, manganese and lithium (often added as a scavenging element) are particularly troublesome. In vacuum, lithium should be avoided or the system should be run at high partial pressures in the order of 1 to 10 torr.

Vacuum acts as a reducing agent as shown in Table 7.5 so that many of the common oxides present break down without the use of a reducing gas such as hydrogen (Figs. 29.3, 29.4; Chapter 29). For example, iron and copper oxides break down at relatively low combinations of pressure and temperature, while other oxides (such as those of aluminum, calcium, magnesium and silicon) will not break down at all under any reasonable combination of pressure and temperature.

TABLE 7.5 | Vacuum level vs. dew point [5]

Vacuum level		Dew point °C (°F)	Water vapor (ppm)
mm Hg	microns		
1×10^{1}	1×10^{4}	+11 (+52)	13,200
1	1×10^{3}	-17 (+1)	1,320
1×10^{-1}	1×10^{2}	-40 (-40)	132
1×10^{-2}	1×10^{1}	-58 (-73)	13.2
1×10^{-3}	1	-74 (-102)	1.32

Which gases can we use?

Argon, hydrogen and nitrogen are the most common partial-pressure gases. Argon is often preferred as it tends to "sweep" the hot zone – that is, the heavy molecule tends to reduce evaporation as compared to nitrogen or hydrogen. Specialized applications such as those in the electronics industry may use helium or even neon (if an ionizing gas is needed). Gases with a *minimum* purity of 99.99% and a dew point of -60°C (-75°F) or lower should be specified.

Certain cautions are in order. For example, nitrogen may react with certain stainless steels or titanium-bearing materials, resulting in surface nitriding. In the case of hydrogen, the normally near-neutral vacuum atmosphere can be sharply shifted to a reducing atmosphere to prevent oxidation of sensitive process work or for furnace/fixture bake-out/cleanup cycles. Embrittlement by hydrogen is a concern for certain materials (e.g., Ti, Ta).

Measurement and Control

It is critical to know the exact pressure, flow and type of gas being injected into the vacuum furnace so that the process being run is under control. Thermocouple gauges typically found on vacuum furnaces are affected by gas species (since they are calibrated for air). It is not uncommon to believe, for example, that you are running an argon partial pressure at 1 torr when in reality you are running at 0.4 torr; or with hydrogen (or helium) that you are at 10 torr when you are really at 1 torr. Absolute-pressure gauges such as the MKS Baratron® should be used to determine precise partial-pressure values.

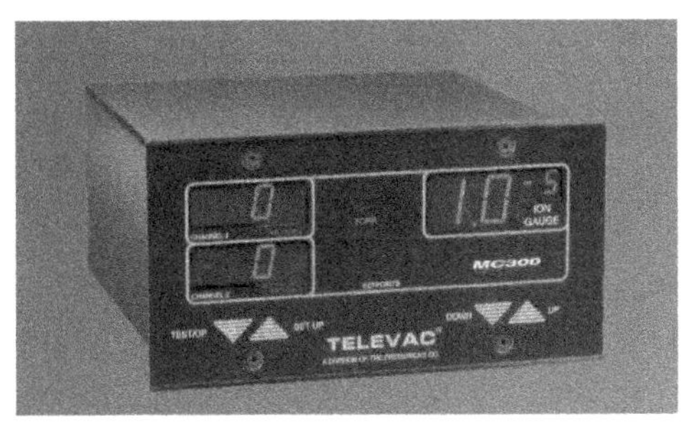

FIGURE 7.1 | Two-channel vacuum and partial-pressure controller (courtesy of Televac, a Fredericks Company)

For flow accuracy, flow meters should have a micrometer needle valve installed in the downstream line and a pressure regulator upstream set to the pressure calibration of the flow meter. On many units, the gas is pulsed in using a solenoid valve and setpoint control on the vacuum gauge, akin to continuous flow with a needle valve installed. Also, it is extremely important to inject the partial-pressure gas directly into the hot zone so that the gas does not short-circuit the work area.

Summing Up

Partial-pressure atmospheres are required in many heat-treating and brazing operations to achieve the results we expect. Introduction of the partial-pressure gas into the furnace hot zone at one or more locations and controlling the partial-pressure injection gas stream as a continuous flow rather than trying to operate at a specific pressure are critical considerations. The choice of partial-pressure gas is also important from both a cost and quality standpoint.

REFERENCES

1. *Practical Vacuum Systems Design*, The Boeing Company
2. *Training Manual, Vacuum Brazing and Heat Treating*, VAC AERO International, Inc.
3. Herring, D.H., "Using Partial Pressure in Vacuum Furnaces," *Industrial Heating*, November 2005
4. *The Nature of Vacuum*, SECO/WARWICK Corporation
5. *Vacuum Furnace Training Manual*, Ipsen
6. Jones, William R., "Partial Pressure Vacuum Processing – Parts I and II," *Industrial Heating*, September/October 1997
7. Fabian, Roger, ed., *Vacuum Technology: Practical Heat Treating and Brazing*, ASM International, 1993
8. William R. Jones, Solar Atmospheres, private correspondence
9. Richard L. Houghton Jr., Hayes Heat Treating, private correspondence
10. Heidi Reimann McKenna, The Fredericks Company – Televac, private correspondence

CHAPTER 8

HOT ZONE CONSTRUCTION

The first commercial vacuum furnace was sold to industry in 1929. In these early years, vacuum furnaces were hot-wall retort designs; that is, alloy retorts placed inside atmosphere furnaces in which a vacuum was created on the retort interior. By the late 1950s, with vacuum furnaces gaining wider acceptance (particularly within the commercial heat-treatment industry), larger sizes were in demand, prompting furnace manufacturers to consider alternative designs. The early 1960s saw the introduction of the first all-graphite-felt hot zone with graphite (cloth) heating elements. This was followed a few years later by a hot zone construction consisting of a molybdenum hot face backed with Kaowool® insulation and graphite (tubular) heating elements. These early designs suffered from a combination of ills (e.g., leaky vacuum vessels, poor element life, workload contamination and contact carburization concerns) as the industry struggled to understand how vacuum furnaces needed to be built.

Construction Methods

The first all-metallic hot zone was introduced in the mid-1960s. The hot zone consisted of two to three molybdenum shields backed by two to three stainless steel shields and molybdenum heating elements. Other combinations soon followed. These designs contained no Kaowool or graphite, and by this time vacuum-vessel technology had improved with excellent vacuum tightness and good leak rates. Although all-metal hot zones appealed to metallurgists concerned with such things as (gaseous and contact) carburization (from the carbon monoxide and carbon dioxide generated when air infiltrated into the hot zone) and alloy depletion in the parts being treated, it wasn't long before the price and availability of molybdenum, (relatively) short life and high power usage had customers looking for alternatives.

Several updated versions of the all-graphite hot zone were introduced in the 1970s. One such design combined graphite-felt insulation with solid graphite elements. Another used a graphite-foil hot face backed by graphite felt then Kaowool and had molybdenum elements in the form of 360-degree bands. These designs had the insulating characteristics of a graphite furnace without the power losses associated with an all-metallic design. The Kaowool was replaced by the early 1980s, and graphite-board insulation became popular.

Today, most vacuum furnaces contain graphite-felt or graphite-board hot zones with graphoil hot faces and solid graphite (bands or hexagonal shaped) heating elements. All-metallic heating chambers are also popular in certain applications as are all fiber-lined units.

Insulation Types

The most common insulation designs and materials (Figs. 8.2-8.5) can, in general, be classified as:

- All metallic (radiation shields or shield pack)
- Combination (inner metallic shields separated or backed by ceramic or graphite insulation)
- All graphite (board, fiber, carbon/carbon composite)
- All ceramic fiber

It is important that the hot zone support structure be designed to prevent distortion of the insulation, which would cause warpage, cracking or gaps through which radiant energy can leak. The structure must be simple and allow a fastening system that avoids undue conductive heat losses while holding the assembly rigid. Hot zone superstructures can be as simple as steel expanded metal mesh or as complex as solid stainless steel enclosures, the latter having the advantage of no rusting and no subsequent outgassing. The critical factor is to help ensure proper temperature uniformity in the workload area and minimize heat loss to the shell.

Another important factor in hot zone design is thermal expansion and contraction, especially in today's high-pressure gas-quench (HPGQ) designs. The expansion rates and temperatures must be taken into careful consideration in the design stage to allow for proper clearances around element supports, nozzles or restraint systems so that the insulation remains flat with minimal buckling or cracking.

FIGURE 8.1 | Examples of hot zone designs [3]

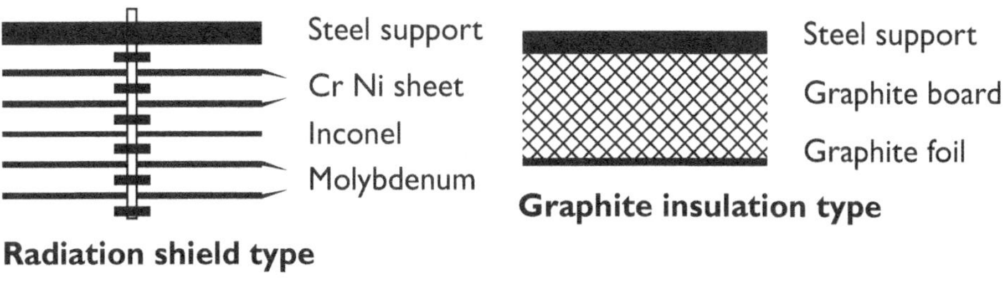

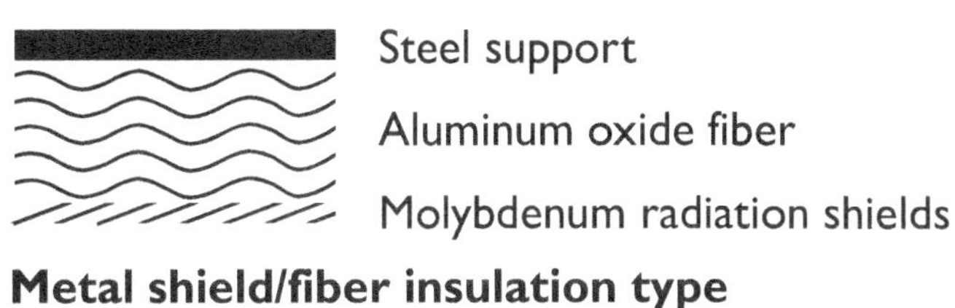

Radiation shields can be manufactured from:

- Tungsten or tantalum having a maximum operating temperature of 2400°C (4350°F)
- Molybdenum having a maximum operating temperature of 1700°C (3100°F)
- Stainless steel or nickel alloys having a maximum operating temperature of 1150°C (2100°F) with 980°C (1800°F) being a common limit

Most all-metallic designs consist of a combination of materials. For example, three molybdenum shields backed by two stainless steel shields would be typical for 1150°C (2400°F) operation. Radiation shields are made with (relatively) expensive materials and are labor-intensive to construct. When compared to other types of insulation, their heat losses are high and become higher with loss of emissivity (reflectivity) due to the gradual oxidation and contamination of the shields. Properly designed, all-metallic hot zones have two distinct advantages: surface area is small (relative to fiber insulation) so that absorbed and desorbed gases are reduced, facilitating pump-down; and heat storage is low, promoting faster cooling.

Metallic shielding is known to have specific properties (Tables 8.1, 8.2) that coexist well in a vacuum environment, including:

- Cleanliness – Refractory metals will not flake off particles that could contaminate the work or pumping system.
- Heat absorption – Refractory metals are reflective to the radiant energy, assisting in heat transfer.
- Outgassing – Refractory metals do not absorb gases as do the other materials and thus avoid prolonged pump-down times.

❖ Low heat storage – Refractory metals do not hold temperatures as long as the other materials and, therefore, allow for faster cooling.

TABLE 8.1 | Typical properties of molybdenum, tantalum and tungsten shield materials [2]

		Molybdenum	Tungsten	Tantalum
Property	Atomic number	42	74	73
	Atomic weight	95.95	183.86	180.95
	Atomic volume	9.41	9.53	10.90
	Lattice type	Body-centered cube	Body-centered cube	Body-centered cube
	Lattice constant, 20°C, A	3.1468	3.1585	3.3026
	Isotope (Natural)	92, 94, 95, 96, 97, 98, 100	180, 182, 183, 184 186	181
Mass	Density at 20° C gm/cc	10.2	19.3	16.6
	Density at 20° C lb/in³	0.368	0.697	0.600
Thermal properties	Melting point, °C	2610	3410	2996
	Boiling point, °C	5560	5900	6100
	Linear coefficient of expansion per °C	4.9×10^{-6}	4.3×10^{-6}	6.5×10^{-6}
	Thermal conductivity at 20°C, cal/cm²/cm°C/sec.	0.35	0.40	0.130
	Specific heat, cal/g/°C; 20°C	0.061	0.032	0.036
Electrical properties	Conductivity, % IACS	30%	31%	13%
	Resistivity, microhms-cm; 20°C	5.7	5.5	13.5
	Temperature coefficient of resistivity per °C (0-100°C)	0.0046	0.0046	0.0038
Mechanical properties	Tensile strength at room temperature, psi	100,000-200,000	100,000-500,000	35,000-70,000
	Tensile strength at 500°C, psi	35,000-65,000	75,000-200,000	25,000-45,000
	Tensile strength at 1000°C, psi	20,000-30,000	50,000-75,000	13,000-17,000
	Young's Modulus of Elasticity, lb/in²			
	Room temperature	46×10^6	59×10^6	27×10^6
	500°C	41×10^6	55×10^6	25×10^6
	1000°C	39×10^6	50×10^6	22×10^6
Spectral emissivity	(Wavelength approx. 0.65)	0.37 (1000°C)	0.45 (900°C)	0.46 (900°C)
Working temperature		1600°C	2800°C	2400°C
Recrystallizing temperature		900-1200°C	1200-1400°C	No recrystallization
Stress-relieving temperature		800°C	1100°C	850°C
Metallography	Etchant	Hot H_2O_2; 6% sol	HF-NH; F sol	Alk.$K_3FE(CN)$ sol
	Polishing	Alumina — rough to finish		

TABLE 8.2 | Radiant shield data for molybdenum [2]

Data element	Data value					
Furnace temperature, °C (°F)	1315 (2400)	1315 (2400)	1315 (2400)	1315 (2400)	1200 (2192)	1200 (2192)
Cold-shell temperature, °C (°F)	65 (150)	65 (150)	65 (150)	65 (150)	65 (150)	65 (150)
Number of shields (1-10)	3	4	5	6	1	2
Average shield emissivity factor (0-10)	0.60	0.60	0.60	0.60	0.60	0.60
Cold-shell emissivity factor (0.9 typical)	0.70	0.70	0.70	0.70	0.90	0.90
Computated shield temperature, °C (°F)						
#1 Shield	1230 (2247)	1250 (2282)	1263 (2305)	1271 (2320)	976 (1789)	1080 (1976)
#2 Shield	1080 (1976)	1142 (2087)	1177 (2151)	1201 (2194)		825 (1517)
#3 Shield	850 (1663)	1000 (1832)	1074 (1966)	1119 (2047)		
#4 Shield		784 (1444)	939 (1723)	1020 (1869)		
#5 Shield			734 (1354)	891 (1636)		
#6 Shield				694 (1282)		
Computated heat loss, kW/m² (kW/ft²)	43.0 (4.0)	33.4 (3.1)	28.0 (2.6)	23.7 (2.2)	78.5 (7.3)	46.3 (4.3)
Note: At 38°C (100°F), the computated heat losses are as follows:						
Computated heat loss, kW/m² (kW/ft²)	43.0 (4.3)	25.8 (2.4)	59.2 (5.5)	35.5 (3.3)	105.4 (9.8)	63.5 (5.9)

Combination (or so-called sandwich insulation pack) designs are composed of one or more radiation shields, typically with ceramic-wool insulation between or behind them. Combinations of graphite-fiber sheets and ceramic-wool insulation are also used. These versions are cheaper to buy and maintain but adsorb higher levels of water vapor and gases (due to the very large surface area of the insulation wool). Their heat losses are considerably lower than those of radiation shields. Advantages of this style include low cost, good maintainability and good insulation value. Disadvantages include a tendency for the blanket to shrink, leaving voids, which allow heat loss; dusting of the material, particularly after devitrification; and a strong tendency toward adsorption of water vapor. The systems must be supported by hangers, which project through the insulation adding to potential heat-loss problems. The maximum operating temperature of this design is approximately 1150°C (2100°F).

Graphite-fiber insulation, especially in the form of fiberboard, has very low adsorption rates, ensuring fast pump-down speeds and reduced outgassing compared to ceramic fiber. The speed at which graphite-lined hot zones reach their ultimate vacuum and life depends strongly on the purity of the graphite. Designs typically cost more than combination insulation. Advantages include ease of installation and extended life. The maximum operating temperature is usually 1800°C (3275°F), but it is dependent on the number of insulating layers. In some applications, such as brazing, a sacrificial layer (typically of graphoil or carbon/carbon composite material) is used to protect the insulation beneath.

TABLE 8.3 | Insulation ratings [1, 6]

Function (condition)	Insulation type		
	All metallic	Combination	All graphite
Outgassing	1	3	2
Heat storage	1	1	1
Maintainability	3	2	1
Cooling	2	2	1
Heat loss	3	1	2
Temperature uniformity	1	1	1
Cost	3	2	1

Key:

1 = Most acceptable
2 = Acceptable
3 = Least acceptable

It is also important to recognize, especially when doing a reline, that not all graphite-based material is the same. For example, carbon felt, which is fired at elevated temperature to produce graphite felt, has porous cells. The higher the porosity, the greater the fiber surface area, thus more air and moisture retention when chambers containing low-purity material are exposed to air. This means more outgassing and longer vacuum pump-down and degassing on heating in vacuum. The higher the firing temperatures, the better the purity and, in general, the more expensive the material.

All ceramic-fiber-lined batch and multi-chamber designs are also used in many low- and high-temperature applications. Heat loss is extremely low in these designs, and the furnaces can often be opened at temperature without damaging the insulation. For equal material thickness, Kaowool, for example, is a better insulator than graphite by perhaps as much as 20%. Vacuum furnaces with ceramic-fiber insulation can be supplied with electric elements, and certain manufacturers produce gas-fired models.

FIGURE 8.2a | Typical all-graphite hot zone: graphite board, graphite felt, curved graphite elements [5]

FIGURE 8.2b | Typical all-graphite hot zone: graphite insulation, graphite-foil hot face, curved graphite elements [5]

FIGURE 8.3 | Typical all-metal hot zone: three molybdenum shields, two stainless steel shields, molybdenum heating elements, ceramic nozzles [5]

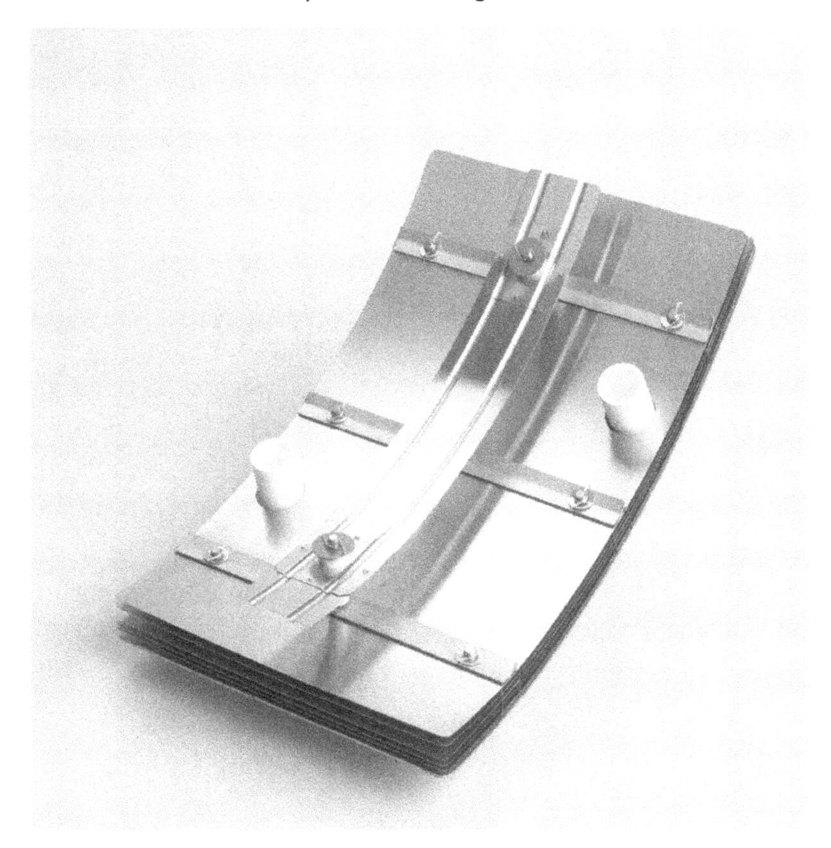

FIGURE 8.4a | Typical combination metal and ceramic-fiber hot zone: molybdenum hot face, ceramic insulation, molybdenum heating elements, molybdenum gas nozzles [5]

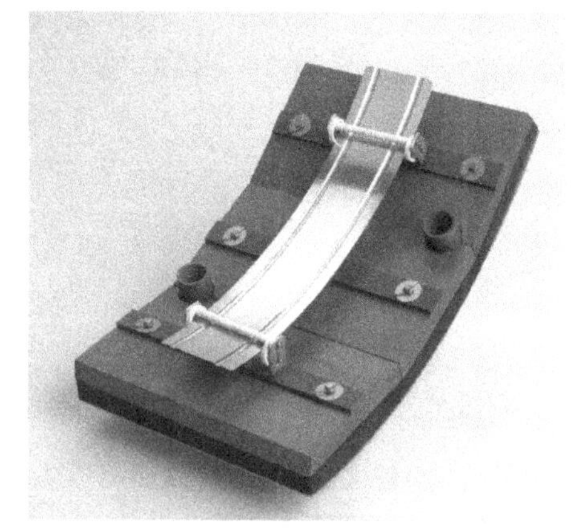

FIGURE 8.4b | Typical combination metal and graphite hot zone: graphite board, graphite felt, curved molybdenum heating elements [5]

FIGURE 8.5 | Typical all-ceramic-fiber-lined hot zone, hot-wall construction (courtesy of Surface Combustion Inc.)

Process Assurance

The process demands on vacuum furnaces are numerous – from hardening to brazing to sintering with a plethora of other thermal processes, from ultrahigh temperatures to temperatures barely above ambient, from ultrahigh vacuum to slightly above atmospheric pressure. Irrespective of how the hot zone is insulated, the goal is to conserve the loss of thermal energy to the walls of the vessel and to protect parts being processed.

Hot Zone Maintenance

As with any piece of equipment, proper maintenance at regular intervals is essential for long service life and trouble-free operation. Maintenance considerations begin with good operating (process) practices. Start by eliminating all sources of part discoloration due to air (oxygen), dirt (debris, oil, cleaning compounds) and water vapor. Simple maintenance methods yield surprisingly good results. Cleaning dirty door seals, inspecting then cleaning/re-greasing or replacing worn or cracked O-rings, avoiding broken thermocouples, and making sure fittings properly seat and seal are just a few of the steps that can be taken to minimize air infiltration and lengthen the life of hot zone components.

CONTAMINATION

In addition to these problems, vaporization of volatile elements when heated under vacuum will eventually contaminate the furnace internals with undesirable residues. To minimize this condition, techniques such as lowering ultimate vacuum levels, introducing inert gas and operating in partial-pressure ranges may be used. Dirt or debris in the form of metallic residues can cause electrical short-circuits in hot zone heating elements. Over time, a considerable amount of contaminants can build up, often with no discernible change in operating pressures until disaster strikes (Fig. 8.6). Regular bake-out cycles must be performed to burn off contaminants before they begin affecting load quality or create electrical problems. Depending on the cleanliness of the work processed previously, a bake-out cycle may also be required immediately prior to processing a critical workload or materials that are particularly prone to contamination.

FIGURE 8.6 | Debris buildup in a hot zone

SEE FULL COLOR IMAGE ON PAGE A

For example, partial-pressure systems should be set up in such a way as to avoid vaporization of elements such as chromium. It is equally important to limit the heating rate, typically no greater than 13.8°C/minute (25°F/minute). Otherwise, the result is a dramatically shortened hot zone life. Another good practice is to be aware of low melting-point eutectics that can damage the interior. For example, molybdenum has a melting point of 2610°C (4730°F), but in contact with nickel it results in a nickel-molybdenum eutectic at about 1265°C (2310°F). Air leaks are a major reason for the destruction of any hot zones and loss of efficiency in metallic-lined units. Brazing is another process that should be carefully considered. Oftentimes, having a sacrificial layer has avoided costly replacement. Oil and even cleaning chemicals can outgas and create problems.

SAFETY

There are a number of safety issues that must be considered when maintaining vacuum furnaces. Standard safety practices must be adhered to in order to avoid injury, burns and electric shock (c.f. NFPA 86). In addition to these, there are several special considerations specific to vacuum furnace equipment. Maintenance of furnace chamber internals should only be conducted using approved confined-space entry and electrical lockout procedures. Lockout procedures to prevent furnace operation must be in place before entering any furnace chamber.

OUTGASSING

Whenever the door to a vacuum-furnace chamber is open, humidity from the air (even in the dry desert) will enter the chamber and water vapor will condense on the chamber walls and be absorbed into the hot zone materials. When the chamber is subsequently evacuated (before heating) and the furnace internals are exposed to this lower-pressure vacuum, outgassing of the entrapped moisture will occur. If sufficient moisture has been entrapped (such as in very humid environments), the outgassing effect will slow the pump-down process and may even give the appearance of a malfunction in the pumping system. Eventually, the outgassed moisture will be pulled out of the chamber by the pumping system, and evacuation rates will improve.

This same effect will be apparent when oily or contaminated workloads are placed in the furnace. It may be more pronounced in furnaces with ceramic-felt or carbon-based hot zone insulation materials. To minimize the outgassing effect, it is important to keep the chamber door closed whenever possible. Ideally, the chamber should also be kept at least partially evacuated whenever the furnace is not in use. Maintaining the recommended temperature of the coolant entering the chamber cooling jacket is also important and recommended to be at least 6°C (10°F) over plant ambient temperature. Condensation of moisture is pronounced on cooler surfaces.

Final Thoughts

In operation, it is not uncommon to find either physical damage to the hot zone due to loading or contact with the insulation or gaps in joints or seams. Missing sections of insulation or puncture holes penetrating through layers of graphoil or carbon felt are common. It is important to remember that most insulation packs are fairly thin and work as designed because the insulation is intact. For best results, repairs should be instituted and training provided to avoid this type of damage.

REFERENCES

1. Jones, William R., Solar Atmospheres, technical and editorial review
2. "Designing with Refractory Metals," white paper, Rembar
3. Pritchard, Jeff, "Hot-Zone Design for Vacuum Furnaces," *Industrial Heating*, September 2007
4. Totten, George E., ed., *Heat Treatment of Steel*, 2nd Edition, Chapter 7, "Vacuum Heat Processing," 2006
5. Vacuum Furnace Systems (VFS) Corporation, product literature
6. Metalsky, William J., "Hot Zone Considerations for Vacuum Furnace Processing," *Industrial Heating*, 1993
7. *The Nature of Vacuum*, SECO/WARWICK Corporation
8. *Maintenance Procedures for Vacuum Furnaces*, VAC AERO International, Inc.
9. Moyer, Michael, "Keeping it Bright," ASM International Vacuum Maintenance Seminar, Anaheim, Calif., 2009

CHAPTER 9

HEATING ELEMENTS

Almost all high-temperature vacuum furnaces are electrically heated. Resistance heating elements are constructed from metal or graphite in a variety of styles. In general, one of the following materials is used:

- Stainless steel alloys – 300-series alloys (e.g., 304L, 316L) can be used for heating elements up to (approximately) 760°C (1400°F).
- Nickel/chromium and iron-aluminum-based alloys – These typically operate up to temperatures of 900°C (1650°F) and exhibit good-to-excellent oxidation resistance, making them useful for a number of applications, including hot-wall-type furnaces.
- Inconel® and other nickel alloys – Depending on material and vacuum level, they can be used up to 1150°C (2100°F). Above 800°C (1475°F), there is a risk of evaporation of chromium from these materials.
- Silicon carbide (SiC) – These elements have a maximum operating temperature of 1090°C (2000°F). There is a risk of evaporation of silicon at high temperatures and low vacuum levels of less than 0.133 mbar (100 microns). Silicon carbide is a somewhat brittle material even at ambient temperature.
- Molybdenum – With a maximum operating temperature of 1700°C (3100°F), molybdenum becomes brittle at high temperature and is sensitive to changes in emissivity brought about by exposure to oxygen or water vapor.
- Graphite – These elements can be used up to 2000°C (3630°F). Graphite is sensitive to exposure to oxygen or

water vapor, resulting in reduction in material thickness. The strength of graphite increases with temperature. However, graphite has limited flexibility and can break if moved.

- Tantalum – Elements made of tantalum can be used up to temperatures of 2400°C (4350°F). Tantalum is a strong getter material and becomes brittle in service, brought about by exposure to oxygen or water vapor. It is also sensitive to changes in emissivity.
- Tungsten – Elements made of tungsten have the highest duty temperature, typically 2800°C (5075°F). Tungsten becomes brittle in service, brought about by exposure to oxygen or water vapor and is sensitive to changes in emissivity.

Note: The above element ratings are downgraded from their upper operating limits (Table 9.1).

The choice of a heating-element material depends largely on operating temperature. For low-temperature operations such as aluminum brazing or vacuum tempering, inexpensive stainless steel or nickel-chromium alloys can be used for the heating elements. For higher-temperature general heat-treating applications such as hardening or brazing, molybdenum or graphite are popular choices for element materials. For specialized heat-treating applications above about 1482°C (2700°F), refractory metals such as tantalum or tungsten are popular choices, although graphite is also used. Other processes, such as low-pressure vacuum carburizing, use graphite or silicon-carbide elements.

For many years, molybdenum elements (Fig. 9.1) were used almost exclusively in vacuum furnaces for aerospace heat-treating and brazing applications. There was a widespread concern that contamination from graphite elements could react with certain gases, such as hydrogen, during partial-pressure or quenching operations and contaminate certain materials. In addition, early graphite-element designs were cumbersome and limited choices to certain simple shapes. Furthermore, connections between element segments and electrical feed-throughs were prone to failure. With advances in materials, design and manufacturing techniques for graphite-based electrical products, the popularity of graphite heating elements now exceeds that of molybdenum elements in general heat-treating and brazing furnaces. The most widely used graphite-element design incorporates either lightweight, curved bands (Fig. 9.2) or segmented bars (Fig. 9.3).

FIGURE 9.1 | Curved molybdenum heating elements (courtesy of VAC AERO International)

FIGURE 9.2 | Curved graphite heating elements (courtesy of Solar Manufacturing)

FIGURE 9.3 | Segmented graphite heating elements (courtesy of Ipsen)

FIGURE 9.4 | Flat nickel-strip heating elements (courtesy of Ipsen)

Refractory metals (Mo, W, Ta) and molybdenum in particular are popular choices for heating elements. In sheet form, the watt density of radiating surface is low compared with cylindrical rod, allowing lower operating temperatures. The trade-off is in mechanical strength. As such, good supports and restraining systems are necessary. All refractory-metal heating elements also undergo changes in electrical resistance, so the design of the power system must control current during the onset of heating to avoid damage to the elements. Also, the heating rate must be limited and controlled. Metallic elements are typically available in strip, wide band, coil (wire), ribbon or rod form.

Graphite is an excellent choice for heating elements because it is lightweight, strong at temperature and stronger at higher temperature, and has a high melting point and a low vapor pressure. In addition, graphite exhibits low contact resistance at internal connections and power feed-throughs; has excellent thermal-shock properties; is not degraded by constant heating and cooling; and has a low heat-expansion coefficient. Graphite has the ability to take very high current density. Therefore, very fast ramp-up times can be achieved. Graphite elements can operate in very corrosive or aggressive atmospheres without significant degradation. The low resistivity of graphite means it requires high-current power supplies and correspondingly large feed-throughs and cables. Graphite also acts as a getter to oxygen although it is attacked and consumed in the process (forming CO and CO_2 gases). Graphite heating elements can be supplied in rod, tube, bar, plate, circular shapes or cloth form.

Carbon/carbon composite (C/C) materials can also be used as heating elements and can be made into very thin sections, typically as thin as 1 mm (0.04 inch) thick, due to their fibrous grain structure. C/C elements have higher resistance than graphite elements, allowing lower-current, higher-voltage power supplies to be used. They also have extremely low thermal conductivity, reducing heat loss.

Silicon-carbide heating elements are typically supplied in bar form (electrically heated units) or tubes (gas-fired units).

Heating elements used in vacuum furnaces are typically resistance-type and do not require the oxidation-resistant properties of their atmosphere furnace cousins. To improve operating efficiency, however, elements used in vacuum run at higher temperatures and often operate at a higher watt density (watts/mm^2 or watts/in^2) since the transfer of radiant heat energy is a T^4 relationship, which follows the Stefan-Boltzmann Law (Equation 9.1). This states that the total energy radiated per unit surface area of a black body in unit time, j* is directly proportional to the fourth power of the black body's thermodynamic (or absolute) temperature, T.

9.1) $j^* = \sigma T^4$

where σ is a constant of proportionality = $5.6704 \times 10^{-8}\ J\ s^{-1}\ m^{-2}\ K^{-4}$

It is also important that the heating elements used have a low vapor pressure (Table 9.1) to ensure long life at elevated temperature.

Depending on the design of the hot zone (cylindrical or rectangular), heating elements can be placed circumferentially in a 360-degree pattern on just the two sidewalls, on the top and bottom as well as the sidewalls, or with the use of end elements on all six sides.

TABLE 9.1 | Characteristics of heating elements used in vacuum furnaces [2]

Material	Melting point	Upper operating limit	Vapor pressure	
	°C (°F)	°C (°F)	@1600°C (2910°F)	@1800°C (3270°F)
Graphite	3700 (6700)	2500 (4530)	10^{-13}	10^{-10}
Iron	1535 (2800)		10^{-1}	1
Molybdenum	2625 (4760)	1700 (3100)	10^{-8}	10^{-6}
Tantalum	2996 (5425)	2500 (4530)	10^{-11}	10^{-9}
Tungsten	3410 (6170)	2800 (5070)	10^{-13}	10^{-11}

Other Considerations

The use of low voltage in the vacuum chamber prevents short-circuiting of the heating elements due to ionization from residual gases within the chamber.

Uniformity of temperature is also of great importance to heat-treatment results. The construction of the heating system should be such that temperature uniformity in the load during heating is optimal. It should be better than ±5.5°C (±10°F) after temperature equalization. This is realized with single- or multiple-temperature control zones and a continuously adjustable supply of heating power for each zone.

In the lower-temperature range below 850°C (1550°F), the radiant heat transfer is low and can be increased by convection-assisted heating. For this purpose, after evacuation, the furnace is backfilled with an inert gas up to an operating pressure of 1-2 bar, and a built-in convection fan circulates the gas around the heating elements and the load. In this way, the time to heat different loads, especially those with large cross-section parts to moderate temperatures, can be reduced by as much as 30-40%. At the same time, the temperature uniformity during convection-assisted heating is much better, resulting in less distortion of the heat-treated part.

Maintenance

Graphite heating elements should not be exposed to air until the element temperature is less than 150°C (300°F). Elements that have been attacked by air exhibit a classic "sugar cube" surface appearance and often lose cross-sectional area, causing resistance change and (over) heating issues. In general, graphite elements cannot be repaired if broken, damaged or severely attacked by oxygen or other contaminants. A new element section must replace damaged sections. Elements can be wiped off and cleaned of debris using a Shop-Vac®.

Molybdenum and other metallic-type heating elements can be repaired as long as no more than three repairs should be attempted on any one heating element (Figs. 9.5, 9.6).

FIGURE 9.5 | Replacing a molybdenum element section at or near a standoff (per U.S. Patent No. 6,023,487; February 2000)

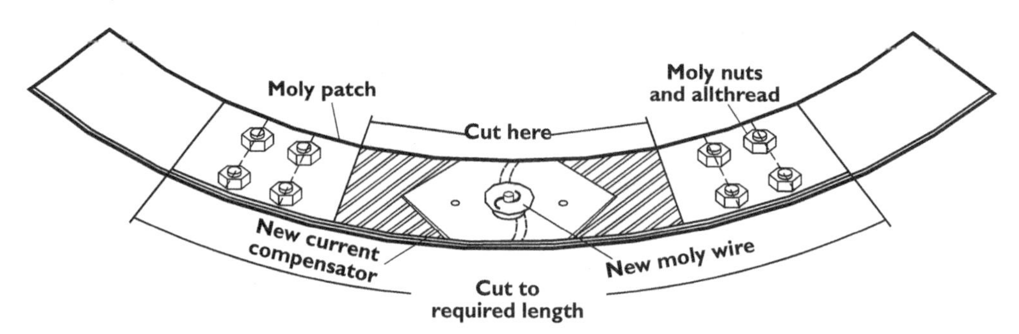

FIGURE 9.6 | Replacing a molybdenum element damaged between standoffs (courtesy of Vacuum Furnace Systems Corp.)

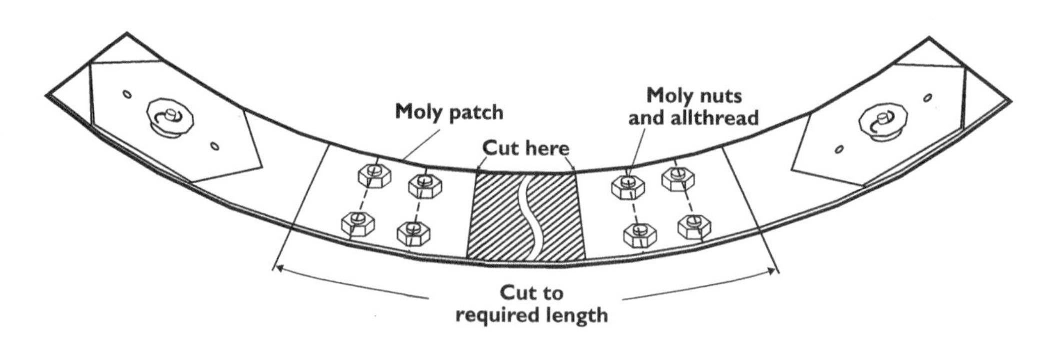

REFERENCES

1. Pritchard, Jeff, "Hot-Zone Design for Vacuum Furnaces," *Industrial Heating*, September 2007
2. Byrnes, Edward R. and Roger C. Anderson, *Heat Treating in Vacuum Furnaces*, 1983
3. Craig, Richard A., "Vacuum Furnace Maintenance," Vacuum Heat Treating SME Technology and Applications Conference, February 1999
4. William R. Jones, U.S. Patent No. 6,023,487

CHAPTER 10

PARTIAL PRESSURE, MEAN FREE PATH AND RELATED TOPICS

One of our goals in vacuum-furnace processing is to minimize both alloy depletion from the part surface and subsequent hot zone contamination. Many of the materials we run are processed at temperatures and pressures at which individual elements can volatilize (i.e. leave the part surface). Partial-pressure systems (Fig. 10.1) are designed to prevent this from happening by establishing a combination of pressure-temperature-time that minimizes the vaporization of the more volatile alloy constituents.

What is partial pressure?

We know from our school days that all gases are made up of molecules that are in constant (and random) motion. The Kinetic Theory of Gases tells us that a gas exerts a pressure on the walls of its enclosure because of the impact of these molecules on the vessel walls. These impacts tend to be elastic in nature, hurling the gas molecules back from the wall with the same speed with which they struck. The direction of the rebound, however, is independent of the direction of the impact, the angular direction of which can be expressed by Knudsen's cosine law (the intensity of the reflected molecular beam is proportional to the cosine of the angle between the molecule's path and the direction normal to the surface).

In simplest terms, therefore, the partial pressure of a gas introduced into a vacuum furnace is the force exerted by the gas (or gases) constrained in the vacuum vessel. If only a single gas is present, the partial pressure of the system is the same as the total pressure. Air, meanwhile, is a good example of a multi-gas system. The atmospheric pressure of air is 760 torr (760 mm Hg) at sea level, while it is only 483 torr (Table 10.1) at an altitude of 3,657 meters (12,000 feet).

TABLE 10.1 | Partial pressure of individual gases present in air [4]

Constituent of air	Partial pressure at sea level, torr	Partial pressure at 3,657 meters (12,000 feet), torr
Nitrogen (79%)	600.4	381.6
Oxygen (20.5%)	155.8	99.0
Other gases (0.5%)	3.8	2.4
Total pressure	760	483

In vacuum systems, partial pressure typically indicates the operation of a vacuum furnace at or above 0.10 torr (100 microns). The chamber is usually evacuated to a higher vacuum level first, commonly between 10^{-3} and 10^{-5} torr (0.1-0.01 microns). Then an inert gas is introduced at a controlled rate to a fixed partial-pressure range and controlled within this range. The high-vacuum portion of the pumping system is usually isolated and bypass circuitry employed using the mechanical pump so that a continuous flow of gas is introduced equal to the pumping capacity (throughput) at the required operating pressure.

FIGURE 10.1 | Dual (two-gas) partial-pressure circuitry on the side of a vertical vacuum furnace (courtesy of VAC AERO International)

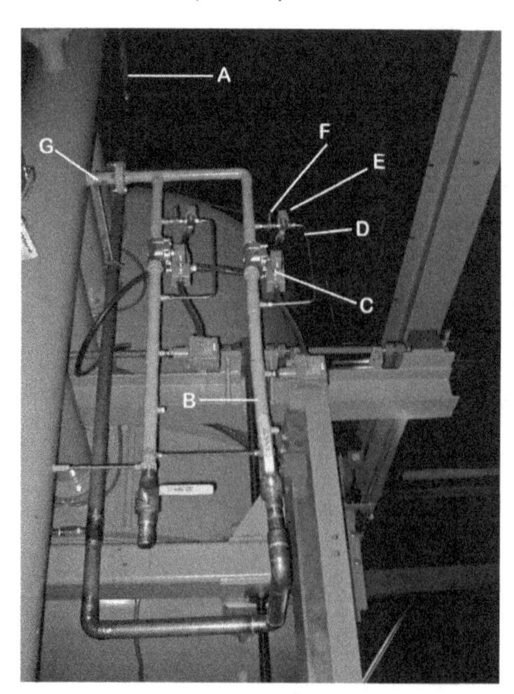

Key
A: Incoming gas supply line
B: Backfill line No. 1
C: Quench solenoid
D: Partial-pressure line
E: Partial-pressure solenoid valve
F: Partial-pressure (micrometer) needle valve
G: Inlet into furnace

Mean Free Path

We touched upon a related topic in Chapter 1 – that of the mean free path of a gas. The mean free path is a function of the gas pressure in the vacuum vessel (Tables 10.2, 10.3).

TABLE 10.2 | Estimated mean free path of gases as a function of pressure, cm (inches)

Gas species	Symbol	Mean free path @ 26.6 Pa (0.2 torr)	Mean free path @ 2.67 Pa (0.02 torr)
Helium	He	8.5×10^{-2} (3.34×10^{-2})	8.0×10^{-1} (3.15×10^{-1})
Hydrogen	H_2	5.5×10^{-2} (2.17×10^{-2})	4.5×10^{-1} (1.78×10^{-1})
Oxygen	O_2	3.0×10^{-2} (1.18×10^{-2})	2.4×10^{-1} (9.45×10^{-2})
Nitrogen	N_2	2.6×10^{-2} (1.02×10^{-2})	2.3×10^{-1} (9.06×10^{-2})
Carbon dioxide and water vapor	CO_2 / H_2O	1.1×10^{-2} (4.33×10^{-3})	1.6×10^{-1} (6.30×10^{-2})
Mercury and chlorine	Hg / Cl	5.0×10^{-3} (1.96×10^{-3})	1.2×10^{-1} (4.72×10^{-2})
Ethanol	C_2H_5OH	9.0×10^{-3} (3.54×10^{-3})	9.5×10^{-2} (3.74×10^{-2})

Note: Values are approximate.

The mean free path is the average distance a molecule travels between successive collisions with other molecules in a gaseous state (Table 10.4). The mean free path establishes the type of gas flow that will occur in the system. The mean free path is given by the following formula (Equation 10.1):

10.1) $$L_x = 0.0085 \frac{n}{p}\sqrt{\frac{T}{M}}$$

where:

- L_x = Mean free path (cm^2)
- n = velocity (poise)
- p = pressure (microns)
- T = temperature (°K)
- M = molecular weight (amu)

TABLE 10.3 | Mean free path of air molecules in 1 cm^3 (0.06 in^3) at various pressures [4]

Pressure, torr	Number of molecules
760 (atmospheric)	3×10^{19}
1×10^{-3}	4×10^{13}
1×10^{-9}	4×10^{7}

TABLE 10.4 | Mean free path at various pressures [4]

Pressure, torr	Mean free path, cm (inches)
760 (atmospheric)	6×10^{-6} (2.36×10^{-6})
1	5×10^{-3} (1.96×10^{-3})
1×10^{-3}	5 (1.96)
1×10^{-9}	5×10^{6} (1.96×10^{6})

Knowing the mean free path allows one to know, or at least estimate, the type of flow that is occurring (Table 10.5). A useful formula for the mean free path in air at room temperature is shown in Equation 10.2:

10.2) $L = 5/p$

where:

L = Mean free path (cm^2)

p = Pressure (microns)

TABLE 10.5 | Average mean free path vs. type of flow [4]

Average mean free path, cm [a]	Type of flow
$L_a/a < 0.01$	Viscous
$L_a/a > 1.00$	Molecular
$0.01 < L_a/a < 1.00$	Transitional

Note:

[a] where:

L_a = Average mean free path (cm^2) based on the average pressure in a flow system

a = Dimension that is characteristic of the opening through which the gas is flowing or the dimensions of the vacuum enclosure

Practical Partial-Pressure Ranges for Common Steels

What follows is a general guideline for partial-pressure settings used in industry for processing some of the more common materials. From a practical standpoint, there are two process considerations for determining partial pressure.

The first is the metal-oxide reduction partial pressure. The partial pressure of oxygen at a given temperature determines the direction of the reaction and consequently whether the part is "bright" or "discolored" (oxidized). These values are typically in the range of 10^{-6} torr to 10^{-2} torr. This is why materials such as titanium alloys and superalloys must be processed at extremely low vacuum levels. The second consideration is the vaporization of metal at high temperature and hard vacuum. The metal solid-to-vapor partial pressures require higher pressures to avoid alloy depletion. These higher pressures often produce sufficient dilutions of contaminants to drive the reaction to be reducing.

What is often overlooked or misunderstood is that higher levels of partial pressure "dilute" any oxygen or water-vapor partial pressure and can produce oxide-free "bright" parts at higher pressures. (This dilution also occurs when a retort is purged with nitrogen or argon to achieve clean processing.) The oxygen partial pressure is reduced by dilution rather than by vacuum. In addition, it can't be overstated that the oxidation on parts from exposure to the atmosphere and moisture absorbed by the furnace lining when the door is open is critical in the processing. Oxidation occurs on heat-up, and when the temperature is high enough to reverse the oxidation reaction, the parts will clean up. This is why it is harder to bright temper than to bright harden. In batch vacuum furnaces, combination hardening and tempering cycles are used to take advantage of not removing the parts from the furnace. The same parts will often discolor if tempered in the same furnace after they have been removed and the furnace is exposed to air.

Also, a thorough understanding of the required component properties and material characteristics (e.g., alloy composition, grain size, hardenability response) is needed to design the final vacuum heat-treat cycles and select the final partial-pressure settings.

For example, stainless steels, tool steels and more exotic alloys are run in a vacuum furnace that will benefit from the use of partial-pressure atmospheres. Chromium present in these materials evaporates noticeably at temperatures and pressures within normal heat-treatment ranges. Processed above 990°C (1800°F), chromium will vaporize as a function of both vacuum level and time – a vacuum level of no better than 1 x 10^{-4} torr (0.1 microns) being typical. Thus, the practical operating vacuum level for most materials is significantly above the equilibrium vapor pressure (Table 10.6). It is also helpful at times to know the temperature at which individual elements exceed a critical (10^{-6}g/cm^2-s) vaporization rate (Table 10.7).

In addition, heat treaters often observe a greenish discoloration (chromium oxide) on the interior of their vacuum furnaces, which is the result of chromium vapor reacting with air leaking into the hot zone. Otherwise, the evaporation deposit is bright and mirror-like. To avoid this, an operating partial pressure between 1-5 torr (1,000-5,000 microns) is typical for many chromium-bearing parts.

TABLE 10.6 | Equilibrium vapor pressure of various elements [6]

Element	Temperature, °C (°F), at a vacuum level of 10^{-4} mm Hg (0.1 micron)	Temperature, °C (°F), at a vacuum level of 10^{-3} mm Hg (1 micron)	Temperature, °C (°F), at a vacuum level of 10^{-2} mm Hg (10 microns)	Temperature, °C (°F), at a vacuum level of 10^{-1} mm Hg (100 microns)	Temperature, °C (°F), at a vacuum level of 760 mm Hg (760,000 microns)
Carbon	2290 (4150)	2470 (4480)	2680 (4858)	2925 (5299)	4825 (8721)
Chromium	990 (1818)	1090 (1994)	1205 (2201)	1340 (2448)	2480 (4500)
Cobalt	1360 (2484)	1495 (2721)	1650 (3000)	1830 (3331)	
Iron	1195 (2183)	1310 (2390)	1450 (2637)	1600 (2916)	2735 (4955)
Manganese	790 (1456)	880 (1612)	980 (1796)	1020 (1868)	2150 (3904)
Molybdenum	2100 (3803)	2295 (4163)	2530 (4591)	3010 (5448)	5570 (10,056)
Silicon	1115 (2041)	1220 (2233)	1340 (2449)	1485 (2705)	2290 (4149)
Tungsten	2765 (5013)	3015 (5461)	3310 (5988)		5575 (10,701)
Vanadium	1585 (2887)	1725 (3137)	1890 (3430)	2080 (3774)	

TABLE 10.7 | Critical vaporization rate temperatures for common elements in a vacuum environment [7]

Element	Temperature, °C (°F)
Chromium	980 (1796)
Zirconium	1620 (2950)
Platinum	1705 (3100)
Rhodium	1750 (3180)
Molybdenum	2095 (3800)
Carbon	2330 (4220)
Tantalum	2595 (4700)
Tungsten	2705 (4900)

STAINLESS STEELS

Common vacuum-processed stainless steels include the 300 series (austenitic grades), which are annealed or subjected to a stress-relief cycle after hardening by cold working, and the 400 series (martensitic grades), which are vacuum hardened (Fig. 10.2). For either process (except low-carbon and stabilized grades), a rapid gas quench is necessary to prevent carbide precipitation that will lead to a loss of corrosion resistance. Quenching with nitrogen at pressures of 1-6 bar is typical for these alloys, depending on the mass and cross (ruling) section of the workload.

TABLE 10.8 | Practical vacuum heat-treating temperatures and partial-pressure ranges for various 300-series stainless steels

Material	Typical process temperature, °C (°F)	Partial-pressure range, torr (microns) [a, b]	Process or alloy remarks
302	1010 (1850)	1 x 10^{-3} (1)	Annealing
303	1040 (1900)	1 x 10^{-3} (1)	Annealing
304	1010 (1850)	1 x 10^{-3} (1)	Annealing
304L	1010 (1850)	1 x 10^{-3} (1)	Annealing
316	1010 (1850)	1 x 10^{-3} (1)	Annealing
321 [c]	955 (1750)	1 x 10^{-5} (0.01)	Annealing
347 [c]	1010 (1850)	1 x 10^{-5} (0.01)	Annealing

Notes:

[a] For annealing operations performed at 1040°C (1900°F) and above, evacuation to an initial vacuum level below 10^{-4} torr range followed by a partial-pressure (argon) range of 1 torr (1,000 microns) is recommended.

[b] To produce "bright" stainless steel surfaces, vacuum levels in the range of 10^{-4} torr are commonly used.

[c] High vacuum in the range of 10^{-5} torr are typical for these stainless steel grades in order to produce "bright" work.

TABLE 10.9 | Practical vacuum heat-treating temperatures and partial-pressure ranges for various 400-series stainless steels

Material	Typical process temperature, °C (°F)	Partial-pressure range, torr (microns) [a, b]	Process or alloy remarks
410	1010-1065 (1850-1950)	1 x 10^{-1} (100) to 1 (1,000)	Hardening
420	980-1065 (1800-1950)	1 x 10^{-1} (100) to 1 (1,000)	Hardening
440A	1010-1065 (1850-1950)	1 x 10^{-1} (100) to 1 (1,000)	Hardening
440B	1010-1065 (1850-1950)	1 x 10^{-1} (100) to 1 (1,000)	Hardening
440C	1010-1065 (1850-1950)	1 x 10^{-1} (100) to 1 (1,000)	Hardening

Notes:

[a] Preheating is suggested for all grades.

[b] To produce "bright" stainless steel surfaces, vacuum levels in the range of 10^{-4} torr are commonly used.

FIGURE 10.2 | 410 stainless steel housings; hydrogen partial pressure 1 torr (1,000 microns) at 1010°C (1850°F) (courtesy of VAC AERO International)

TABLE 10.10 | Practical vacuum heat-treating temperatures and partial-pressure ranges for various precipitation-hardening steels and stainless steels

Material	Typical process temperature, °C (°F)	Partial-pressure range, torr (microns) [a, b]	Process or alloy remarks
17-4	480 (900)	1×10^{-5} (0.01)	Age hardening
17-7	565 (1050)	1×10^{-5} (0.01)	Age hardening
A286	720 (1325)	1×10^{-5} (0.01)	Age hardening

Notes:

[a] The higher vacuum range is required for aging of most precipitation-hardening alloys, especially if a significant amount of time has elapsed between solution treatment and aging (film and oxide formation may result in discolored parts).

[b] To produce "bright" stainless steel surfaces, vacuum levels in the range of 10^{-5} torr are commonly used.

SUPERALLOYS

Superalloys fall into two general categories: nickel-based and cobalt-based. Most of these alloys are hardened by solution treating in vacuum and then subjected to age hardening. Solution treating involves austenitizing at high temperature followed by gas quenching. In most cases, a very fast quench speed is not required due to the large amount of alloying present, and gas quenching with nitrogen at pressures of 2 bar or less is usually sufficient. This is followed by reheating to an intermediate temperature for extended periods of time. Certain superalloys require other special treatments to develop required properties.

TABLE 10.11 | Practical vacuum heat-treating temperatures and partial-pressure ranges for various superalloys

Material	Typical process temperature, °C (°F)	Partial-pressure range, torr (microns) [a, b]	Process or alloy remarks
Inconel 600	1120 (2050)	1 x 10^{-4} (0.1) to 1 x 10^{-1} (100)	Annealing; nickel-based alloy
Inconel 718	720 (1325)	1 x 10^{-4} (0.1) to 1 x 10^{-3} (1)	Nickel-based alloy
Hastalloy X	1175 (2150)	1 x 10^{-4} (0.1) to 1 x 10^{-1} (100)	Nickel-based alloy

Notes:

[a] For cobalt-based superalloys, partial-pressure settings, when allowed by specification in the 1 x 10^{-4} torr to 1 x 10^{-1} torr (0.1-100 micron) range, are typical.

[b] In typical industry practice, the furnace is pumped down into the 10^{-4} torr range followed by use of a partial pressure of nitrogen in the range of 1 torr (1,000 microns).

FIGURE 10.3 | Knee implants (cobalt-chromium-molybdenum alloy) vacuum heat treated under an argon partial pressure at 1 torr (1,000 microns) to prevent elemental evaporation (courtesy of Solar Atmospheres Inc.)

HIGH-STRENGTH STEELS

High-strength, low-alloy steels in the 41xx and 43xx series are hardened by either oil quenching or high-pressure gas quenching. Modified versions of 4340, up to a critical section size of approximately 65 mm (2.5 inches), can be hardened by gas quenching with nitrogen at pressures in the range of 6-12 bar. 41xx steels having lower hardenability are usually oil quenched if maximum mechanical properties are required.

TABLE 10.12 | Practical vacuum heat-treating temperatures and partial-pressure ranges for various alloy steels

Material	Typical process temperature, °C (°F)	Partial-pressure range, torr (microns) [a, b]	Process or alloy remarks
4140	860 (1575)	1×10^{-2} (10) to 2×10^{-1} (200)	Hardening
4340	815 (1500)	1×10^{-2} (10) to 2×10^{-1} (200)	Hardening
52100	845 (1550)	1×10^{-2} (10) to 2×10^{-1} (200)	Hardening

Notes:

[a] Partial-pressure ranges are not as critical for alloy steels as with other materials.

[b] In typical industry practice, the furnace is pumped down into the 10^{-4} torr range followed by use of a partial pressure of nitrogen in the range of 1 torr (1,000 microns).

TOOL AND HIGH-SPEED STEELS

Most tool and high-speed steels are hardened in vacuum furnaces and gas quenched. The cooling rate is dependent on the grade. Generally, the air-hardening grades (e.g., A, D, S and H series) can be consistently hardened by nitrogen gas quenching at pressures in the 2-10 bar range. Other steels (e.g., O and W series) are typically oil quenched. High-speed steels (e.g., M and T series) usually require high-pressure quenching in the 5-12 bar range to develop full properties.

TABLE 10.13 | Practical vacuum heat-treating temperatures and partial-pressure ranges for various tool steels

Material	Typical process temperature, °C (°F)	Partial pressure, torr (microns) [a]	Process or alloy remarks
A2	970 (1775)	1×10^{-4} (0.01) [c]	Hardening
A6	860 (1575)	1×10^{-4} (0.01) [c]	Hardening
A7	955 (1750)	1×10^{-4} (0.01) [c]	Hardening
D2 [b]	1010 (1850)	1×10^{-4} (0.01) [c]	Hardening
D4	955 (1750)	1×10^{-4} (0.01) [c]	Hardening
D7 [b]	1080 (1975)	1×10^{-4} (0.01) [c]	Hardening
H11 [b]	1010 (1850)	1×10^{-4} (0.01) [c]	Hardening
H12 [b]	1010 (1850)	1×10^{-4} (0.01) [c]	Hardening
H13 [b]	1010 (1850)	1×10^{-4} (0.01) [c]	Hardening
H14	995 (1825)	1×10^{-4} (0.01) [c]	Hardening
H21 [b]	1095 (2000)	1×10^{-4} (0.01) [c]	Hardening
O1	800 (1475)	1×10^{-4} (0.01) [c]	Hardening
O2	800 (1475)	1×10^{-4} (0.01) [c]	Hardening
O6	790 (1450)	1×10^{-4} (0.01) [c]	Hardening
O7	845 (1550)	1×10^{-4} (0.01) [c]	Hardening
S1	940 (1725)	1×10^{-4} (0.01) [c]	Hardening
S2	900 (1650)	1×10^{-4} (0.01) [c]	Hardening
S5	900 (1650)	1×10^{-4} (0.01) [c]	Hardening

Notes:

[a] Partial-pressure values at the high end of the range are typically used in industry for A series and D series.

[b] For hardening operations performed at 1010°C (1850°F) and above, evacuation to an initial vacuum level below 0.05 torr (50 microns) followed by use of a partial pressure of nitrogen in the range of 1×10^{-1} torr (100 microns) to 3.5×10^{-1} torr (350 microns) is typical industry practice.

[c] Vacuum levels in the 10^{-4} torr range are often run (if a diffusion pump is available) or in a partial pressure in the range of 1 torr (1,000 microns) using nitrogen.

TABLE 10.14 | Practical vacuum heat-treating temperatures and partial-pressure ranges for various high-speed steels

Material	Typical process temperature, °C (°F)	Partial-pressure range, torr (microns) [a, b]	Process or alloy remarks
M1	1220 (2225)	1×10^{-2} (10) to 5×10^{-1} (500)	Hardening
M2	1220 (2225)	1×10^{-2} (10) to 5×10^{-1} (500)	Hardening
M3 type 1	1220 (2225)	1×10^{-4} (0.01) [b]	Hardening
M3 type 2	1220 (2225)	1×10^{-4} (0.01) [b]	Hardening
M10	1175 (2150)	1×10^{-4} (0.01) [b]	Hardening
M42	1175 (2150)	1×10^{-4} (0.01) [b]	Hardening
T1	1290 (2350)	1×10^{-4} (0.01) [b]	Hardening
T2	1290 (2350)	1×10^{-4} (0.01) [b]	Hardening
T4	1290 (2350)	1×10^{-4} (0.01) [b]	Hardening
T5	1290 (2350)	1×10^{-4} (0.01) [b]	Hardening
T15	1230 (2250)	1×10^{-4} (0.01) [b]	Hardening

Notes:

[a] For M series and T series, partial pressure 3×10^{-1} torr to 5×10^{-1} torr (300-500 microns) is required to minimize effects of diffusion bonding (part sticking).

[b] Vacuum levels in the 10^{-4} torr range are often run (if a diffusion pump is available) or in a partial pressure in the range of 1 torr (1,000 microns) using nitrogen.

TITANIUM ALLOYS

Vacuum heat treating of titanium alloys is usually limited to age hardening and stress relief. The solution treatments that many titanium alloys require involve fast cooling rates only achievable by liquid quenching. Because of their propensity to react with contaminants, the heat treating of titanium alloys requires tight control of cleanliness and high vacuum levels. No partial-pressure processing is allowed. Titanium alloys will react with residual water, oxygen, hydrogen, nitrogen and carbon dioxide, producing brittle surface conditions such as alpha case, which must be removed before the part goes into service. Pre-cleaning of titanium is also critically important (even fingerprints!) before heat treatment.

OTHER MATERIALS

Many other materials can be vacuum heat treated. Even materials such as beryllium copper and brass (Fig. 10.4) can be heat treated in vacuum or in equipment employing a vacuum purge and positive pressure, + 0.5 psi (0.035 bar), provided the partial pressure suppresses zinc vaporization. (Zinc has a very high vapor pressure and is susceptible to vaporization near atmospheric pressure). Surfaces of these materials may turn from a shiny appearance to a dull finish indicative of some zinc depletion.

TABLE 10.15 | Practical vacuum heat-treating temperatures and partial-pressure ranges for other materials

Material	Typical process temperature, °C (°F)	Partial pressure, torr (microns) [a]	Process or alloy remarks
Ni Span C	400 (750)	1×10^{-4} (0.01)	Hardening
Be-Cu	315 (600)	1×10^{-4} (0.01)	Hardening
AM 350	930 (1710)	1×10^{-4} (0.01)	Hardening

Note:

[a] Vacuum levels in the 10^{-4} to 10^{-5} range are often run.

FIGURE 10.4 | Brass coils annealed under a nitrogen partial pressure at approximately 800 torr (15.5 psi) (courtesy of Solar Atmospheres Inc.)

BRAZING

For vacuum brazing (Fig. 10.5) involving silver, copper and nickel alloys, depletion of the filler-metal alloy can be avoided by raising the pressure in the furnace to a level above the vapor pressure of the alloy at brazing temperature. For example, copper has an equilibrium vapor pressure at 1120°C (2050°F) of 1×10^{-3} torr (0.1 microns), so it is usually run at a partial pressure between 1 and 10 torr (1,000-10,000 microns). By contrast, nickel brazing is normally done in the 1×10^{-4} torr (1 micron) range, because in the 1×10^{-5} to 1×10^{-6} torr (0.01 to 0.001 micron) range you run the risk of losing some of the nickel, which has an equilibrium vapor pressure of 1×10^{-4} torr (0.1 micron) at 1190°C (2175°F).

In brazing, the quality (dew point) of the gas and leak rate of the furnace (typically 10-20 microns per hour or less) are important variables to control.

FIGURE 10.5 | Typical vacuum brazing cycle

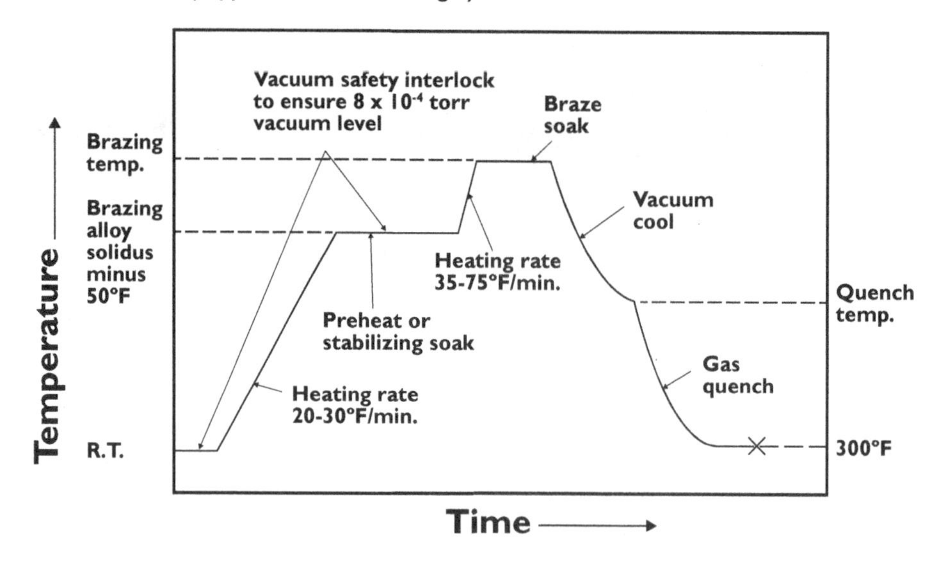

Partial-Pressure Gases

Argon, hydrogen and nitrogen are the most common partial-pressure gases. Argon is often preferred, as it tends to "sweep" the hot zone (i.e. the heavy molecule tends to reduce evaporation as compared to nitrogen or hydrogen). Specialized applications such as those in the electronics industry may use helium or even neon (if an ionizing gas is needed). Gases with a *minimum* purity and a dew point (Table 10.16) should be specified.

Certain cautions are in order. For example, nitrogen may react with certain stainless steels or titanium-bearing materials, resulting in surface nitriding. In the case of hydrogen, the normally near-neutral vacuum atmosphere can be sharply shifted to a reducing atmosphere to prevent oxidation of sensitive process work or for furnace/fixture bake-out/cleanup cycles. Embrittlement by hydrogen is a concern for certain materials (e.g., Ti, Ta).

TABLE 10.16 | Gas purity [5]

Gas	Purity (%) [a]	Oxygen content, maximum ppm	Maximum dew point, °C (°F)
Nitrogen	99.995	10 [b]	-67 (-89)
Argon	99.995	5	- 51 (-60)
Helium	99.995	5	-48 (-58)
Hydrogen	99.995	10 [c]	-68 (-90)

Notes:

[a] Most industrial gases are supplied at higher purities depending upon the supply source (cylinders, dewars, bulk storage, etc.).

[b] Typical oxygen content is 2 ppm.

[c] Typical dew point range is -68°C (-90°F) or approximately 5 ppm.

Measurement and Control

It is critical to know the exact pressure, flow and type of gas being injected into the vacuum furnace (Fig. 10.6) so that the process being run is under control. Thermocouple gauges typically found on vacuum furnaces are affected by gas species (since they are calibrated for air). It is not uncommon to believe, for example, that you are running an argon partial pressure at 1 torr when in reality you are running at 0.4 torr; or with hydrogen (or helium) that you are at 10 torr when you are really at 1 torr. Absolute-pressure gauges should be used to determine precise partial-pressure values.

For flow accuracy (as stated in Chapter 7), flow meters should have a micrometer needle valve installed in the *downstream* line and a pressure regulator upstream set to the pressure calibration of the flow meter. On many units, the gas is pulsed in using a solenoid valve and setpoint control on the vacuum gauge akin to continuous flow with a needle valve installed. It is also extremely important to inject the partial-pressure gas directly into the hot zone so that the gas does not short-circuit the work area.

FIGURE 10.6 | Partial-pressure control (broad white band) (courtesy of VAC AERO International)

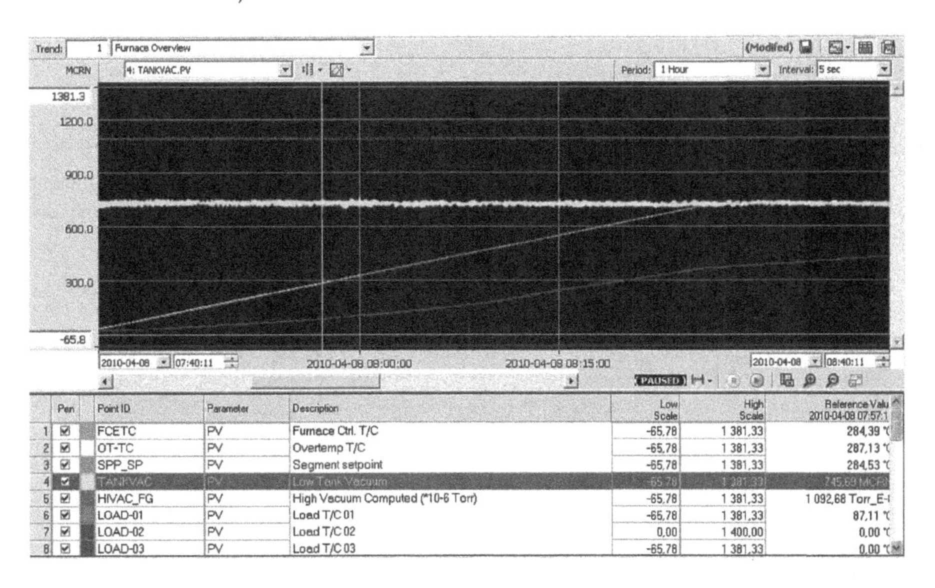

Equipment Limitations

It is worth noting that there are some constraints to the partial-pressure setpoints due to furnace design. Some semi-continuous furnaces may require limits on the partial-pressure setpoints since, for example, load-transfer doors must equalize at around 0.5 torr (500 microns). If a partial pressure greater than this is used, it can affect the function of the furnace's load-transfer system.

Another constraint on a single-chamber furnace may be the ability to measure the partial pressure if it exceeds 1 torr (1,000 microns) for, say, a copper brazing application. Many of the gauges found on these type of furnaces are calibrated in the range of 1 x 10^{-3} to 1 torr (1-1,000 microns) so that when you try to control in a setpoint range of, say, 0.9-1.2 torr (900-1,200 microns), the gauge will flash "999" (i.e. you are out of the calibrated range). Many companies in a copper brazing application, therefore, end up using setpoints of 0.7-0.9 torr (700-900 microns) so that values are recorded on the gauge or use a capacitance manometer to accurately set and control the pressure in the range of 1.3-6.7 mbar (1-5 torr).

In Conclusion

Partial-pressure atmospheres are required in many heat-treating and brazing operations to achieve the results we expect. They are necessary from both a cost and quality standpoint and are easily controlled using today's modern technology.

REFERENCES

1. Richard Houghton, Hayes Heat Treating, private correspondence
2. Norman Sousa, The Sousa Corporation, private correspondence
3. Herring, D.H., "Using Partial Pressure in Vacuum Furnaces," *Industrial Heating*, November 2005
4. *The Nature of Vacuum*, SECO/WARWICK Corporation
5. O'Hanlon, John F., *A User's Guide to Vacuum Terminology*, John Wiley & Sons, 1980
6. Ipsen, Harold N., *Fundamentals of Vacuum Heat Treating*, 3rd Edition, 1974
7. AMS 2769 Rev B (12/2009), "Heat Treatment of Parts in a Vacuum"
8. Jones, William R., "Partial Pressure Vacuum Processing – Parts I and II," *Industrial Heating*, September/October 1997
9. Danielson, Phil, "Understanding Pressure and Measurement," *R&D Magazine*, February 2003
10. Fabian, Roger, ed., *Vacuum Technology: Practical Heat Treating and Brazing*, ASM International, 1993
11. *Training Manual, Vacuum Brazing and Heat Treating*, VAC AERO International Inc.
12. Dan Kay, Kay & Associates, private correspondence

CHAPTER 11

VACUUM VALVES, PENETRATIONS, FEED-THROUGHS AND FLANGES

Valves intended for vacuum service are subjected to a variety of special conditions (Fig. 11.1), ranging from high- and ultrahigh-vacuum levels to low, high and ultrahigh pressures; differentials in pressure; and differentials in temperature as well as variable frequencies of mechanical operation. They can be supplied in a number of configurations: ball valves, gate valves, butterfly valves, needle valves, isolation valves, pressure-relief valves and control valves just to name a few. The type of valve in use is typically identified by its design or function, and each type can be actuated in a variety of ways (manually, electro-magnetically, pneumatically, electro-pneumatically or even via electric motor). Position indicators and limit switches located on the valves are common.

FIGURE 11.1 | Plethora of valves, seals, penetrations, feed-throughs and flanges on a modular quench chamber subject to a wide variety of process conditions, including transitions from negative to positive pressure (courtesy of ALD-Holcroft)

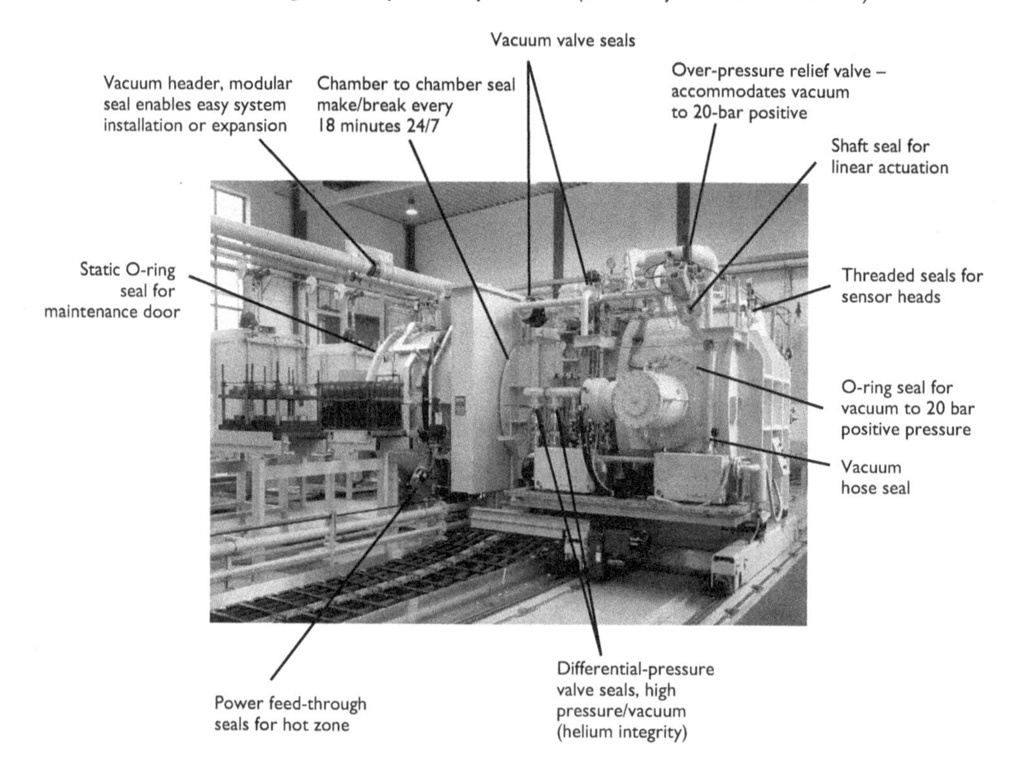

Here's a summary of the most common types of valves and how they function.

Ball Valves

APPLICATION

Ball valves (Fig. 11.2) offer straight-through, unimpeded flow with a minimal valve-body footprint. Because of their unique design, ball valves are less sensitive than other vacuum valves to particulate contamination and, therefore, are especially useful in "dirty" vacuum applications. For example, ball valves are almost always used to isolate scrubbers and traps downstream of mechanical pumping systems.

HOW THEY WORK

Ball valves are made of a body, stem, ball and end caps. The ball is sealed within the body by end caps, creating a vacuum-tight central cavity. The valve is opened and closed by turning the stem 90 degrees back and forth (1/4 turn). The sealing seat for the ball wipes the ball clean as it is opened and closed. This last feature provides the "self-cleaning" action that makes this valve fairly robust to particle-rich effluent streams. Pneumatic actuators are available.

FIGURE 11.2 | Pneumatically operated ball valve (courtesy of A&N Corporation)

Gate Valves

APPLICATION

Gate valves (Fig. 11.3) provide straight-through, unimpeded flow in relatively large diameters. These valves are characterized by their low flow resistance and compact size. They are ideal for space-constrained applications that require maximum flow conductance. Typical inside-diameter sizes range from 25-300 mm (1-12 inches) with port terminations normally available in multiple flange families (ISO, CF and ASA). These valves can come with either manual or pneumatic actuators. For ultrahigh-vacuum applications, a copper bonnet seal version with CF port terminations is typical.

HOW THEY WORK

A central carriage or gate is raised and lowered by the actuator within the body. In the open position, the gate retracts completely from the tube aperture allowing unrestricted flow. In the closed position, a sealing ring is compressed against the surface on the inside of one of the body ports. Depending on the direction of movement of the gate, a distinction is made between rebound valves, shuttle valves and rotary-vane valves. Typically, they can seal against a differential pressure in the range of 1-10 torr (1.3-13.3 mbar).

FIGURE 11.3 | Gate valve (courtesy of A&N Corporation)

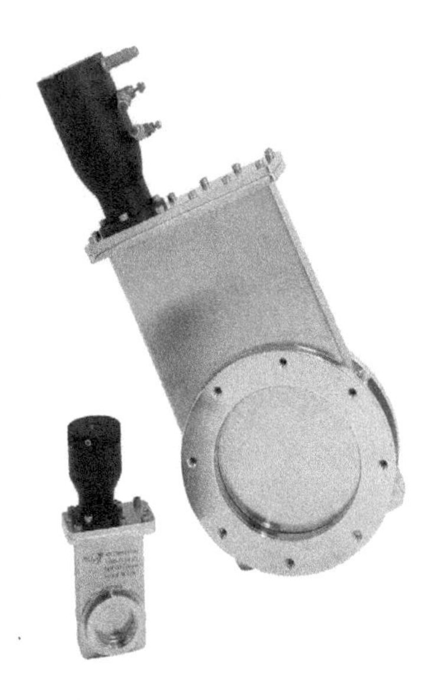

Poppet and Angle Valves

APPLICATION

Poppet valves (Fig. 11.4) are typically used for diffusion-pump vacuum sealing and pumping ports ranging in size from 152-889 mm (6-35 inches) and fitted to a suitable diffusion pump normally of equal size. These valves can operate to the 10^{-7} torr range but more typically to the 10^{-4} to 10^{-6} torr range, and they are sealed with Buna N or Viton O-ring seals for easy assembly for cleaning and service if required.

HOW THEY WORK

Poppet valves are made up of a body, actuator and internal poppet. The actuator (manual or pneumatic) transports the poppet linearly. When the poppet is moved toward the top of the valve, the internal body cavity is open to the system and flow is unimpeded (maximum conductance). Once the actuator moves the poppet to the bottom of the valve, a sealing ring is compressed onto the sealing surface of the lower body of the valve. This action creates a vacuum-tight seal and stops the flow.

FIGURE 11.4 | Poppet valve (courtesy of Solar Manufacturing)

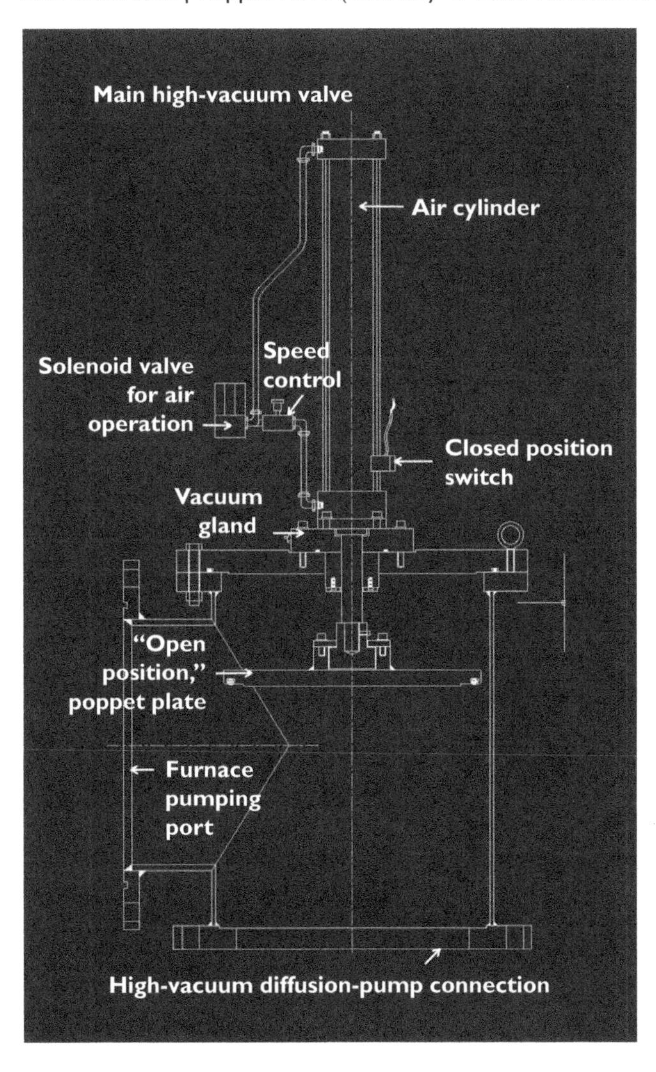

Butterfly or Plate Valves

APPLICATION

Butterfly or plate valves (Fig. 11.5) are commonly used to isolate the pumping system of a vacuum furnace from the main vessel. They are subjected to repeated cyclical operation, and their simple design ensures a high degree of reliability.

HOW THEY WORK

Butterfly or plate valves use a sealing plate that is swung open by some sort of lever or tilted open by means of a simple rotary motion, with the valve plate remaining in the valve opening. Plate valves, due to their design, are used to close very large openings.

FIGURE 11.5 | Butterfly valve (courtesy of C.I. Hayes)

Other Types of Valves

Valves used on vacuum furnaces have many special purposes, including:

- Vent valves release pressure that has built up inside a vacuum vessel in a slow and controlled manner. These valves or vent systems often include a muffler to reduce noise levels.
- Needle valves allow small and measured amounts of gas to be added into a vacuum system in a highly repeatable and precise way. Micrometer needle valves are useful in many partial-pressure systems.
- Pressure or flow valves are for manual or automatic pressure or flow regulation.
- Dump or fast-acting valves are designed to close rapidly in the event of a malfunction.
- Pressure-relief and differential-pressure valves open and close automatically within a given pressure range.

Lubrication of any valve depends on whether you are dealing with the internal (vacuum) side or the external (pressure) side. Internal lubricants must be suitable for the required pressure and temperature ranges or, if possible, avoided entirely in high- and ultrahigh-vacuum applications.

Feed-throughs

When electrical power, optical signals or mechanical movement must be transmitted from outside the vacuum vessel to inside the vacuum vessel, feed-throughs are usually involved. They are classified in the following general types.

ROTARY FEED-THROUGHS

Continuous motion and high-speed rotational movement are best conveyed through rotary feed-throughs. Application examples include fans mounted inside the heating chamber and seals for rotational part movement (Fig. 11.6).

FIGURE 11.6 | Rotary seal assembly (courtesy of Pfeiffer Vacuum)

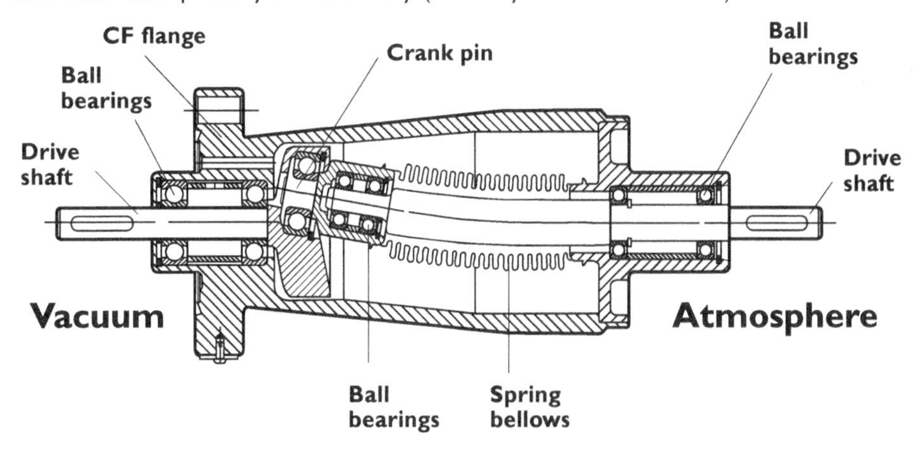

ELECTRICAL FEED-THROUGHS

Electrical feed-throughs come in all shapes, sizes and applications – from the need to pass thermocouples into the chamber to high-voltage and/or high-current power feed-throughs. The choice of electrical feed-throughs (Figs. 11.7-11.9) depends to a great extent on the amount of current and voltage that must pass through the connection. Feed-throughs with glass-to-metal seals are best suited for high-voltage and low-current devices. Ceramic insulation offers the greatest stability (mechanical, thermal).

FIGURE 11.7 | Ultrahigh-voltage ceramic-to-metal feed-through (courtesy of Pfeiffer Vacuum)

FIGURE 11.8 | Vacuum furnace (interior view) power feed-through (courtesy of C.I. Hayes)

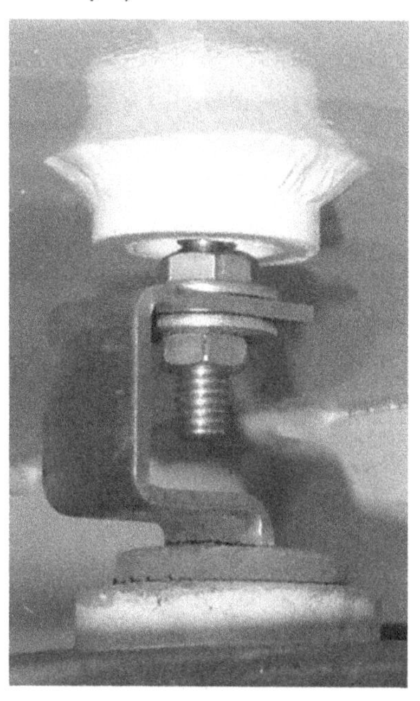

FIGURE 11.9 | Interior view showing vacuum furnace with internal electrical insulator arrangement (courtesy of C.I. Hayes)

MECHANICAL FEED-THROUGHS

Gases as well as liquids (e.g., hydraulic fluid) can be introduced into the vacuum vessel by means of either permanently welded connection points or via flange-type pipe connections.

Site Glasses

These are windows into the vacuum process and are normally positioned on the vessel so that a particular event (e.g., a load transfer from the heating chamber to the quench) can be observed. Site glasses typically consist of safety glass held in place by a flange and sealed vacuum tight by means of an "O" ring. Sight ports can also be provided with an arrangement whereby half the window is see-through glass and half is a light designed to illuminate the area under observation (Fig. 11.10).

FIGURE 11.10 | Combination site port and light illumination (courtesy of J.G. Papailias Co.)

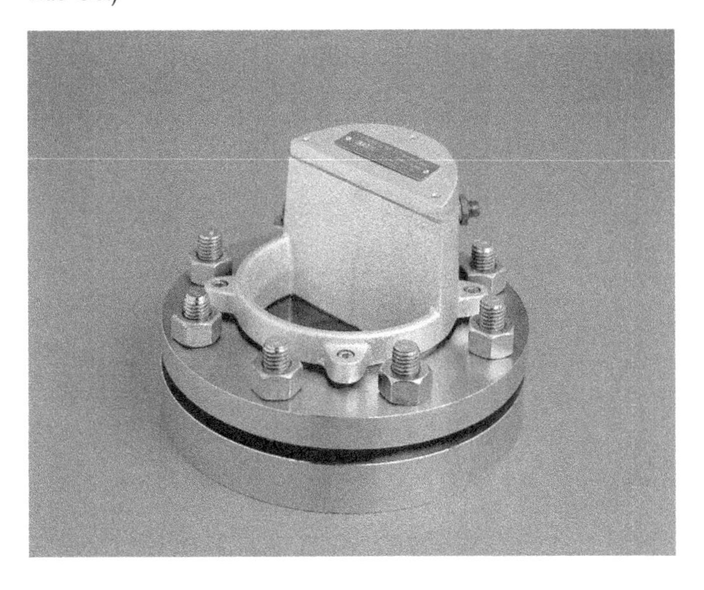

Often when observing directly into the heating chamber, an optically dense baffle of some type is provided to block line-of-sight radiation from damaging the seals holding the glass in place when not making observations.

Flanges

Two types of connections are found on vacuum equipment: detachable (flange-type) and non-detachable (connection-type). The latter are typically attached to the vacuum vessel by brazing, welding or soldering. Mechanical strength and resistance to changes in temperature and pressure are key requirements.

Detachable joints typically involve some type of fixed flange and clamping arrangement. Since leakage is a major concern, stainless steel-reinforced hoses are generally preferred over thick-walled rubber or thermoplastics.

In Conclusion

Without specialized valves, feed-throughs, flanges and penetrations, modern vacuum equipment would be plagued with both operational problems and leaks. These devices allow our vacuum systems to operate in a range that permits heat treating of today's demanding products.

REFERENCES

1. *The Vacuum Technology Book*, Volume I, Pfeiffer Vacuum
2. A&N Corporation
3. George Papailias, J.G. Papailias Co., private correspondence

CHAPTER 12

LEAK RATES, LEAK DETECTION AND LEAK REPAIR

A common problem experienced by almost every vacuum user is that, over time, leaks develop that are both damaging to product quality and to furnace internal components. In extreme cases, the problem is obvious – the furnace will not pump down and/or the hot zone (or heating elements) shows clear signs of oxidation. Small, more common leaks, however, often go undetected because the pumping system can overcome any air infiltration. But even small leaks can cause continuous and sometimes catastrophic damage. Thus, routine leak checking should become a part of any good vacuum-furnace maintenance program.

What is a vacuum leak?

A leak is an opening such as a crack or a hole that allows a substance to be admitted to or to escape from a confined space. A vacuum-system leak allows air to enter into the vacuum vessel. Suspect areas on vacuum furnaces include threaded and brazed joints, fittings that have been improperly sealed or installed, and damage (e.g., cut, worn, melted or dirty O-ring seals, especially around doors). Areas where maintenance was last performed and components that rotate or reciprocate (e.g., vacuum blowers and/or other motor-shaft seals) are other prime leak sites.

Due to the small atomic and molecular radii of the elements involved (Table 12.1), even extremely small holes can yield appreciable leak rates. These atoms are so small that leaks no larger than those caused by slight porosity in welds can cause serious trouble in high-vacuum systems.

TABLE 12.1 | Atomic and molecular radii [2]

Constituent	Symbol	Atomic radius (calculated), picometers	Atomic radius (empirical), picometers	Molecular radius, picometers
Hydrogen	H	42	50	
Helium	He	31		
Nitrogen	N	56	65	
Fluorine	F	42	50	
Carbon	C	67	70	
Chlorine	Cl	79	100	
Argon	Ar	71		
Carbon monoxide	CO			156
Carbon dioxide	CO_2			161
Freon 22	$CHClF_2$			173

What other types of leaks are there?

Failing to achieve a predetermined and acceptable leak rate, as averaged over a one-hour time period, is not necessarily an indication of a failed leak-rate test. You must determine if you are dealing with real leaks or other types of anomalies (e.g., high gas loads, outgassing, etc.).

Vacuum-system gas load results from sources such as leaks (real and internal), surface conditions (outgassing and virtual leaks) and system materials (diffusion and permeation).

FIGURE 12.1 | Gas-load sources [5]

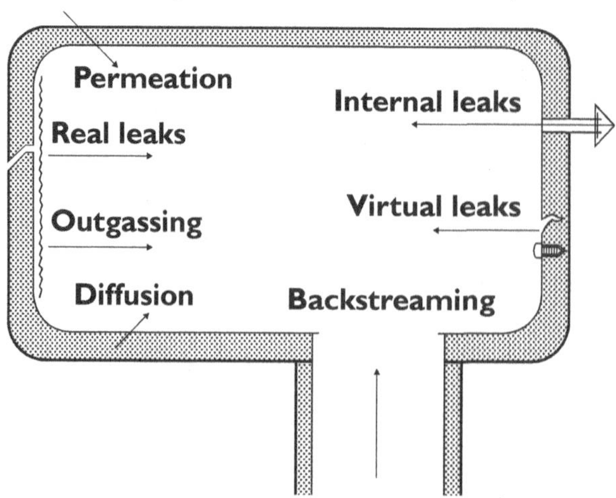

The gas load (Fig. 12.1) is simply the rate gas enters the system volume from such areas as:

- External and internal leaks
- Outgassing (i.e. gas evolving from surfaces) and/or virtual leaks

- Materials by diffusion and permeation (i.e. gas emanating from or passing through the material in question)
- Process gas flows

At a known temperature, the gas load, Q, is the amount of mass (gas) entering the system volume per unit time. Q is expressed in units of pressure times volume per unit time such as torr-liters/second, atmosphere-cubic centimeter/second, mbar-liters/second or Pascal-cubic meter/hour. Each potential leak source can be further broken down.

Sources for real leaks (Figs. 12.2, 12.3) include:

- Connections between leak detector and system
- The last component worked on
- Components that have often leaked in the past
- Seals where motion occurs along seal surface
- Sliding seals (e.g., non-bellows valve shaft seals)
- Rotating seals (e.g., fan or fixture drive seals)
- Seals on chamber doors

Sources for outgassing or virtual leaks (Figs. 12.2, 12.4) include:

- Residual solvents following wet cleaning or preventative maintenance
- Liquid leaks (e.g., cooling fluids)
- Trapped volumes of gas or liquid
- Trapped space under non-vented/non-sealed hardware
- Gases or solvents in spaces with poor conductance
- High vapor-pressure materials or products
- Porous materials exposed to liquid or atmosphere

By contrast to other leaks, a virtual leak is a source of gas molecules that is physically trapped within the chamber. As the pressure within the chamber drops during the pump-down cycle, these molecules are released in a small but steady flow into the vacuum vessel. Water vapor is a good example (which is why one should not leave a vacuum furnace door open any longer than necessary and why one should pump down the furnace as soon as practical). Water vapor present inside the furnace will slowly desorb and reabsorb on another area. Since virtual leaks tend to result in spikes of pressure, gauges and leak detectors should easily be able to sense them as non-linear signals.

Virtual leaks are most prevalent in four general areas: gaps, cracks, surface contact and trapped (gas) pockets. Good design and maintenance/repair practices can help avoid their creation.

FIGURE 12.2 | Response of outgassing and real leaks to (mechanical pump) pump-down [3]

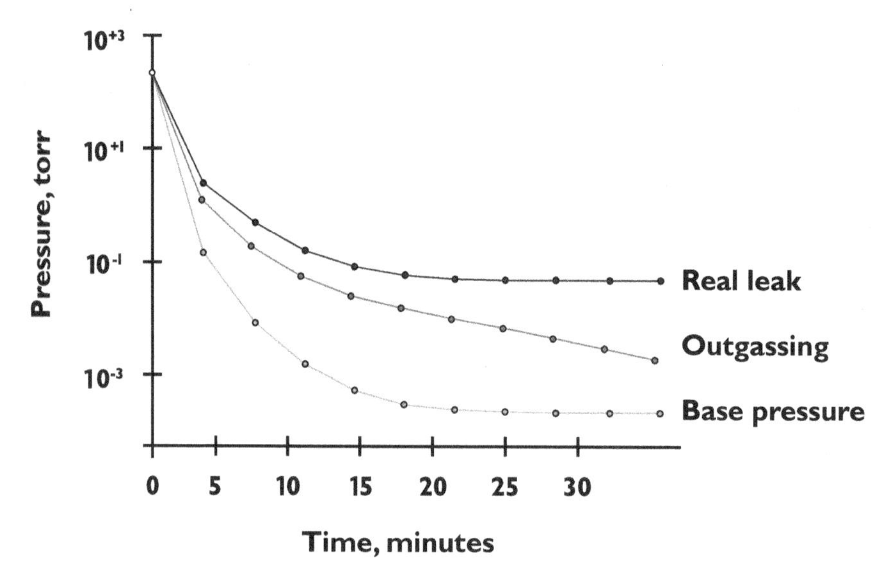

Note: The slope of the lines represents the pumping speed of the system.

Possible sources for internal leaks are:

- Process-gas delivery valves
- Vent gas valves
- Seals between adjacent internal volumes
- Transfer chamber to process chamber
- Load lock to transfer chamber
- Gas purge or ballast set incorrectly

FIGURE 12.3 | Rate of rise due to a real leak [4]

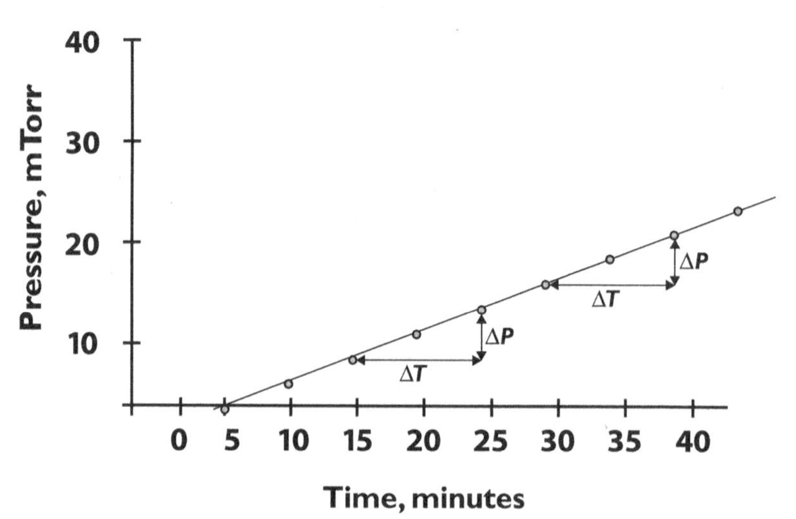

FIGURE 12.4 | Rate of rise due to an outgassing or virtual leak [4]

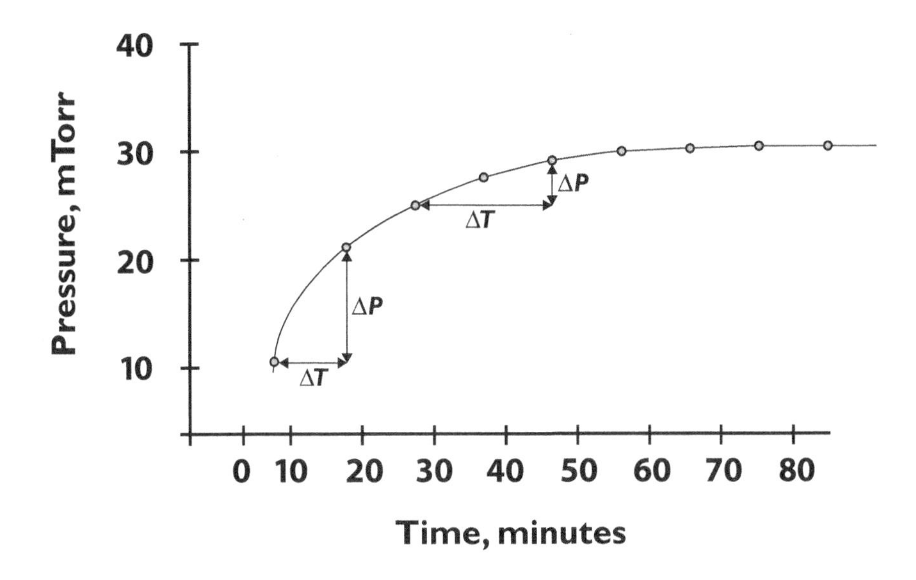

What is leak testing?

Leak testing consists of leak checks that should be done on a weekly or, in some applications, daily basis to ensure that the vacuum system is maintaining its integrity. Leak detection (or finding a leak) should be performed once it is determined that the vacuum system has lost integrity. Keeping a written record or log is an important tool because changes over time can detect problems in advance.

The vacuum level as indicated by vacuum-gauge readings is not always a true indication of the actual conditions within a vacuum furnace. It is possible to have two identical furnaces operating at the same pressure but producing entirely different results from a similar heat treatment. This is due to the relative tightness of each furnace. Most furnaces are equipped with pumping systems sufficient to overcome even reasonably significant leaks. On the furnace with the higher leak rate, air can be continuously infiltrating into the furnace, resulting in higher residual oxygen content than in a tight furnace. The higher oxygen content will adversely affect the heat-treating results, particularly at operating temperatures under 760°C (1400°F).

In general, leak testing involves measuring the amount of leakage (vacuum degradation) over time in the vacuum environment. In most shops, a leak-up rate test is performed for a period of one hour, although shorter times may be used. (Caution: Use of a 1-, 5- or 10-minute leak-up rate test may give erroneously high leak-rate values.) Remember, the leak-up rate of a vacuum system is a function of actual (real) leaks, internal (through) leaks, outgassing and virtual leaks in the chamber or vacuum system (Table 12.2).

TABLE 12.2 | Potential sources of vacuum-system leaks [5]

Type	Potential source
Actual (real) leaks	• Compression seals • Welds and brazed joints • Shaft seals and bellows on valves • Flexible connectors in piping • Threaded joints on vacuum gauge and plugs • Static gasket seals on sight ports, feed-throughs, manifolds and air-operated cylinder glands • Connection between leak detector and system • The last component worked on • Components that have often leaked in the past • Seals where motion occurs along seal surface • Sliding seals such as non-bellows valve-shaft seals • Rotating seals such as fixture drive seals • Seals on chamber doors
Internal (through) leaks	• Process-gas delivery system • Vent-gas exit system • Seals between adjacent internal volumes • Improper gas ballast, gas purge, partial-pressure control systems and backfill valves
Outgassing/ virtual leaks	• Residual solvents • Residual liquids (water, cooling fluids) in blind holes and restricted passageways • Pockets of trapped gases or liquids • High-vapor-pressure materials • Porous materials and condensed by-products from processes already run

Why do I need to leak test?

Everything leaks. And although a leak may be extremely small, it still may pose a problem. Leaks can be inherent in the material; created during the manufacturing process; introduced during maintenance or repair; or can occur over time due to wear, fatigue or stress. The source of a leak may often be revealed by answering the question, What was the last area worked on or modified (e.g., thermocouples, front-door O rings)?

However, the question that really matters is: Can the system tolerate the leak? In other words, can the process and equipment survive and be unaffected by the leak? The answer is almost always no.

Principles of Leak Testing

There are three basic types of flow-through leaks in components and vacuum systems (Table 12.3), which can be defined as follows:

- Turbulent flow – Flow in a gross leak situation (10^{-1} cc/sec and larger). Gross leaks (Table 12.3) occur, for example, in unsoldered joints, open welds, open valves and plugs left off piping. Typical methods used to find these leaks include soap solution, water immersion, ultrasonic detection and pressure decay.

- Laminar flow – Flow in small or fine leaks (10^{-1} cc/sec to 10^{-6} cc/sec). In this case, the leak is proportional to the differential pressure (i.e. doubling the pressure across the leak doubles the apparent leak rate). Generally, a change in detector gas will not affect the leak rate. Typical methods used to find these leaks include soap solution, water immersion, ultrasonic detection, pressure decay, halide torch, halogen leak detection and helium mass spectrometry.
- Molecular flow – Flow through extremely fine leaks (10^{-7} cc/sec or smaller). These occur when the mean free path of the molecule is greater than the diameter of the hole. An increase in pressure differential has no affect on the leak rate. Typical methods used to find these leaks include helium mass spectrometers, radiographic gauges and hydrogen-sensitive leak detectors.

TABLE 12.3 | Leak-rate bubble equivalents [6]

Leak rate, cc/sec	Equivalent flow (approximate)	Bubble equivalent (approximate) [a]
10^{-1}	1 cc/10 sec	steady stream
10^{-2}	1 cc/100 sec	10 per sec
10^{-3}	3 cc/hour	1 per sec
10^{-4}	1 cc/3 hours	1 in 10 sec
10^{-5}	1 cc/24 hours	
10^{-6}	1 cc/2 weeks	
10^{-7}	3 cc/1 year	
10^{-8}	1 cc/3 years	
10^{-9}	1 cc/30 years	
10^{-10}	1 cc/300 years	
10^{-11}	1 cc/3,000 years	smallest detectable leak by He mass spectrometer

Note:

[a] Assuming a bubble of 1 mm³ (6.1 x 10^{-5} in³) in volume and 1.241 mm in diameter

Both viscous and molecular flow conditions can exist in a leak situation. Near atmospheric pressure, the flow is viscous. The lower in pressure one goes, the more the flow conditions become molecular in nature. The type that is dominant will depend on the volume (quantity) of leakage.

1. If the leakage is over 10^{-5} torr-liters/second, viscous flow will predominate and the leak rates will be:

a.) Proportional to the difference between the square of the detector (tracer) gas pressure on opposite ends of the leak

b.) Inversely proportional to the detector-gas viscosity

2. If the leakage is under 10^{-7} torr-liters/second, molecular flow will predominate and the leak rates will be:

   a.) Proportional to the difference in pressure across the leak

   b.) Inversely proportional to the square root of the molecular weight of the gas

3. If the leakage is between 10^{-5} and 10^{-7} torr-liters/second, flow conditions will be in the transition region and the effect on leak rate is difficult to predict.

The most influential changes to the system pressure depend on:

- ❖ The maximum detector-gas partial pressure (after the leak is covered by the detector gas)
- ❖ The time required for the detector gas to reach its maximum partial pressure
- ❖ The time required for the partial pressure of the detector gas to drop from maximum to minimum
- ❖ The rate of increase of the detector-gas partial pressure in the vacuum system

How to Ensure the Most Accurate Leak-Test Value

For best results, leak testing should be done in a clean, cold, empty and outgassed chamber. To accomplish this once a vacuum furnace has been put into service, a furnace burnout cycle should be conducted. The purpose of a burnout is to condition the hot zone. It is important to note, however, that a burnout cycle should never be conducted in a badly leaking vacuum furnace. If oxygen is present during a burnout cycle, in the case of metallic shields, these can be damaged or destroyed by oxidation (i.e. loss of the ability to re-radiate). If the hot zone is graphite construction, the graphite can be vaporized, compromising thermal integrity. In either hot zone construction, there is a danger of metalizing the heating-element electrical insulators, which can result in short-circuits or heating-element failures. So, before a burnout cycle is run, pump on the empty furnace for approximately four hours. If the vacuum level fails to lower to acceptable limits, a leak check to find and seal gross leaks needs to be done before attempting a burnout cycle.

A furnace burnout typically involves heating the *empty* furnace to a temperature approximately 38°C (100°F) above the maximum process temperature (but less than the rated maximum operating temperature of the furnace) followed by a vacuum cool. If you are running a standard furnace at temperatures below 1090°C (2000°F), the burnout temperature should still be in the 1090-1260°C (2000-2300°F) range. Fixtures, baskets, parts and work thermocouples are removed from the furnace. A normal burnout cycle is run at the highest vacuum level possible and, in some cases (for furnaces equipped only with mechanical pumps), in partial-pressure sweep gas (e.g., nitrogen, argon, helium or hydrogen) to aid in the clean-up cycle.

A typical set of burnout instructions might be as follows (for a standard graphite or metal-lined hot zone):

1. Ramp the furnace at 10°C (20°F) to a setpoint of 38°C (100°F) over the maximum normal daily cycle operating temperatures. Do not exceed the recommended maximum operating temperature of the furnace.
2. During the ramp portion of the cycle, run a partial pressure of 1,000-2,000 microns (1-2 torr), typically with either nitrogen or argon. If a hydrogen partial pressure is used, appropriate safety considerations must be taken and the furnace temperature monitored very carefully to prevent a runaway condition due to the high heat-transfer characteristics of hydrogen.
3. Once the furnace reaches burnout temperature, the furnace should be allowed to remain in this condition for approximately two hours. At the end of the two-hour soak, the partial-pressure event should be stopped and the furnace allowed to pump down to its lowest vacuum level.
4. The furnace is then soaked at the burnout temperature for an additional one hour in the hard vacuum condition.
5. The heat is turned off and the furnace is allowed to vacuum cool to between 260-425°C (500-800°F) so that vaporization of previously heated materials can take place. *Note: To obtain the best possible leak rate, the furnace must be allowed to vacuum cool to below 50°C (120°F).*
6. Backfill and cool to room temperature. If desired, open door, inspect hot zone and elements, and check element-to-ground resistance per manufacturer's specifications.
7. When finished, the furnace can be leak tested and then placed back into production.

The leak-rate test is normally performed immediately after the bake-out cycle without opening the furnace to atmosphere. Pump the system down to normal vacuum operating pressure (or less), then isolate the pumping system from the furnace chamber. The vacuum level is recorded after 30 minutes and again after 60 minutes. The leak rate can then be calculated in microns per hour and compared to previous values or the original equipment manufacturer's requirements. Vacuum furnaces should not have leak rates exceeding 10 microns per hour at a pressure of 100 microns or less. For older furnaces, leak rates of 20-25 microns per hour are not unusual. For vacuum oil-quench furnaces, the quench tanks will typically have leak rates of 100 microns or less. These leak rates ensure that the volume of impurities that may be leaking into the furnace is sufficiently low so as not to cause any significant detrimental effects to the materials being processed.

A furnace exhibiting a leak rate greater than these limits should not be used for production until the leak is repaired. In this case, the normal procedure is to backfill the furnace with nitrogen, but do not open the chamber to atmosphere. All thermocouple fittings and other vacuum feed-throughs should be tightened. The furnace can then be re-tested for leak rate as before. Failure of the second leak-rate test is an indication that the furnace requires more extensive maintenance, possibly including helium leak checking.

Leak Testing Explained

There are many methods used to measure leak rate (Table 12.4), and there are various types of detector gases or leak-detecting (tracing) agents (Table 12.5). The choice of tracing agent usually depends on the measurement method used. Selection of the best method for a specific application requires consideration of economics, accuracy, tolerance to environmental conditions, leak-rate requirements and equipment limitations. Helium is the preferred detector gas in most heat-treat applications.

To perform a leak-up rate check, pump the furnace down to ultimate pressure with the heat turned off and the furnace cold (ambient temperature or below). This will require a minimum of one hour and in many instances up to four hours. Record the vacuum level and the time. Next, isolate the furnace from the pumping system by closing the vacuum valve(s) to the chamber. Allow at least one hour to obtain an accurate leak-up rate. (Note: This step is often shortened to only a few minutes, but this is poor practice and should be avoided.) Record the time and vacuum level. The leak-up rate is the difference in the vacuum levels divided by the elapsed time and is expressed in microns/hour (mbar/hour).

TABLE 12.4 | Different methods of leak detection [6]

Method	Minimum detectable leak, cc/sec	Probe material	Remarks
Water immersion	10^{-3}	Air, nitrogen, helium	Low equipment cost; wet product; operator judgment
Soap (liquid) application	10^{-3}	Air, nitrogen, helium	Low equipment cost; wet product; operator judgment
Dye penetrant	10^{-4}	Liquid	Low cost; may be destructive; time consuming
Pressure change – thermocouple gauge	10^{-5}	Acetone, alcohol, hydrogen	Low cost; may be destructive; residue vapors
Pressure change – ionization gauge	10^{-7}	Gaseous hydrocarbons, hydrogen, helium, oxygen, solvents	Operator judgment
Rate of rise	10^{-5}	Atmosphere	Qualitative (real or virtual leaks); time consuming
Ultrasonic	10^{-2}	Air, various gases	Remote detection; sound system; affected by background "noise"
Halogens	10^{-6}	Freon, chlorine, fluorine	Ventilation required; good for refrigerant-filled systems and devices; low/moderate cost; some judgment needed
Radioscopic	10^{-10}	Krypton – 85	Good for small sealed parts; special precautions required (radiation); destructive; high equipment cost
Helium mass spectrometer	10^{-6} to 10^{-10}	Helium	Versatile and fast; nondestructive; no operator judgment; moderate/high equipment cost

TABLE 12.5 | Detector-gas characteristics [7]

Detector gas	Molecular weight	Thermal conductivity	Viscosity	Leakage ratio
Air	29	24	1.69	
Hydrogen	2	173	0.83	3.8
Helium	4	142	1.78	2.7
Neon	20	45.5	2.94	1.2
Argon	40	16.6	2.08	0.85
Carbon Dioxide	44	14.6	1.35	0.81
Propane	44	15.3	0.77	0.81

Test Methods

There are three general categories of leak-detection procedures:

- Effect-of-leak types – pressure decay (differential, increase) and vacuum decay
- Amount-of-leak types – mass flow (inside/out, outside/in, accumulation), carrier gas, helium mass spectrometers
- Traditional types – immersion, sniffing

In some cases, residual-gas analyzers (RGA) are used as pseudo leak detectors, but these instruments suffer from sensitivity issues, are complicated to use and are typically applied only as an analytical tool to supplement helium mass spectrometers. The most common procedures for detecting leaks in vacuum furnaces are mass spectrometers and, in some instances, the solvent procedures.

TOTAL PRESSURE MEASUREMENT

It is often possible to use the existing vacuum gauges on a system as a basic leak-detection system. However, the change in gauge indication is subject to a number of conditions that can cause the gauge to either increase or decrease in its value. These conditions include: the change in leak rate when the leak is covered by the detector gas; the change in gauge response to the detector gas (compared to its response to air); and the difference in pumping speed of the detector gas (compared to the speed for air).

For thermal-conductivity gauges, the most effective technique is to use two detector gases – one that causes an increase in gauge response and one that causes a decrease. The procedure is to alternate the gases and measure the degree of gauge response. This difference can be as high as a factor of five. Pirani gauges are preferred over thermocouple-type for this type of leak detection. For ionization gauges, the technique is similar, with hot-cathode gauges preferred over cathode models.

SOLVENTS

The use of a solvent test method is simple but effective for locating gross leaks (i.e. leaks considered in the intermediate and large size ranges). If a thermocouple gauge is connected on the pump side of the system such that it reads manifold pressure and the system can be evacuated to a range of at least 200 microns (0.2 torr), then a solvent such as alcohol (preferred) or acetone can be carefully sprayed on a suspect area and any change in vacuum level inside the chamber, based on the vacuum-gauge reading, noted.

In using this method, one must be careful to allow enough time (up to 20 seconds) for a pressure increase to occur. This procedure is more sensitive at lower pressures. Solvent checking is typically used to enable the system to be evacuated into the range where a mass spectrometer instrument can be used to check for smaller leaks. Be sure to observe all required safety precautions when using hazardous solvents, including proper ventilation and spill containment. Remember also that these solvents may remove paint!

If a leak is located, a temporary sealant such as Glyptal® red alkyd lacquer (Glyptal Inc.), Kinseal clear vacuum sealant (Kinney Vacuum Div., Tuthill Corp.), vacuum seal putty or wax can be used to patch the area and allow leak checking to continue. A common mistake is to neglect to permanently fix the problem after testing is completed. A misconception is to seal such a leak in a vacuum furnace running at 2 bar or above, allowing the gas pressure to break the seal. In these instances, the component must be taken apart and fixed.

RESIDUAL-GAS ANALYZERS

An RGA can be used to find both small and large leaks in a vacuum system as well as differentiate between water vapor, nitrogen and air leaks. It is a very powerful, albeit expensive, tool since it uses a quadrapole mass analyzer, which requires separate dry pumping along with a turbomolecular pump and throttling manifolds to connect it to the vacuum system. Once the mass scan is run, it can be compared with later scans to see which peaks have increased. The RGA cannot measure the leak in a quantitative way since it is very difficult to determine the speed of the leak. RGAs locate leaks by applying any gas to a small area on the outside of the vessel and looking for an increase in partial pressure of that particular gas species.

HELIUM MASS SPECTROMETERS

A helium mass spectrometer (Fig. 12.5) is a highly accurate instrument for locating leaks in hard-to-reach areas. In some instances, it is necessary to "bag" or isolate a specific area on the furnace and inject helium into the contained space. This dynamic, nondestructive technique is sensitive enough to check parts that have moving seals or those that may leak only during the transition from pressure to vacuum (or vice versa).

FIGURE 12.5 | Portable helium mass spectrometer with remote-control capability (courtesy of Agilent Technologies) [2]

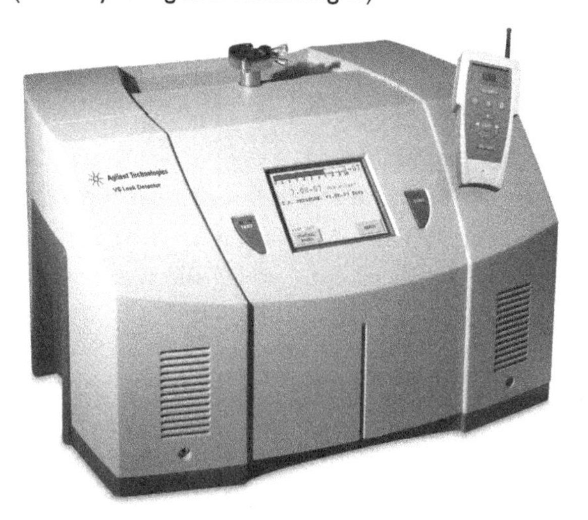

A mass spectrometer can detect extremely small amounts of helium. When the gas enters the spectrometer tube, it is ionized and accelerated. These high-speed, charged particles are then exposed to a magnetic field perpendicular to their direction of motion. What results is a force perpendicular to both the velocity vector and magnetic field. This force causes the particles to follow a curved path, the radius of which depends on the mass of the particle, allowing separation of the particle stream into different ions. A properly positioned collector plate (ion detector) enables the concentration of any gas to be very accurately measured. Every electron given up by the collector plate equates to the presence of one helium ion. The amount of helium collected is then converted to a leak rate.

Helium is the tracer gas of choice because it is inert, nontoxic, relatively inexpensive (in small quantities) and not easily absorbed. Helium also easily flows through small leaks and has only a trace presence in air (usually 5 ppm).

A good hint is to always look for leaks in the last area serviced on the vacuum furnace. The following technique is a practical guide to using a helium leak detector.[1] Be sure that the system you are using is properly calibrated.

Begin by pumping down the chamber (including use of the diffusion pump if available), and be sure you have reached the lowest vacuum level possible. Next, valve in the leak detector and note the background leak indication. It must be less than the standard leak, preferably a decade lower. High background indications suggest gross leaks in the 10^{-4} or 10^{-3} cc/second range, so these leaks need to be repaired in order to get the background down to a level where helium mass spectrometry is effective. Adjust the helium flow on the wand to a low-flow setting (allowing the helium to contact a wet part of the skin) or use a liquid and adjust the flow to several bubbles per second.

Start at the top of the vacuum system and work down, checking the following areas:

1. Power feed-throughs
2. Thermocouples ports – check the sheath gland and pipe threads, and don't forget to bleed helium at the connector to verify that there is no leak down the inside of the sheath
3. Door seals
4. Viewport seals
5. All tapered pipe-thread connections
6. Vacuum/pressure switches (note: remove the switch covers and bleed helium in)
7. Electrical feed-through on quench blower motor (if applicable)
8. Atmospheric glands on internal cooling loops
9. Nitrogen/inert gas bleed valves (this will require opening the upstream connection and introducing helium into the valve)
10. Any Dresser-type couplings
11. Oil drain/fill connection on the diffusion pump (note: flood bottom of the pump to see if there are leaks on the boiler plate)
12. Cold cap glands on the diffusion pump (if applicable)
13. Butterfly valve shafts
14. High-vacuum valve piston-rod seal (may require introducing helium into the lower port of the high-vacuum cylinder)
15. Shaft seals on oil agitators. If agitators have "canned" motors, check electrical feed-through on motor.
16. Oil to water heat exchangers used on oil-quench furnaces. This would require blowing out the water on the tube side and introducing helium, a very involved procedure.
17. Mechanical seal on quench-oil circulation pump
18. Mechanical seal, end plate and oil line leaks on the roughing pump
19. Mechanical seal, end plate and oil fill/drain leaks on the blower

Additional leak checking hints include:

❖ On old furnaces, especially those that have been cooled with untreated water, tube penetrations from the outer to inner shell often corrode through and leak water. Gross leaks will be obvious, but pinhole leaks will allow ice to form on the vacuum side. Since ice has a much lower vapor pressure than liquid water, it delays pumping below 0.067 mbar (50 microns). Very small leaks will not show up as water in the mechanical pump. Gross water leaks will turn the pump fluid milky.

- Helium "drift" is the major problem in locating the site of the actual leak. The problem is the use of too large a flow of helium, which results in a leak indication that is not in the location probed. When you are on top of the leak, the response time will be very quick (in the order of a second or two), and you should be able to repeat the leak indication at least three times successively at the same location.
- If no leaks are found, note the vacuum level and slowly open the needle valve on the main chamber to give a small upscale deflection on the vacuum level. Now introduce a small amount of helium into the metering valve. You should see a large leak indication within 1-2 seconds. Then close the metering valve.
- Probe with helium into the gas-ballast check valve on the mechanical pump. Close the port to seal it off and then quickly open and close the gas-ballast ball valve to introduce a small amount of helium into the pump. It will "back diffuse" into the blower and allow you to see a large leak indication quickly.
- Leaks that remain could be due to water leakage on the furnace shell or gas-cooling finned heat exchangers. The gas-cooling heat exchangers will have to be blown out to eliminate water, and then helium will have to be introduced using a large helium tank with regulator. Water leaks on the diffusion-pump cold cap are also common. Check this by blowing out the cap water line and introducing helium.
- With all leaks repaired, a background leak indication in the low 10^{-9} cc/second range should be achievable. You may have to vent and re-pump several times to get a low enough background.
- At the conclusion of leak check, get in the habit of valving off the leak detector so that it won't be dumped to air on the next roughing cycle.

Mass Spectrometer Leak-Testing Tips

Mass spectrometer leak testing requires that the unit be exposed to helium leaking for only three to four seconds. As with the solvent test method, however, a dwell or lag time between test areas is needed to prevent false readings. So, being too quick (aggressive) in finding leaks using helium is a bad practice. This is especially true since in most applications helium is squirted on the outside of suspected areas with a small blast from a pressurized tank in such a way that the helium will pass through the leak and into the vacuum system before being intercepted by the leak detector. There is a tendency to want to move on before allowing enough time for the helium to migrate from outside to the detector.

In utilizing helium, checking should always begin at the bottom of a pressurized system and at the top of an evacuated system. When dealing with moving or transition seals (e.g., vacuum seals that must also withstand pressurization), it may be necessary to "bag" the area under investigation by using a plastic bag and tape to seal off a component and then injecting helium into the closed area. Remember that you are dealing with total leakage within the enclosed space.

Many modern vacuum furnaces come equipped with software (Fig. 12.6) to make the job of leak-rate testing and detection easier.

FIGURE 12.6 | Automated vacuum leak-rate check software [3]

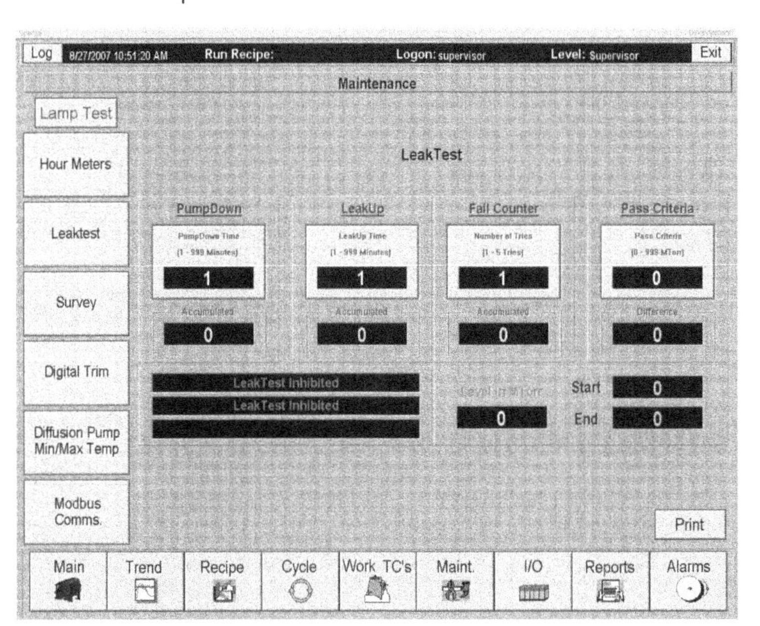

Backfill Lines and More

There is no acceptable minimum leak size. Since the vacuum chamber is not capable of determining how many leaks are present or their sizes, the effect is cumulative. For example, five small leaks can be more harmful than one large leak if the sum of their leak rates exceeds that of the large leak.

When leak testing a vacuum furnace, do not forget to check the gas backfill lines (including all fittings) from the gas supply to the furnace. It is good practice to install vacuum-tight shutoff valves near the source (just inside the building if using a cryogenic system located outside) and at the equipment. These lines are often pressurized and soap tested when first installed. However, that does not guarantee that they do not leak. The gas inside backfill lines often travels at near supersonic velocities, and a pinhole leak in a line will draw in air via a venturi action. Always perform leak-up tests first with the backfill lines closed and then with them open (up to the shutoff valve near the source) to confirm that the lines are not leaking. The backfill lines may then be evacuated using the vacuum-furnace pumping system up to the source and the lines themselves vacuum checked utilizing a helium leak detector.

In addition, clean and accurate vacuum measuring devices are essential for obtaining a meaningful value of leak-up rate. The need for periodic checks and annual calibrations of all vacuum instruments against known standards cannot be overemphasized.

Finally, there are differences between North American and European leak-up rate specifications. In North America, the specs usually involve a fixed leak-up rate and do not take into account the chamber volume. OEM specifications for new equipment vary from 0.003-0.013 mbar/hour (2-10 microns/hour). In Europe, the leak-up rate factors in chamber volume and is expressed in units such as mbar-liter/second.

Final Thoughts

A comprehensive preventative-maintenance program is essential to minimizing downtime from vacuum leaks. Proper care of pumps, replacement of O rings as they age, cleaning of flange sealing surfaces (particularly loading-door seals) and regular inspection of vacuum feed-throughs will help prevent leaks. In addition, daily leak checks or continuous monitoring of vacuum levels during processing can also help to identify potential problems before they develop into major repairs.

REFERENCES

1. Geoff Humberstone, Metallurgical High Vacuum Corporation, editorial review
2. *Practical Vacuum Systems Design,* The Boeing Company
3. *Leak Detection, Applications & Techniques,* Agilent Technologies
4. Grann, James, "Understanding the Difference Between Linear and Non-Linear Leak Rates," Symposium on Vacuum Furnace Maintenance, ASM International, October 2007
5. Herring, D. H., "The Why, When and How of Leak Checking a Vacuum Furnace," *Heat Treating Progress,* September/October 2003
6. *The Nature of Vacuum,* SECO/WARWICK Corporation
7. Brunner, William F., "Vacuum Leak Detection," American Vacuum Society, 1981

CHAPTER 13

CLEANING OF PARTS AND FIXTURES

When vacuum furnaces were first introduced, many in the industry felt that the only acceptable part and fixture cleaning method was solvent vapor degreasing. Over the years, however, environmental and other factors have dictated the use of aqueous systems. It is important, therefore, to understand how each method can successfully get the job done.

Cleaning is the application of time, temperature, chemistry and energy to remove contamination from the surface of a part to a level appropriate for the intended application. In other words, cleaning is simply moving contaminants from where they are not wanted (on the parts) to where they should be (in the waste disposal system). If all four aspects of the cleaning process are not working together, the parts will not be properly cleaned. Vacuum heat treating demands a high level of cleanliness compared to other methods. Contamination left on parts can cause significant problems both in the equipment (Fig. 13.1) and on the parts themselves (Fig. 13.2).

FIGURE 13.1 | Typical interior of a contaminated vacuum-furnace hot zone

SEE FULL COLOR IMAGE ON PAGE A

FIGURE 13.2 | Soft spots found after shot peening on a low-pressure vacuum-carburized gear due to cleaning issues

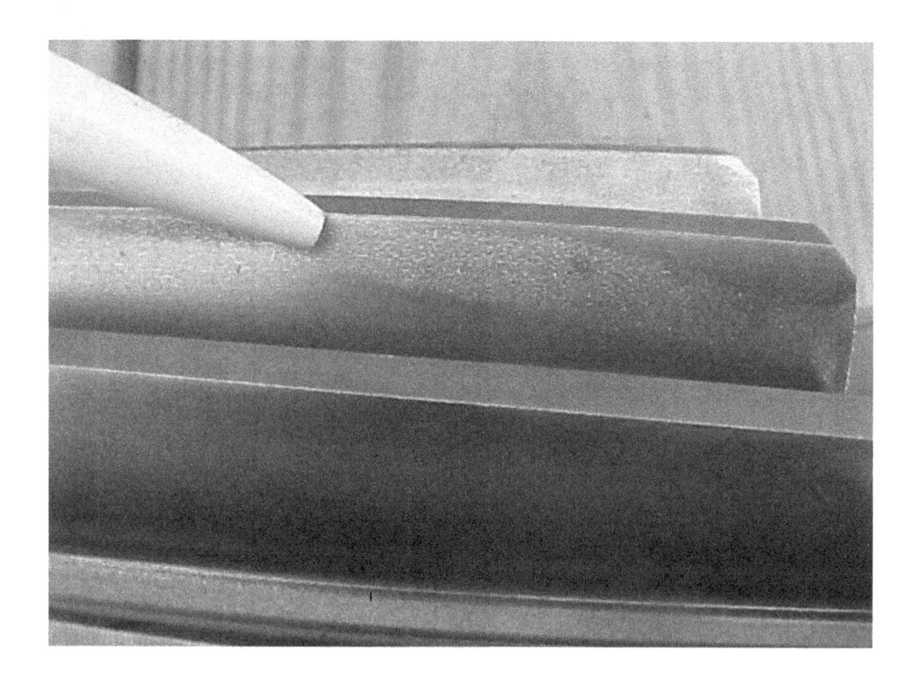

Most vacuum systems are required to operate below 1 torr (1.33 mbar) and as such cannot contain or introduce any contaminants with a vapor pressure (at the process temperature) near the operating vacuum pressure. As is often the case, more than one contaminant is present, so the sum of the vapor pressures of each will be the limiting pressure of the system.

All cleaning systems depend on one or a combination of three basic actions:

- A *physical* action (i.e. a mechanical force, such as spray agitation, dunking, ultrasonics or even hand cleaning) to remove the contaminants from the part surface
- A *thermal* action to improve the activity of the cleaning solution and increase the kinetic energy of the system
- A *chemical* action to allow contaminants to be either desorbed from the part surfaces with the aid of surface-active agents or be dissolved by an action of absorption and dilution

Solvent Cleaning

Cleaning in a solvent offers a level of simplicity and forgiveness not seen in other cleaning methods. Solvent cleaning involves three basic steps: wash, rinse and dry. Washing is where the parts are immersed in or contact the solvent, which is typically boiling to help the removal process. The purpose of rinsing is to bring fresh or clean solvent in contact with the parts. The goal is to dilute the contaminated solvent present on the surface of the parts after washing. It is important to remember that the rinse solvent must be kept clean. Contaminated solvent is a very common problem and will only reintroduce contaminants back onto the surface. The drying step evaporates the solvent and separates the rinse solvent from the parts.

Solvent cleaning has a negative connotation in the heat-treating industry primarily due to environmental (VOC) concerns, safety and cost issues. The emergence of vacuum vapor degreasing in a sealed vessel (Fig. 13.3) offers an attractive alternative. It takes advantage of the best aspects of solvent cleaning – significantly reducing the size and amount of residual contaminants – while meeting the most stringent environmental requirements and avoiding the traditional problems of open degreasing systems.

FIGURE 13.3 | Typical vacuum vapor-degreasing system used for cleaning parts prior to running in vacuum furnace (courtesy of Dr. Donald Gray, Vacuum Processing Systems LLC)

Aqueous Cleaning

Aqueous cleaning (Figs. 13.4, 13.5) is the dominant approach used in the heat-treat industry. Its simplicity, ease of use and overall flexibility is what makes it an attractive process. Aqueous cleaning uses detergents to lift contaminants from the surface of the parts; heat to make the detergents more compatible with the contaminants and to soften them; fluid force to dislodge the contaminants from the parts and to collect the insoluble contaminants in some removal systems; and time to allow the process to take effect. For cleaning parts and fixtures intended for vacuum operation, a clean-water rinse followed by a forced hot-air drying system is highly recommended. Being absolutely dry prior to placing parts or fixtures into a vacuum furnace is mandatory.

Aqueous cleaning is not perfect however. It often leaves a surface residue or "film" on the parts that may interfere with certain processes, such as brazing or case hardening, and thus requires subsequent removal. In general, aqueous cleaners don't dry well, and the solution is often difficult to remove from internal part surfaces such as holes, crevices and recessed areas. Finally, aqueous cleaners evaporate slowly, requiring large amounts of energy to dry parts and have been known to damage certain sensitive parts.

Once cleaned, parts cannot remain in unprotected conditions inside the plant due to a concern over rusting.

FIGURE 13.4 | Typical aqueous parts washer used for cleaning parts prior to running in a vacuum carburizing furnace (courtesy of ALD Thermal Treatment)

FIGURE 13.5 | Aqueous parts washer used for cleaning parts prior to hardening (courtesy of Aichelin USA)

Other Types of Cleaning

Blast cleaning, pressure washing, steam cleaning, abrasive cleaning and other mechanical methods have all been used to clean parts and fixtures prior to vacuum heat treatment. Fluidized beds have also been used for years to remove contaminants on part surfaces. In cases where it is beneficial to remove imperfections – such as stains or surface corrosion, heat discoloration, oxide films, weld marks, scratches and particles of all sizes – electropolishing techniques can be used. These "nontraditional" cleaning approaches have value but present their own unique set of challenges.

Cleaning the Cleaning Equipment

One of the most overlooked aspects of successful parts cleaning is to schedule routine maintenance on the cleaning equipment. Batch chemistry, concentration and pH should be checked and, if necessary, adjusted daily. This not only includes replacing the cleaning solution (or distilling it in the case of solvent systems), but cleaning the parts washer thoroughly before recharging. Steam cleaning and scraping methods are often used on the inside of the tank to ensure that all areas are cleaned.

How long and how clean is clean?

Cleaning time depends to a large extent on the system and the parts. An aqueous-based process typically needs to run for 10-20 minutes, while solvent-based techniques need 5-10 minutes to complete the cleaning process. As a rule, it is more difficult to clean clean parts than dirty ones. A key question is always, how do we know when the parts are clean?

There are a large number of tests to measure cleaning effectiveness. The most common in the heat-treat industry include:

- Visual inspection
- Stereomicroscope (macroscopic, 5X-50X) inspection (Fig. 13.6)
- White-glove inspection (Fig. 13.7)
- Ultraviolet (black) light observation (Fig. 13.8)
- Tape sampling
- Water-break tests
- Surface-tension test fluids (Fig. 13.9)
- Nordtest technique – a quantitative approach using low surface-tension droplets spread onto the part surface; measure of the droplets ability to wet the surface and form a film layer (Table 13.1)
- Gravimetric methods (Fig. 13.10)

TABLE 13.1 | Spreading surface tensions for clean surfaces

Material	Surface tension, mN/mm or dyne/cm
Hot-rolled steel	32-34
Cold-rolled steel	30-32
Stainless steel	26-28
Aluminum	24-26
Bronze	24-26
Copper	22-24
Polyethylene	22-24

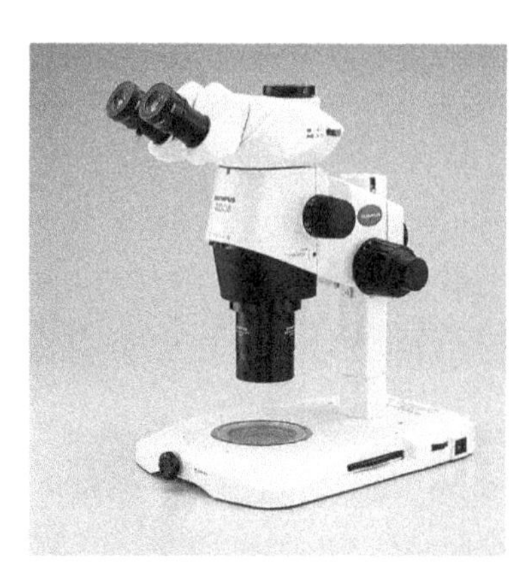

FIGURE 13.6 | Example of stereomicroscope inspection (courtesy of Olympus Corp.)

FIGURE 13.7 | Example of white-glove inspection

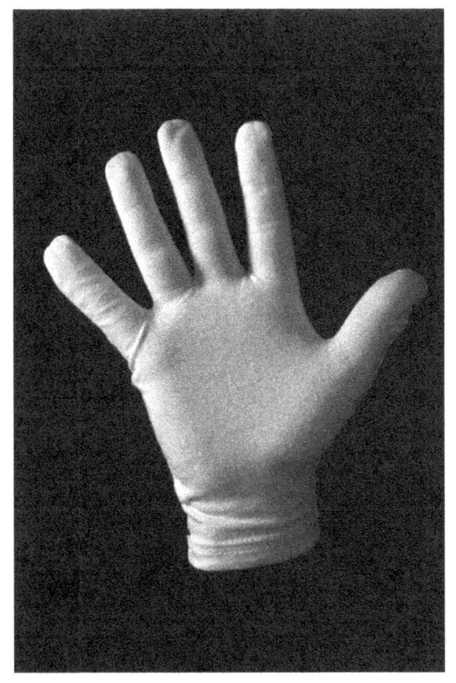

FIGURE 13.8 | Example of ultraviolet (black) light observation

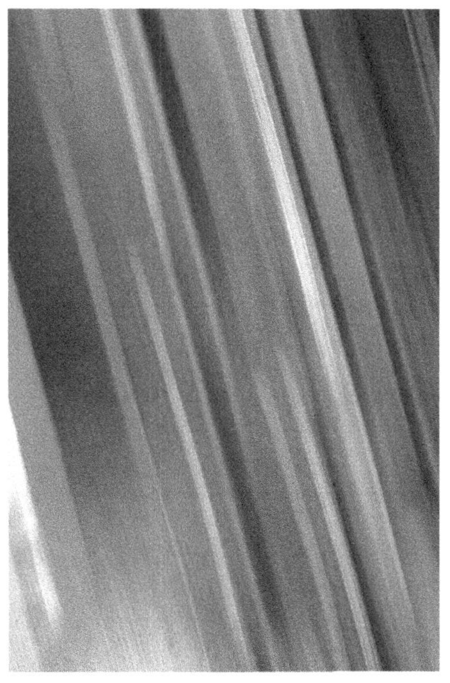

FIGURE 13.9 | Example of surface-tension testing

FIGURE 13.10 | Example of gravimetric method

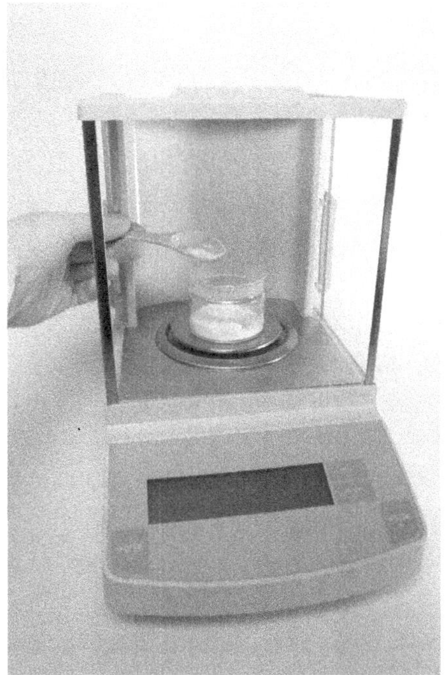

Direct and indirect verification methods for part and fixture cleanliness can be classified into three main categories: gross verification, semi-precise verification and precise verification. The test choice is dependent on the needs of the specific vacuum process.

Gross Verification Tests

Gross verification looks for visible contamination, but it does not quantify them. (For example, dirt is just visible at 0.1 gram/square inch and above.) Methods include the use of stereomicroscopes, nonvolatile residue methods, tape tests, ultraviolet (UV) fluorescence observation, water-break test and "white glove" test.

NONVOLATILE RESIDUE (NVR)

The NVR test requires extraction of contamination from a dirty part into a volatile solvent, evaporating off the solvent and measuring the weight of the remaining residue using an analytical balance. Almost any clean volatile solvent can be used.

TAPE TEST

For polished or lapped parts, a strip of transparent (Scotch®) tape is affixed to the surface in question with firm pressure. The tape is removed and placed on a clean, white sheet of paper. The surface should appear as white as the original sheet of paper.

ULTRAVIOLET (UV) FLUORESCENCE

Fluorescence can provide a visual indication of where contamination remains on a surface since contaminants will fluoresce in the presence of UV light. The intensity of the radiation can also be measured via a registered signal on an instrument, which dictates the degree of contamination on a surface. This form of analysis is useful for locating contamination, not identifying it.

WATER-BREAK TEST

This is a simple test in which the part is immersed in clean water. Upon removal, the water film must continuously cling to the part surface for 30 seconds. If water beads the surface, it is considered to be contaminated with a film of oil or grease.

Semi-precise Verification Tests

Semi-precise verification can be qualitative or quantitative to a moderate level of precision (0.001-0.1 gram/square inch). Methods include the use of contact-angle measurements, the Millipore test and optical microscopy.

MILLIPORE (PATCH) TEST

The Millipore test consists of spraying a representative number of parts with filtered hexane, isopropyl alcohol or trichlorethylene at a pressure of 60-80 psi through a jet nozzle equipped with a 1.2-micron membrane filter. The spray is collected and vacuum filtered onto a clean filter membrane, and the membrane is inspected for contaminants (placed under a microscope to measure, in microns, and count the number of dirt particles remaining). Weighing the membrane pad determines the total contaminant (in milligrams) that has been left behind.

OPTICAL MICROSCOPY

Optical microscopes use a beam of light and lenses to magnify objects. Optical microscopes are ideal for viewing residual oils and greases, flux residues, certain particles and surface anomalies.

Precise Verification Tests

Precise verification is quantitative to extreme levels of accuracy (0.001 gram/square inch down to absolute zero). Methods include the use of Auger electron spectroscopy (AES), carbon coulometry, electron spectroscopy for chemical analysis (ESCA), Fourier transform infrared (FTIR), fluorometer, gas chromatography/mass spectrophotometry (GC/MS), ion chromatography, optically stimulated electron emission (OSEE), particle counting, scanning electron microscope (SEM) and secondary ion mass spectroscopy (SIMS).

AUGER ELECTRON SPECTROSCOPY (AES)

AES is used for compositional analysis or determining which atoms are present on a surface. An argon (or other selected gas) stream directs electrons toward the surface, ionizing surface atoms by causing the removal of an electron from the atom's inner shell. The atom now becomes excited and must release energy to "relax" and return to its original state. This is done by transferring the extra energy to an electron that can leave the atom. The exiting electron is known as the auger electron. The AES method of analysis measures the energy of the auger electron, which is unique to each particular atom. AES is used in the semiconductor field for corrosion and failure analyses and thin-film analyses.

CARBON COULOMETRY

The technique employs in-situ direct oxidation of surface carbon to CO_2, followed by automatic CO_2 coulometric detection.

ELECTRON SPECTROSCOPY FOR CHEMICAL ANALYSIS (ESCA)

ESCA is a spectrophotometric technique in which X-ray bombardment of a surface results in the emission of an electron from a given atom. Knowing the energy of the X-ray and measuring the energy of the emitted electron can determine the binding energy of the electron. ESCA methods reveal chemical structure, bonding and oxidation state. ESCA has the potential to be very useful in identifying organic compounds.

GAS CHROMATOGRAPHY/MASS SPECTROPHOTOMETRY (GC/MS)

GC/MS is used to identify surface contamination by extracting contaminants into solvent and analyzing them. Organic compounds are separated via gas chromatography and are then identified, based on molecular weight, by mass spectrophotometry.

ION CHROMATOGRAPHY

Ion chromatography separates, identifies and quantifies ions. The analysis begins with a sample, typically a water matrix containing ions of interest. A portion is injected into the system and combined with a chemical stream that carries the sample to the analytical column. The analytical column separates the ions of interest in the sample into narrow bands within the stream of the chemicals. The chemical stream then sweeps these groups of ions into the suppressor device, which electrolytically transforms the chemical stream into pure water leaving just the ions of interest in pure water to be swept downstream to the conductivity detector. The detector detects the ions based on their conductivity relative to the water.

Final Thoughts

A number of technical organizations, including ASTM, offer cleaning standards, often based on the type of material to be cleaned. Remember, clean is generally observed, not measured. Cleaning effectiveness is established by answering the question, can we do what we need to do next? Cleaning tests provide quantification of the nature of the part surface so that the influence of the remaining contaminants can be factored into the heat-treatment operation as well as subsequent manufacturing operations.

Cleaning is critical to the success of vacuum heat treating. The more we embrace the fact that parts and fixtures must be absolutely clean (and dry), the better the quality of our parts and the longer the life of our equipment.

It is also important to recognize that both solvent and aqueous cleaning processes can be made to clean almost all parts and fixtures requiring vacuum heat treatment. However, the choice between the two should be made first by the degree of part cleanliness needed followed by other factors. The focus today is on improving physical action (force and volume) in combination with a chemistry choice balanced for the type of cleaning required.

Finally, take a systems approach – consider the manufacturer of the cleaning system and the supplier of the cleaning agent as partners in the long-term success of any cleaning operation.

REFERENCES

1. *Practical Vacuum Systems Design*, The Boeing Company
2. Herring, D.H., "It's Time to Clean Up Our Act!," *Industrial Heating*, January 2008
3. "Choices for Cleaning Verification," *Parts Cleaning Magazine*, 2001
4. Durkee II, John B., *Management of Industrial Cleaning Technology and Processes*, Elsevier, 2006, ISBN 0-080-44888-7
5. Joseph P. Schuttert, Vacuum Processing Systems LLC, private correspondence

CHAPTER 14

DIFFUSION BONDING, EUTECTIC MELTING, OUTGASSING AND RELATED TOPICS

Problems that can occur during processing of parts include diffusion bonding, eutectic melting and unexpected or uncontrolled outgassing. In all cases, part quality may be compromised, so we must understand these phenomena in order to avoid them.

What is diffusion bonding?

Diffusion bonding is a solid-state joining process capable of bonding together a wide range of small and large metal or ceramic part combinations. In instances where our intent was not to join components together, however, diffusion bonding can be an unexpected problem.

In vacuum processing, metal surfaces remain very clean and free of oxides. When these near-perfect surfaces are in contact with each other or other surfaces (e.g., baskets or fixtures), certain elements have a tendency to interact between these surfaces via solid-state diffusion (i.e. interdiffusion of atoms across the interface). The result is that the parts "stick" together or to the baskets or fixtures, the equivalent of being welded on a microscopic level. In some cases, the effect is minor. A slight tapping of the components separates them, and the surface "damage" is inconsequential. In other cases, parts are fused together so strongly that the surfaces have to be literally ripped apart, ruining the components.

In most metals, the presence of oxide layers (e.g., atmospheric gas heat treating) at the surface will avoid diffusion bonding. For some vacuum-processed metals and alloys, their oxide films either dissolve in the bulk of the metal or decompose at the process temperature (e.g., carbon and alloy steels, stainless and tool steels, copper, titanium, tantalum, columbium and zirconium). So, metal-to-metal contact can be readily established at the interface. If the oxide film is chemically stable (as for aluminum alloys), however, the chances for diffusion bonding to take place are limited.

In its simplest form, diffusion bonding involves placing components under load at elevated temperature usually in vacuum (or a very pure protective atmosphere). The loads used are typically below those that would cause macro-deformation of the parent material(s). Temperatures of 50-90% of the absolute melting point (0.5-0.9 T_m, where T_m is the melting point in K) are needed. Times at temperature can range from as little as 1 minute to 60 minutes or more. This depends on the materials being bonded, the joint properties required and the remaining bonding parameters.

An example of diffusion bonding is when loads of saw blades or knives are packed closely together in fixtures so as to minimize distortion during vacuum processing. This pressure, in combination with high temperature, is enough to cause sticking to occur, even though partial pressures of nitrogen are commonly used. In most instances, the parts can be tapped gently and separated. In extreme cases, they bond together and must be scrapped. To negate the effects of diffusion bonding, partial-pressure levels can be set high enough to prevent inter-surface diffusion, or coatings (e.g., magnesium oxide or boron nitride) can be applied to the part surfaces in question to prevent sticking from occurring.

FIGURE 14.1 | The mechanism of diffusion bonding: a.) initial point contact, showing residual oxide contaminant layer; b.) yielding and creep, leading to reduced voids and thinner contaminant layer; c.) final yielding and creep, some voids remain with very thin contaminant layer; d.) continued vacancy diffusion, eliminates oxide layer, leaving a few small voids that ultimately disappear as bonding is complete [1]

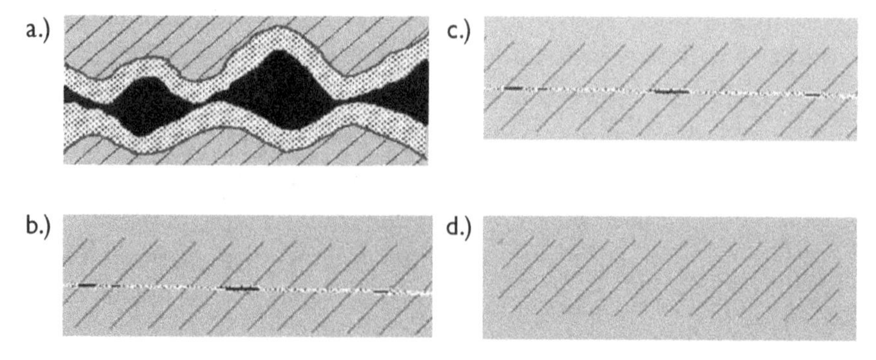

What is eutectic melting?

When two elements of a specific chemical composition melt at a lower temperature than any other composition of those elements, the phenomenon is known as eutectic melting. The temperature at which this occurs is called the eutectic temperature, and the composition at that temperature produces a eutectic point (Fig. 14.2). For example, carbon and nickel react by melting at temperatures as low as 1326°C (2421°F) and cause localized melting, also known as eutectic melting.

FIGURE 14.2 | Phase diagram depicting the eutectic composition, temperature and point [3]

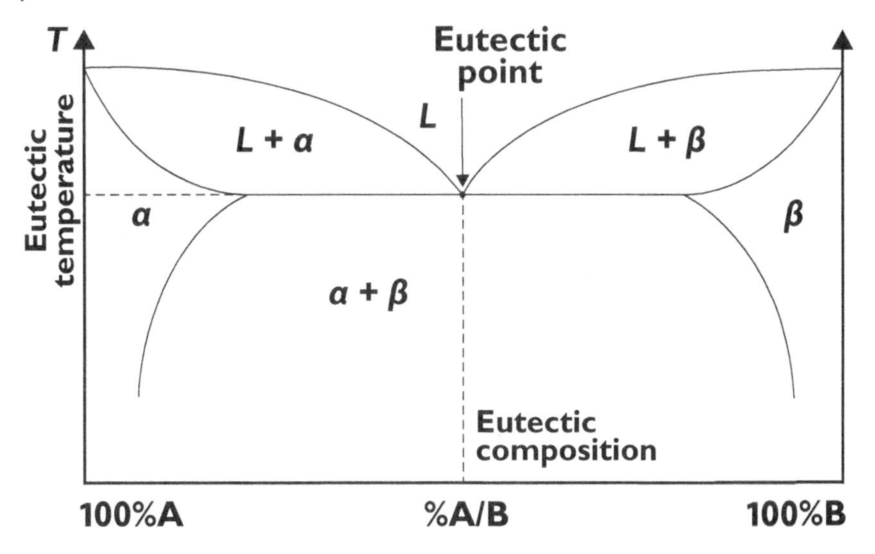

Since nickel is a common element in many steels (particularly stainless steels), these materials should not be allowed to come into contact with graphite hearths or fixturing materials during heat treating. For this reason, graphite hearth rails are usually designed to accommodate ceramic or molybdenum inserts that separate the load from the graphite. Severe eutectic melting reactions can cause extensive damage to workloads and furnace hot zones.

To avoid damage, it is best to understand the alloy system you are using, the eutectic melting temperature or temperature range (Table 14.1) and separate potentially reactive materials from each other with completely non-reactive insulators. There are several choices, including high-purity ceramics such as aluminum oxide or zirconium oxide. These are often available in the form of papers, cloth blanket, plates or other solid shapes upon which the workpieces can be placed.

It should be noted that some ceramic papers and cloths are hydroscopic (i.e. retain moisture) and can release fiber particles that can lead to deleterious effects on vacuum pumping systems. Ceramic materials are also available in the form of stop-off paints, which can be applied liberally to furnace baskets, grids or other fixtures. These coatings must be inspected and touched up regularly as they will deteriorate with use. The same materials can be applied by plasma spray. This process provides a thicker and more durable protective finish, but it is slightly more expensive than the paint-on versions.

Table 14.1 is intended as a guide (only) for heat treatment of these materials. The values listed are believed to be correct, but they have been found to vary and are subject to industrial interpretation. Small quantities of an impurity present in the alloy or furnace parts may cause unexpected melting if there is the possibility of a eutectic reaction between the alloy and the impurity. Table 14.1 is as accurate as possible. In some instances, there is more than one eutectic composition for each binary combination.

TABLE 14.1 | Selected eutectic melting points (courtesy of Solar Atmospheres Inc.)

Alloy (Element-Wt. %)	Melting point, °F	Melting point, °C
Aluminum-Beryllium (Be-0.8)	1191	644
Aluminum-Copper (Cu-33)	1019	548
Aluminum-Copper (Cu-82)	1904	1040
Aluminum-Gold (Au-96)	968	520
Aluminum-Gold (Au-91)	1056	569
Aluminum-Gold (Au-7.5)	1202	650
Aluminum-Iron (Fe-98)	1211	655
Aluminum-Lithium (Li-93)	351	177
Aluminum-Lithium (Li-8)	1040	596
Aluminum-Magnesium (Mg-66)	819	437
Aluminum-Magnesium (Mg- 36)	842	450
Aluminum-Nickel (Ni-5.7)	1182	639
Aluminum-Nickel (Ni-86.4)	2570	1410
Aluminum-Silicon (Si-12.6)	1071	577
Aluminum-Zinc (Zn-94)	718	381
Aluminum-Zirconium (Zr-88)	2426	1330
Aluminum-Zirconium (Zr-76)	2705	1485
Aluminum-Zirconium (Zr-57)	2732	1500
Beryllium-Chromium (Cr-13)	2232	122
Beryllium-Chromium (Cr-44)	2277	1247
Beryllium-Chromium (Cr-92)	2757	1514
Beryllium-Copper (Cu-94.7)	1576	858
Beryllium-Copper (Cu-64)	2190	1199
Beryllium (Fe-22.6)	2219	1215
Beryllium-Nickel (Ni-95)	2102	1150
Beryllium-Nickel (Ni-73)	2257	1236
Beryllium-Nickel (Ni-43)	2449	1338
Beryllium-Silicon (Si-64)	1994	1090

Alloy (Element-Wt. %)	Melting point, °F	Melting point, °C
Carbon-Iron (Fe-22.6)	2107	1153
Carbon-Molybdenum	4001	2205
Carbon-Nickel (C-2.5)	2421	1327
Chromium-Carbon (C-3.6)	2793	1534
Chromium-Carbon (C-10)	3141	1727
Chromium-Cobalt (Co-59)	2489	1365
Chromium- Molybdenum, (min. melt Mo-21)	3308	1820
Chromium-Nickel (Ni-46)	2453	1345
Chromium-Tantalum (Ta-34)	3200	1760
Chromium-Tantalum (Ta-78)	3605	1985
Chromium-Titanium (min. melt, Ti-53)	2570	1410
Chromium-Zirconium (Zr-86)	2430	1332
Chromium-Zirconium (Zr-28)	2898	1526
Cobalt-Carbon (C-2.6)	2498	1320
Cobalt-Molybdenum (Mo-39)	2435	1335
Cobalt-Tantalum (Ta-21)	2336	1280
Cobalt-Tantalum (Ta-70)	2858	1570
Cobalt-Tantalum (Ta-81)	3038	1670
Cobalt-Tin (Sn-34)	2034	1112
Cobalt-Titanium (Ti-73)	1868	1020
Cobalt-Titanium (Ti-20)	2156	1180
Cobalt-Vanadium (V-38)	2278	1248
Cobalt-Tungsten (W-45)	2680	1471
Copper-Magnesium (Mg-66)	909	487
Copper-Magnesium (Mg-36)	1026	552
Copper-Magnesium (Mg-9)	1389	726
Copper Silicon	1476	802
Copper-Titanium (Ti-12)	1607	875
Copper-Titanium (Ti-50)	1796	980
Iron-Carbon (C-4.2)	2107	1153
Iron-Molybdenum (Mo-37)	2640	1449
Iron-Niobium (Nb-19)	2503	1373
Iron-Niobium (Nb-75)	2552	1400
Iron-Niobium (Nb-55)	2795	1535
Iron-Silicon (Si-20 & Si-53)	2197	1203
Iron-Silicon (Si-53)	2205	1207
Iron-Silicon (Si-20 & Si-53)	2214	1212
Iron-Tin (Sn-49)	2066	1130
Iron-Tantalum (Ta-20)	2606	1430
Iron-Ta (Ta-85)	2898	1592
Iron-Titanium (Ti-67)	1985	1085
Iron-Titanium (Ti-14)	2352	1289
Magnesium-Nickel (Mg-24)	946	508

Alloy (Element-Wt. %)	Melting point, °F	Melting point, °C
Magnesium-Nickel (Mg-91)	2006	1097
Magnesium-Silicon (Si-1.34)	1180	638
Magnesium-Silicon (Si-57)	1334	946
Magnesium-Tin (Sn-98)	399	204
Magnesium-Tin (Sn-37)	1042	561
Magnesium-Zinc (Zn-53)	644	340
Magnesium-Zinc (Zn-97)	687	364
Manganese-Titanium (Ti-59)	2156	1180
Manganese-Titanium (Ti-9)	2246	1230
Molybdenum-Nickel (56-Mo)	2388	1309
Molybdenum-Silicon (Si-94)	2577	1414
Molybdenum-Silicon (Si-26)	3452	1900
Molybdenum-Silicon (Si-9.5)	3668	2020
Nickel-Carbon (C-0.6)	2421	1327
Nickel-Niobium (Nb-23)	2347	1286
Nickel-Niobium (Nb-51)	2147	1175
Nickel-Silicon (Si-29)	1767	964
Nickel-Silicon (Si-38)	1774	968
Nickel-Silicon (Si-12)	2089	1143
Nickel-Silicon (Si-16)	2210	1210
Nickel-Tantalum (Ta-63)	2462	1350
Molybdenum-Silicon (Si-94)	2577	1414
Molybdenum-Silicon (Si-26)	3452	1900
Nickel-Tantalum (Ta-39)	2516	1380
Nickel-Tin (Sn-33)	2066	1130
Nickel-Titanium (Ti-73)	1728	942
Nickel-Titanium (Ti-34)	2044	1118
Nickel-Titanium (Ti-12)	2379	1304
Magnesium-Tin (Sn-98)	399	204
Magnesium-Tin (Sn-37)	1042	561
Magnesium-Zinc (Zn-53)	644	340
Magnesium-Zinc (Zn-97)	687	364
Manganese-Titanium (Ti-59)	2156	1180
Manganese-Titanium (Ti-9)	2246	1230
Molybdenum-Nickel (56-Mo)	2388	1309
Molybdenum-Silicon (Si-94)	2577	1414
Molybdenum-Silicon (Si-26)	3452	1900
Molybdenum-Silicon (Si-9.5)	3668	2020
Nickel-Carbon (C-0.6)	2421	1327
Nickel-Niobium (Nb-23)	2347	1286
Nickel-Niobium (Nb-51)	2147	1175
Nickel-Silicon (Si-29)	1767	964
Nickel-Silicon (Si-38)	1774	968

Alloy (Element-Wt. %)	Melting point, °F	Melting point, °C
Nickel-Silicon (Si-12)	2089	1143
Nickel-Silicon (Si-16)	2210	1210
Nickel-Tantalum (Ta-63)	2462	1350
Nickel-Tantalum (Ta-39)	2516	1380
Nickel-Tin (Sn-33)	2066	1130
Nickel-Titanium (Ti-73)	1728	942
Nickel-Titanium (Ti-34)	2044	1118
Nickel-Titanium (Ti-12)	2379	1304
Nickel-Vanadium (V-47)	2196	1202
Silicon-Tantalum (Ta-5)	2543	1395
Silicon-Titanium (Ti-15 & Ti-92)	2426	1330
Silicon-Zinc (Zn-99.98)	787	419
Silver-Aluminum (Al-26.5)	1053	567
Silver-Beryllium (Be-0.03)	1616	880
Silver-Calcium (Ca-60)	878	470
Silver-Calcium (Ca-6)	1211	655
Silver-Calcium (Ca-19.5)	1013	545
Silver-Copper (Cu-28)	1434	779
Silver-Magnesium (Mg-51.4)	882	472
Silver-Tin (Sn-96.5)	430	221
Tantalum-Carbon (C-2)	5149	2843
Tantalum-Carbon (C-9.5)	6233	3445
Tungsten-Carbon (W-1.3)	4919	2715
Tungsten-Carbon (W-3.4)	5000	2760
Tungsten-Carbon (W-4.4)	5000	2760

What is outgassing?

In simplest terms, outgassing is the evolution of gases from a material when subjected to a vacuum environment. Outgassing can also be due to the release of low-vapor-pressure contaminants present in the vacuum system itself. Sources of these contaminants include water vapor, volatile liquids, dirt, grease, oxides and gases adsorbed on or in surfaces and pores. These contaminants outgas at different rates at different vacuum levels and temperatures. To minimize outgassing, cleanliness is essential, especially of workpieces, part baskets and fixtures introduced into the furnace. It's also a good idea to limit the time that an open furnace is exposed to the factory environment.

The rate of outgassing increases with temperature, up to the flash point of the material involved. Outgassing is often made obvious during the process cycle by a large spike or rise in pressure during heating. It can also be detected by comparing successive leak-rate values after long pump-down cycles. If the leak rate improves (decreases) with each successive pump-down, then outgassing is suspected. If the rate remains essentially the same, then an actual (real) leak may be considered the primary cause.

In most cases, both leaks and outgassing are present simultaneously. Bear in mind that real leaks are linear and outgassing is not (Fig. 14.3). Note that the slope of the lines shown in Figure 14.2 represents the pumping speed of the system. For example, a four-hour leak-rate test, with measurements taken at 30-minute intervals, will generate a linear rate. The outgassing curve will deviate from linearity, yielding the true leak-up rate.

Since the outgassing rate decreases with time, if pumping speed is assumed to be constant and independent of pressure, the time when a decrease in pressure can be expected in an outgassing system is given by Equation 14.1:

14.1) $t = (P_o/P)^{1/n}$

where:

t is time, in hours
P_o is the initial pressure of the system, in torr
P is the ultimate pressure of the system, in torr
n is a constant whose value is between 1.2 and 1.4

Possible sources for outgassing (or virtual leaks) include:

- Residual solvents or liquids following PM or wet clean
- Liquid leaks such as cooling fluids
- Trapped volumes of gas or liquid
- Trapped space under non-vented/non-sealed hardware
- Gases or solvents into spaces with poor conductance
- High vapor-pressure materials or products
- Porous materials exposed to liquid or atmosphere

OUTGASSING CYCLE (FOR SEVERE OUTGASSING PROBLEMS)

The following procedure has been reportedly used to overcome issues with severe vacuum-furnace outgassing or when re-commissioning a unit that has been idle for a long period of time.

1. Confirm that the problem is indeed outgassing (as opposed to, for example, issues with leak rate, condensation from water jackets, etc.).
2. Change the mechanical-pump oil and ballast the mechanical pump.
3. Pump down the heating chamber (heat off) to under 0.13 mbar (100 microns). Backfill the chamber with argon (preferred) or nitrogen to 339 mbar (-10 inches Hg) and again pump to under 0.13 mbar (100

microns). Repeat the backfill/pump-down step a second time. *Note: This procedure helps eliminate a great deal of moisture (water vapor) that may have accumulated in the chamber while it sat open (especially in high humidity environments).*

4. Turn heat on at 0.13 mbar (100 microns).
5. Turn heat off if the vacuum-furnace pressure exceeds 0.67 mbar (500 microns), and leave heat off until vacuum level returns to 0.13 mbar (100 microns) or lower.
6. Repeat steps 4 and 5 as many times as necessary until the temperature reaches 120°C (250°F). Heat at a ramp rate not to exceed 8°C/minute (15°F/minute). Soak for one hour at 120°C (250°F) while continuing to pump on the heating chamber. Record the lowest vacuum level achieved.
7. Heat up to 230°C (450°F) at a ramp rate not to exceed 8°C/minute (15°F/minute). Repeat steps 4 and 5 as many times as necessary. Soak for one hour at 230°C (450°F) while continuing to pump on the heating chamber. Record the lowest vacuum level achieved.
8. Heat to 400°C (750°F) at a ramp rate not to exceed 8°C/minute (15°F/minute). Repeat steps 4 and 5 as many times as necessary. Soak for one hour at 400°C (750°F) while continuing to pump on the heating chamber. Record the lowest vacuum level achieved.
9. Heat to 565°C (1050°F) at a ramp rate not to exceed 13°C/minute (25°F/minute). Repeat steps 4 and 5 as many times as necessary, except the vacuum-level limit is now 1.33 mbar (1,000 microns). Soak for a half hour at 565°C (1050°F) while continuing to pump on the heating chamber. Record the lowest vacuum level achieved.
10. Heat to 730°C (1350°F) at a ramp rate not to exceed 13°C/minute (25°F/minute). Repeat steps 4 and 5 as many times as necessary, except the vacuum-level limit is now 1.33 mbar (1,000 microns). Soak for a half hour at 730°C (1350°F) while continuing to pump on the heating chamber. Record the lowest vacuum level achieved.
11. Proceed to process operating temperature plus 10°C (18°F). Caution should be observed to stay below the maximum operating temperature of the furnace. Significant outgassing – that from which the vacuum pumping system is not able to hold at least 1.33 mbar (1,000 microns) – should not occur above 730°C (1350°F). Soak at this temperature for a minimum of four hours.
12. Vacuum cool overnight.
13. Check the mechanical pump oil and change if necessary. Ballast the pump for a minimum of 20 minutes.
14. Production cycles can now begin.

FIGURE 14.3 | Response of outgassing and real leaks to pump-down (mechanical pump) [4]

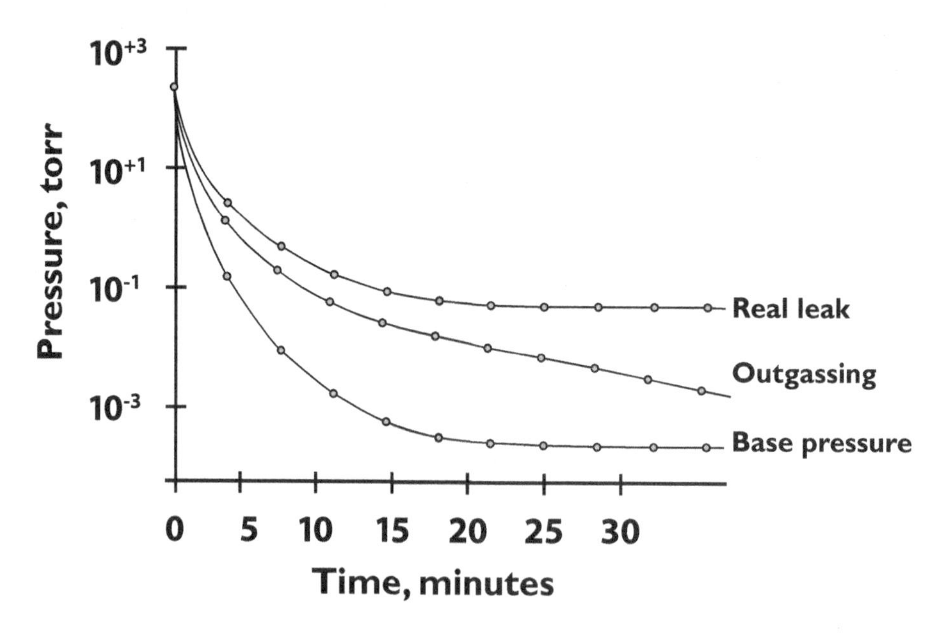

Summing Up

Diffusion bonding, eutectic melting and outgassing are subjects that we tend to ignore until their effects cause us to notice. Taking a proactive approach by understanding these phenomena makes all the difference when trying to avoid their negative consequences.

REFERENCES

1. Dunkerton, Sue, "Diffusion Bonding," TWI Knowledge Summary, TWI Ltd., 2001
2. Kazakov, N.F., *Diffusion Bonding of Materials*, Pergamon Press, 1985
3. Wikipedia, eutectic system
4. *Critical Melting Points for Metals and Alloys with other metallurgical reference data*, Solar Atmospheres, 2011

CHAPTER 15

HEAT EXCHANGER DESIGN AND MAINTENANCE

Most heat exchangers used in vacuum furnaces are essentially fin cooling units (Fig. 15.1), which depend on the surface area of their coolers as well as the temperature of the incoming water supply to achieve a given cooling rate. Factors to consider when designing these units include the total surface area as well as the type and design of the individual elements that make up the surface since there can be many types of finned tubes with differences in shape, arrangement and relative dimensions between the fins and the tubes.

FIGURE 15.1 | Typical vacuum-furnace heat exchanger arrangement (courtesy of Surface Combustion Inc.)

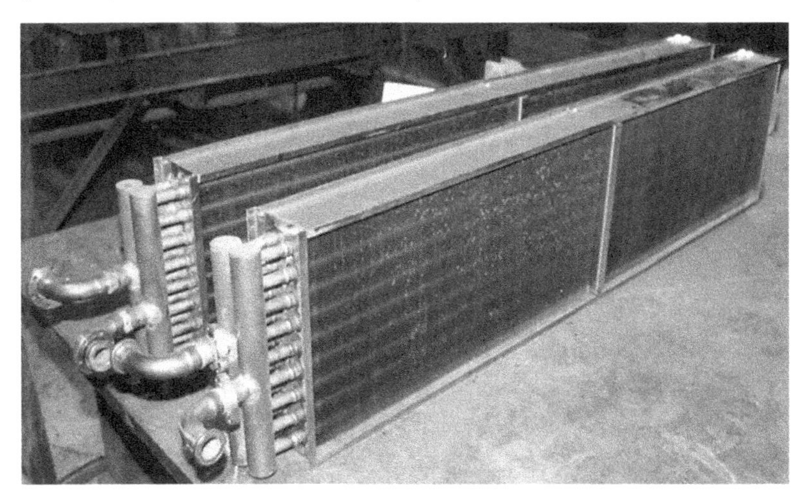

Design Considerations

The objective in the design of most heat exchangers is to transfer a given amount of heat to or from a selected fluid (either gas or liquid) while minimizing three things: cost, overall size/weight and (fluid) pressure drop. Heat transfer takes place by convection and conduction. For any given surface-element design, as velocity increases (which reduces the relative size and cost), there is an increase in pressure drop. Thus, the critical criteria for performance are a balance of these elements.

The independent variables in heat exchanger design are:

- Fluid properties (e.g., viscosity, density)
- Total amount of heat that must be transferred (e.g., Joules/second, BTU/hour)
- Total flow of the principal fluid (e.g., gas)
- Log mean temperature difference between the principal (e.g., gas) and secondary (e.g., water) fluids
- Coefficient of heat transfer between the principal (e.g., gas) and secondary (e.g., water) fluids
- Allowable temperature rise (of the principal fluid)
- Fixed pressure drop of the principal fluid (gas side) or the flow area of that fluid

Notice that fluid velocity does not appear in the list above because it is considered a dependent variable being a function of total flow, face area and the ratio of free area to face area. Depth in the direction of flow is also not fixed.

Other design considerations include:

- When baffles are provided, the system directs the shell fluid from axial flow to top-to-bottom flow or side-to-side flow with the effect that the heat-transfer coefficient is higher than for undisturbed flow along the axes of the tubes.
- Patterns of tube layout influence turbulence and, therefore, heat-transfer coefficient. (For example, triangular pitch gives greater turbulence than square pitch, and under comparable conditions of flow and tube size, the heat-transfer coefficient for triangular pitch is roughly 25% greater than for square pitch.)
- The closer the baffle spacing, the greater the number of times the shell fluid is to change its direction, resulting in greater turbulence.
- Shell-side coefficient is also affected by tube size, clearance and fluid-flow characteristics.
- Shell-side flow area varies across the bundle diameter with the different number of tube clearances in each longitudinal row of tubes. That's why

there is no true shell-side flow area by which the mass velocity of the shell fluid can be computed. The correlation obtained for fluids flowing in tubes is obviously not applicable to fluids flowing over the tube bundles punctuated with segmental baffles.

Some of the terms used in heat exchanger specifics can be confusing – words such as rating, design and selection. These can be explained as follows:

- Rating defines as the computational process in which the inlet flow rates and temperatures, fluid properties and heat exchanger parameters are taken as input, and the outlet temperatures and thermal duty (if the exchanger length is specified) or the required length of the heat exchanger are calculated as output.
- Design defines as the process of determining how the exchanger must perform given heat duty and respect limitations on shell-side and tube-side pressure drop.
- Selection is simply choosing a heat exchanger from among a number of existing designs.

Overall, you want the (theoretical) heat-transfer coefficient, U, to be as high as possible. The U value represents the lack of resistance to transferring heat between the two fluids (gas and water). The more resistance there is to heat moving from one stream to the other, the greater the temperature difference needed to get the process temperature you want. This means unnecessary waste of heating or cooling.

Internal vs. External Styles

In the external heat exchanger quench-loop design (Fig. 15.2), the blower housing, heat exchanger housing and most quench piping are located outside the vacuum heating chamber. The biggest advantage of this design is easier maintenance access to the blower and heat exchanger. The gas circulation pattern (Fig. 15.3) is such that pressurized gas is directed by the external blower into the hot zone, where it exits either through a series of gas nozzles typically located in a 360-degree pattern around the work or through a bung opening(s). In this way, the load gives up its heat, and the hot gas exits the hot zone and is directed through a heat exchanger, where it is cooled.

FIGURE 15.2 | Vacuum furnace with external heat exchanger (courtesy of VAC AERO International)

FIGURE 15.3 | Gas circulation pattern for an external heat exchanger system (courtesy of VAC AERO International)

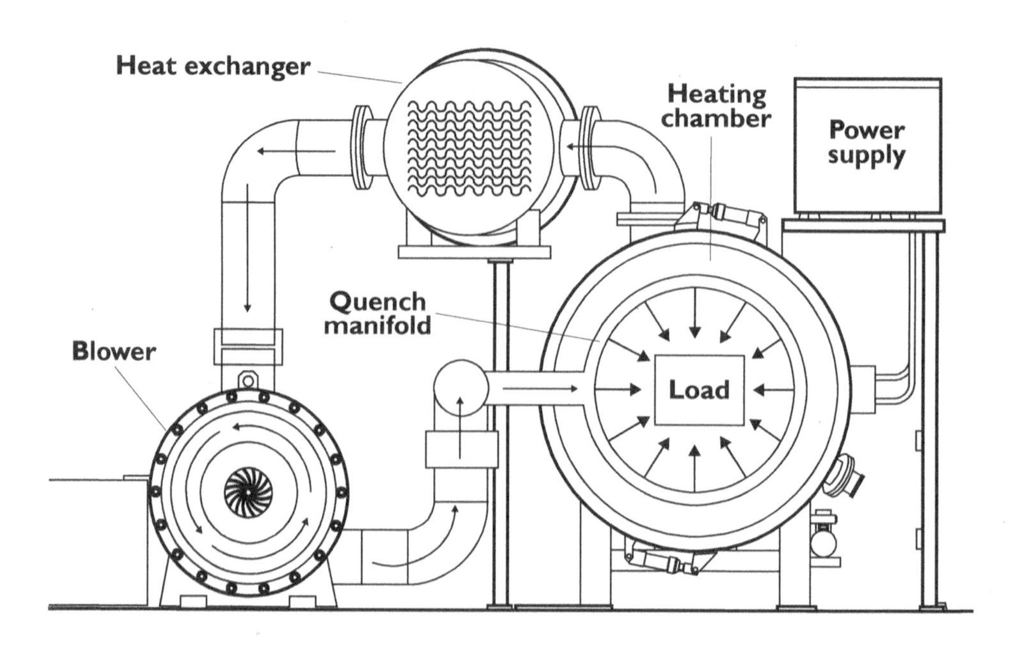

In the internal heat exchanger quench-loop design (Fig. 15.4), the quench blower and heat exchanger are contained within the chamber. The heat exchanger is most often located either in front of the fan (Fig. 15.5) or on the sidewalls of the vessel. The biggest advantage of this design is the close proximity of the cooling system to the hot gas stream to maximize heat transfer. Another advantage is the compact furnace design. Internal designs also eliminate the need for separate housings for the quench blower and heat exchangers. Its chief drawback is the close proximity of the quench blower motor (including drive shaft and bearings) to heat emanating from the furnace hot zone. If a failure of a motor component or heat exchanger does occur, it may be necessary to remove the entire hot zone to gain access for repairs.

FIGURE 15.4 | Vacuum furnace with internal heat exchanger: a.) front view; b.) rear view (courtesy of Ipsen Inc.)

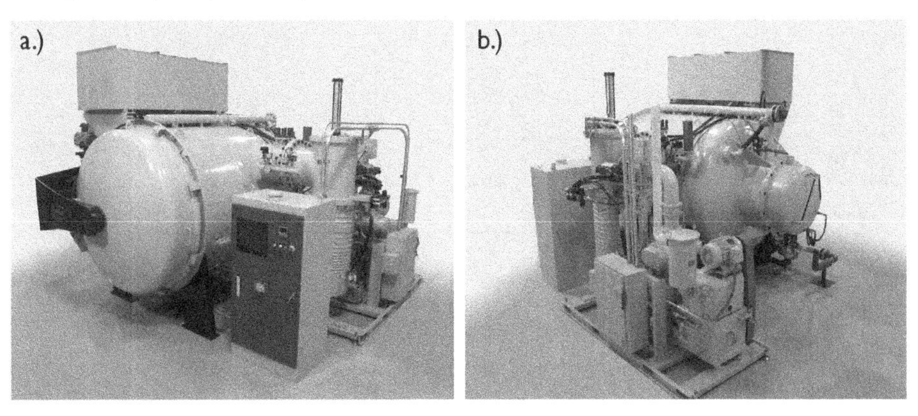

FIGURE 15.5 | Gas circulation pattern for an internal heat exchanger system (courtesy of Ipsen Inc.)

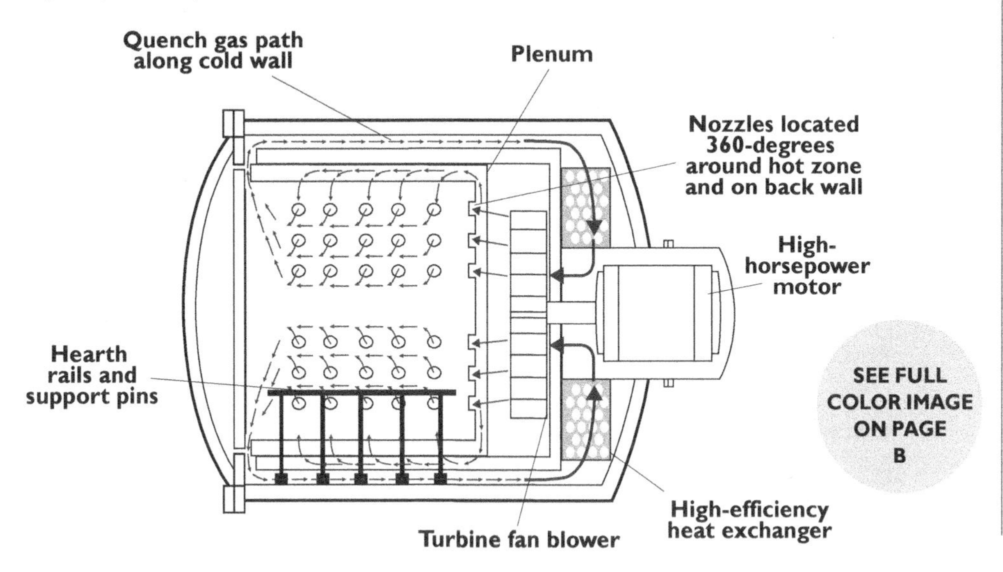

SEE FULL COLOR IMAGE ON PAGE B

Gas quenching follows the final thermal soak in vacuum. In normal operation, the furnace is quickly backfilled with inert gas to slightly below (sub-atmospheric) or above positive pressure. The gas – driven by a fan or blower – is continuously circulated, flowing at high velocity over the workload and through a gas-to-water heat exchanger. The typical gases used for gas quenching are nitrogen, argon and helium. Quench rates are enhanced through the use of cooling gas at greater than atmospheric pressures (Fig. 15.6). The advantage of higher-pressure cooling is a denser gas with increased mass flow and, therefore, greater thermal conductivity, all of which add up to improved cooling rates. In addition, gas blowers and heat exchangers operate at better efficiency at increased pressure.

FIGURE 15.6 | Typical cooling time vs. gas pressure [5]

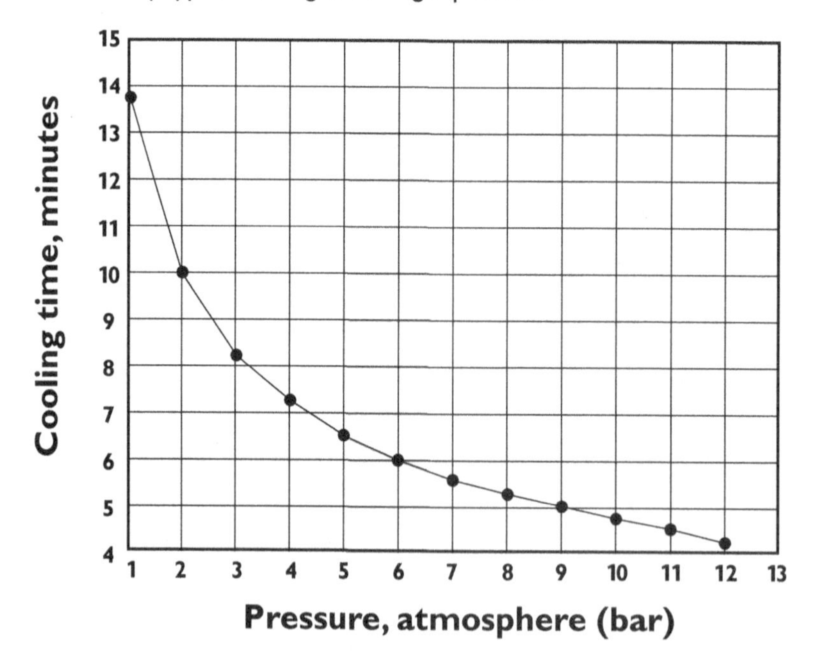

Maintenance Considerations

Heat exchangers should be scheduled for an annual maintenance check to ensure that the coils and finned tubing are clean and that the tubes are not plugged. In reality, however, heat exchangers are seldom checked or maintained, going years without inspection unless a cooling or pump-down problem exists. The most common problems that do occur include:

WATER LEAKS – PUMP-DOWN ISSUES

If the vacuum furnace fails to pump down properly, one of several possibilities is that a pinhole leak exists in the heat exchanger tubing or brazed joints. To test the heat exchanger in place, the water is removed from the heat exchanger and a helium leak check is performed while the furnace is cold. It is often helpful, before introducing helium, to heat the heat exchanger up to several hundred degrees with the ends open so as to eliminate trace amounts of water and possibly open up the pinhole leak. Appropriate cautions must be observed. If a leak is detected, the heat exchanger must be pulled from the unit, pressurized while placed in a tank of water and bubble tested to locate the leak source. Repair typically involves re-brazing.

DIRTY FINS – LOSS OF COOLING EFFICIENCY

Many times heat exchanger fins are covered or coated with dirt (Fig. 15.7) or a film-like residue from improperly cleaned workloads, or they have small parts embedded in the fins. This lowers their overall cooling efficiency and cooling cycle's length in response. Maintenance includes blasting the surfaces with clean, dry compressed air or, in some instances, removal followed by steam cleaning. A high degree of care must be taken not to use too high a pressure so as to bend the copper heat exchanger fins. (If this occurs, they must be straightened, which is a labor-intensive operation.)

FIGURE 15.7 | Dirty heat exchanger (courtesy of Solar Atmospheres Inc.)

FLOW RESTRICTIONS – LOSS OF COOLING EFFICIENCY AND (WATER) TEMPERATURE INCREASE

An increase in water temperature and/or pressure is a clear indication that blockage exists in the heat exchanger system. This is most often accompanied by a loss of cooling efficiency (i.e. longer workload cooling cycles, lower hardness or improper part microstructure). This situation should be corrected as soon as it is discovered to prevent additional problems.

In rare instances, the flow restriction is such that steam forms inside the heat exchanger tubing and creates a dangerous situation often accompanied by loud noises similar to sounds heard in steam pipes under similar circumstances. The vacuum furnace should be immediately shut down, the heat exchanger removed, the restriction(s) found and removed, and the unit pressure tested and reinstalled.

Cleaning

Cleaning of a blocked heat exchanger typically involves the following procedures.

- Circulate chloride-free cleaning solution through the shell at a good velocity to remove sludge or other soft deposits.
- Remove soft deposits by circulating hot fresh water through the shell.
- Remove hard scale, mineral and other hard deposits preferably by mechanical means (e.g., brushes). Chemical cleaning, although possible, may cause damage to the tubes and must be performed by a qualified source familiar with the chemicals being used.

In the event that the heat exchanger tubes are to be steam cleaned, extreme precautions must be taken to ensure that damage to the tubes (e.g., hammering) does not occur. It is important that the shell side remains in operation or flooded and equipped with a relief valve. This will prevent shell liquid from evaporating and possibly increase chloride concentration to an unsafe level and cause stress corrosion cracking of the tubes. Also, if the shell fluid is heated, it is possible for the pressure to rise above design conditions and cause leaks.

Summing Up

Heat exchangers are one of those components in vacuum systems that are critical to the overall cooling performance of the furnace. They are often taken for granted, which is why they have been singled out for discussion here.

REFERENCES

1. Craig Moller, Ipsen Inc., technical and editorial review
2. *Design Data: Heat Exchangers,* General Electric, 1959
3. Than, Su Thet Mon, Kin Aung Lin and Mi Sandar Mon, *Heat Exchanger Design,* World Academy of Science Engineering and Technology 46, 2008, p. 604-611
4. Fabian, Roger, *Vacuum Technology: Practical Heat Treating and Brazing,* ASM International, 1993, p. 110
5. *One Minute Mentor,* ASM International, 2007
6. Ralph Poor, Surface Combustion Inc., private correspondence

CHAPTER 16

WATER COOLING SYSTEMS

All vacuum furnaces require some form of cooling, either by water or a suitable fluid (e.g., ethylene glycol) with high heat-transfer characteristics. Cooling systems can be designed as either open systems (discharging to a sewer) or closed (recirculation) types. There are a number of factors to consider when selecting a vacuum-furnace cooling system, including type, treatment methods and system maintenance. The importance of each of these factors cannot be overstated, and neglecting any one can have disastrous consequences on the performance and life of the vacuum furnace and cooling system.

Cooling System Choices

Most plant systems fall into one of three general categories (or combinations thereof): single pass, multiple pass or recirculation. The water source is usually municipal, well or river, all with high levels of dissolved oxygen. This leads to early rusting of the vacuum vessel, reducing its life expectancy by as much as two-thirds.

In a single-pass system, the cooling/process water is used only once and then discharged from the system. Today, it must often be treated before being discharged into the sewer system and is subject to a number of EPA, federal, state and local codes. Many consider this the least expensive type of system, and it is perhaps the most common (although, in this writer's opinion, not the best). In areas where the water has high hardness levels or large amounts of total dissolved solids, the rate at which rust or scale will form and other problems occur increases, thus necessitating frequent cleaning or flushing of the cooling system and associated piping.

Recirculation systems are those in which the cooling medium is reused and recycled. Three basic types are additive systems, open systems and closed systems.

An additive system uses make-up water to maintain a specific temperature or temperature range for the cooling system. Higher-temperature water is discharged from the system to help maintain temperature control. The advantage over a single-pass system is that the amount of water used is minimized. A typical system requires a recirculation pump, sump tank and automatic water-temperature control system.

Open recirculation systems utilize evaporative cooling via a cooling tower, evaporative condenser, sump tank or spray pond to remove large amounts of heat with small amounts of water loss. One concern with this type of system is the tendency to concentrate contaminants, and it must be dealt with by regulating the amount of water that is bled off the system and by monitoring the condition of the water present in the system. The issue of dissolved oxygen in the water is also a great concern.

Closed recirculation systems with water/water, water/liquid or water/air heat exchangers have become popular methods of controlled cooling, especially when water-treatment costs are to be minimized or zero-discharge situations are required. These will be further detailed below.

Water Requirements

In order to have satisfactory performance over an extended period, the following requirements are needed (as a minimum).

- Hardness: 7 grains/gallon (maximum)
- Calcium carbonate level: 3-100 ppm
- pH: 7.0-8.0
- Suspended solids: < 10 ppm
- Conductivity: 300 micro mho/cm

Hardness is a measure of the dissolved salts (e.g., magnesium, calcium) present in the water, typically as ions. A high ion count is also responsible for an increase in conductivity. Controlling the pH is important. Low pH values (<7.0) indicate acidity that will contribute to corrosive attack, while high pH values (>8.0) indicate alkalinity, which can contribute to the precipitation of calcium carbonate ($CaCO_3$) as scale. Other important considerations include biological conditions (e.g., algae) and dissolved gases (e.g., oxygen).

A typical set of water specifications for horizontal vacuum-furnace systems requires specifying the following:

- Maximum inlet water temperature, typically 30-38°C (85-100°F)
- Minimum inlet water temperature to the furnace above ambient dew point (to avoid condensation)
- Minimum flow rate (as required by the particular system involved)
- Maximum inlet water pressure, typically 345-415 kPa (50-60 psig)
- Minimum pressure drop through the system, typically 205 kPa (30 psig)

Signs of Trouble

In addition to spiraling utility bills, other tell-tale signs that a recirculation water system is needed include:

- ❖ Hot surfaces (e.g., near the bottom or top of the vessel)
  - Inadequate cooling from low flow or scaling
  - Damage by corrosion inhibitors
  - Increase in pressure at furnace inlet indicates scaling problem
- ❖ Freezing tower
  - Possible fouled plate and frame heat exchanger
  - Clogged strainer or filter
  - Temperature controls set too low
- ❖ Discoloration, odors or foam in the water
  - Root cause – inadequate water treatment, possibly bacteria or algae
  - Corrective action – flushing and cleaning of all piping

A dedicated water system can often extend the life of major components (e.g., shell, power feed-throughs, pumps, heat exchanger) and ensure that the right temperature and pressure of cooling water is delivered where and when needed. Ideally, the furnace should be placed on a closed water system from initial start-up, but it is never too late to implement a closed water system.

Cooling Requirements

In general, the following criteria are applied to water systems for vacuum furnaces:

- ❖ Large low-pressure reservoir
  - Large thermal mass required during quench
  - Low pressure – must *not* put back-pressure on the furnace shell, typically under 69 kPa (10 psig) maximum
- ❖ Large furnace return water pipes
  - Eliminate restrictions in return water lines
  - Make return pipes as short as possible
- ❖ Maximum elevation of about 9 meters (30 feet) for return piping
- ❖ Balancing valves or pressure regulators may be needed in multiple furnace installations

As an example, a typical single-chamber vacuum furnace (Fig. 16.1) has the following water requirements.

- ❖ Overall system requirement: 345-415 kPa (50-60 psi) water at 30-38°C (85-100°F) at 1,090 lpm (288 gpm)
- ❖ Ancillary equipment cooling: 129 lpm (34 gpm)

- Mechanical pump, fan, blower, feed-throughs, power terminals
- Prefers high pressure and cool water

❖ Diffusion pump requirements: 15 lpm (4 gpm)
- Flows from 1.9-15 lpm (0.5-4 gpm)
- High pressure required: 205 kPa (30 psi) or greater
- Inlet temperature: 15-38°C (60-100°F)

❖ Heat exchanger: quench 757 lpm (200 gpm)
- Typically provided with solenoid that opens only during quench
- High flow, not pressure-sensitive

❖ Furnace jacket and heads: 190 lpm (50 gpm)
- High flow, low pressure: less than 103 kPa (15 psi)

Quenching a typical load (Fig. 16.2) from 1205°C (2200°F) results in different water-system flow requirements over the length of the cycle (Fig. 16.3).

FIGURE 16.1 | Typical horizontal vacuum-furnace water schematic

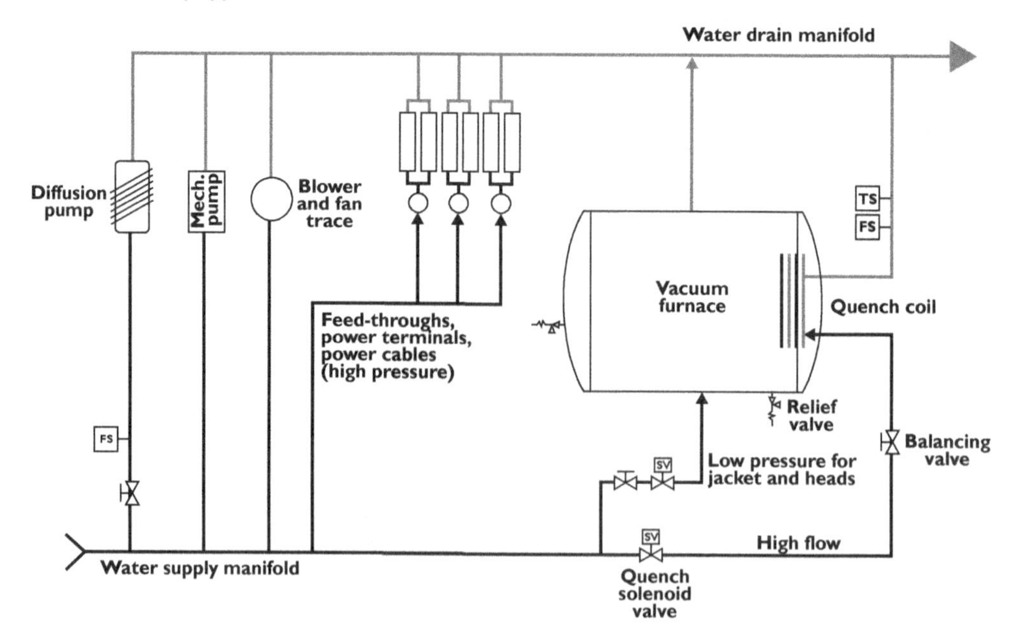

FIGURE 16.2 | Typical workload cooling rate

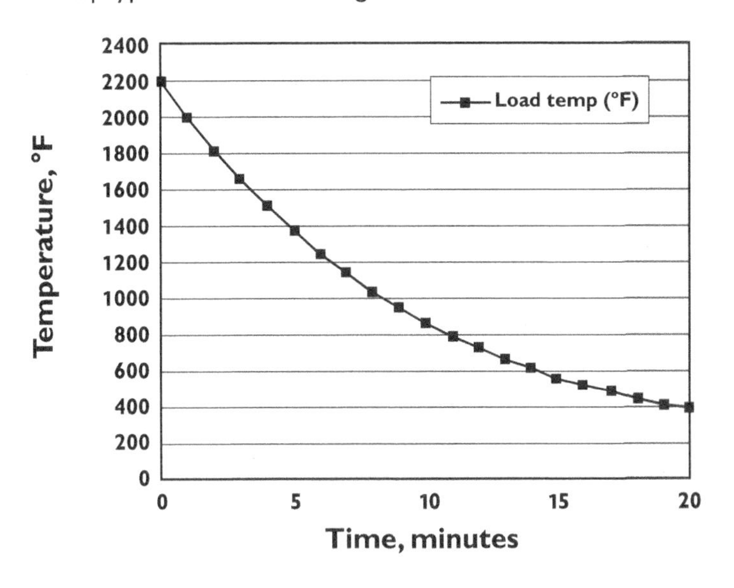

FIGURE 16.3 | Typical cooling-system response on quenching a load from 1205°C (2200°F)

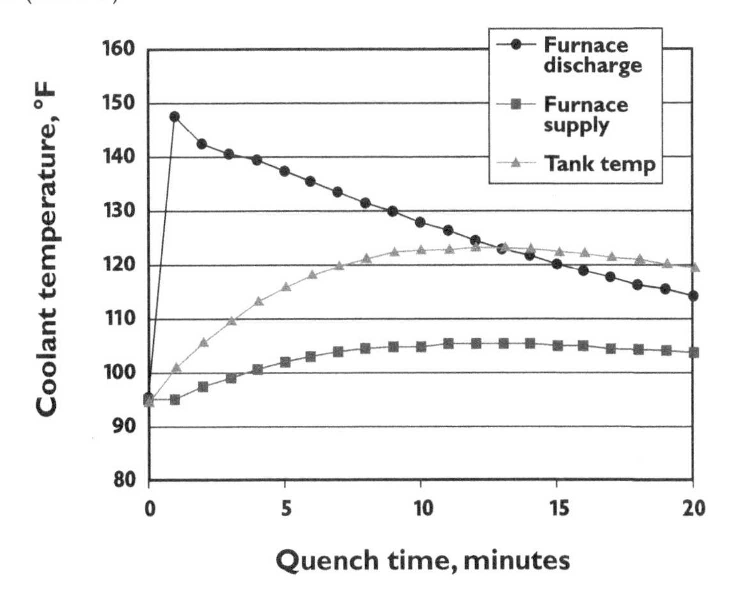

Types of Cooling Systems

TOWER, PLATE AND FRAME

Systems using tower, plate and frame cooling arrangements (Fig. 16.4) provide 29-35°C (85-95°F) water in a closed cooling loop for the vacuum furnace. Low water usage reduces operating cost, and glycol is not required. These systems are considered to be a moderately priced capital investment.

Limitations to these systems are that they require water treatment and filtration, are best suited for warmer-climate operation (they are susceptible to freezing in northern climates) and require regular annual maintenance (e.g., PHEs, strainers). In addition, the towers act as air scrubbers.

FIGURE 16.4 | Schematic of tower, plate and frame cooling system [3]

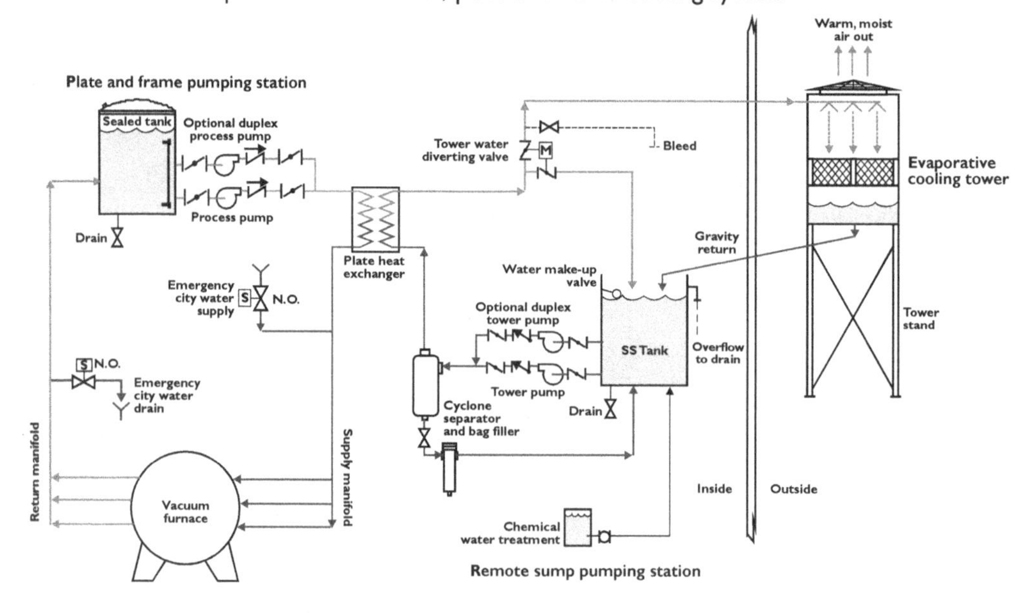

FIGURE 16.5 | Typical 757-lpm (200-gpm) tower, plate and frame cooling system with 100-mm (4-inch) piping (courtesy of Dry Coolers Inc.) [3]

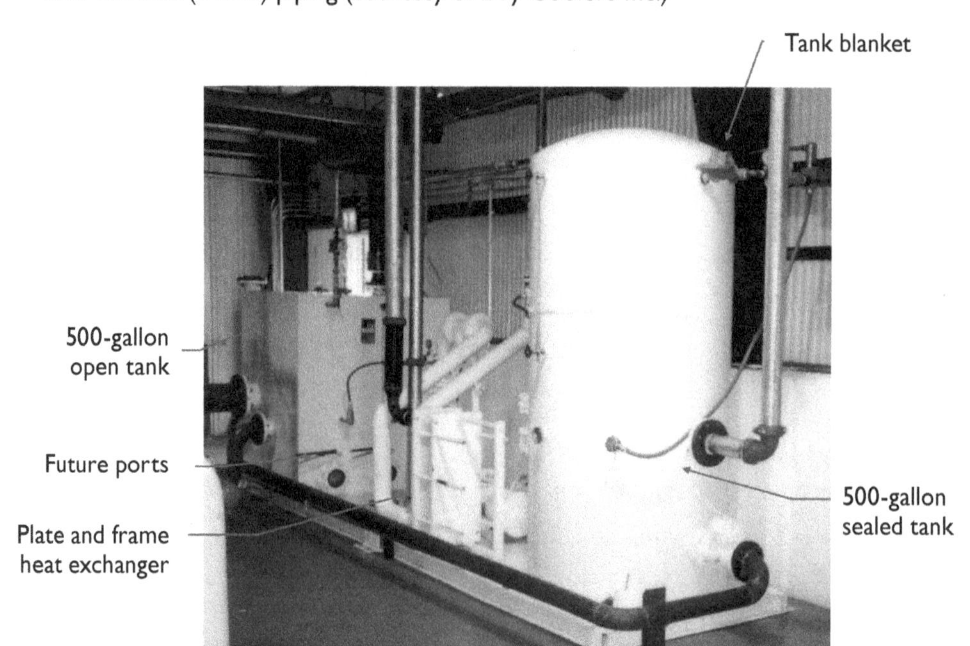

CLOSED-CIRCUIT EVAPORATIVE COOLING SYSTEMS

Evaporative cooling uses a closed-loop system with glycol antifreeze (Fig. 16.6) to deliver lower system operating temperatures, typically in the 29°C (85°F) range or lower if required.

AIR-COOLED HEAT-EXCHANGE SYSTEM

Closed-loop systems can be combined with air cooling (Fig. 16.7) for zero water usage since there is no evaporation and no environmental emissions. They offer the lowest maintenance of any system, creating minimal scale or corrosion buildup. In-line or bypass filtration and chemical treatment is still necessary. They require the highest capital expenditure of any system, but (in general) they are highly reliable.

FIGURE 16.6 | Typical evaporative cooling-system schematic [3]

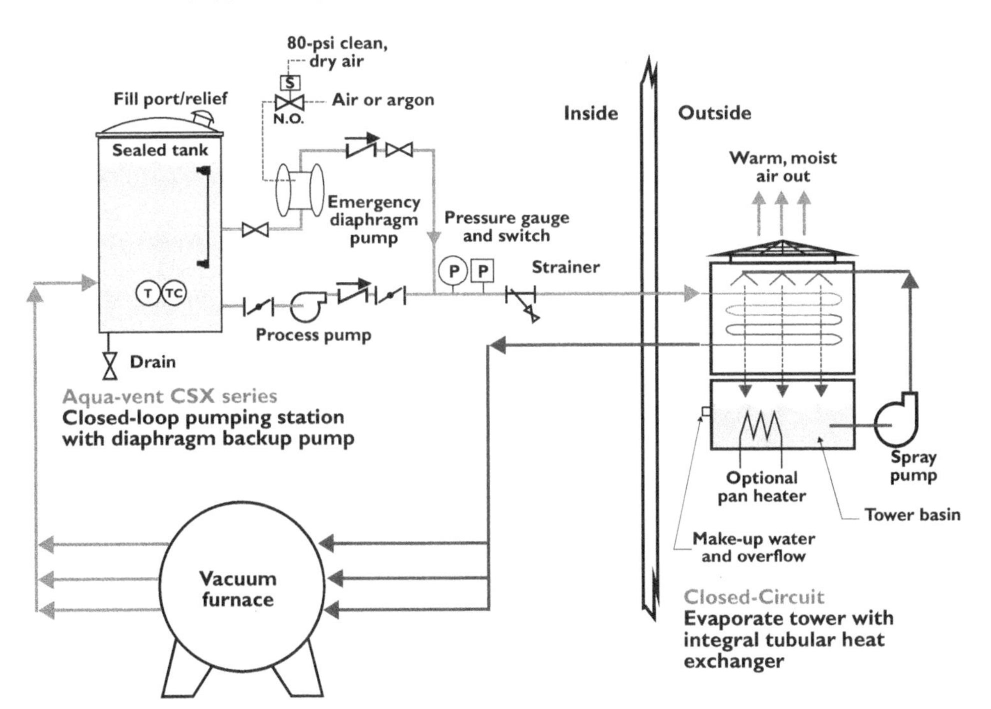

FIGURE 16.7 | Typical evaporative system with air-cooled heat exchanger [3]

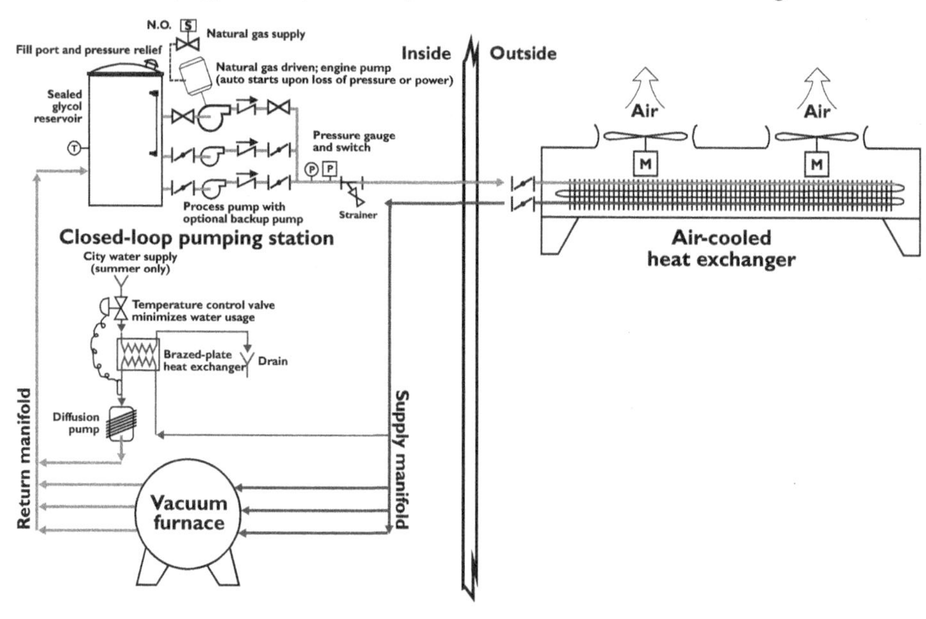

FIGURE 16.8 | Typical air-cooling system components (courtesy of Dry Coolers Inc.)

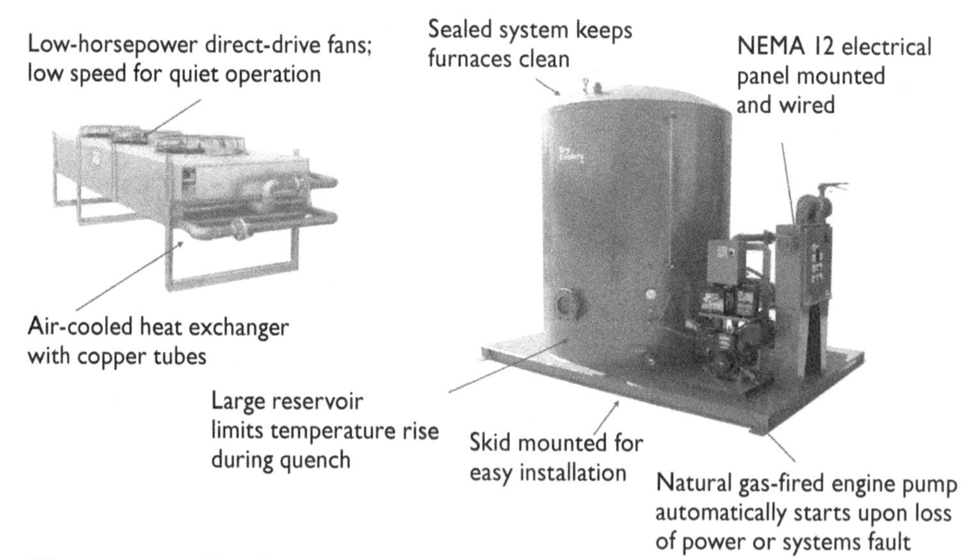

Emergency Backup

Emergency backup is intended to provide a continuous cooling flow to the vacuum furnace during a loss of power. Backup options include: a normally closed solenoid valve system designed to activate city water to the system in the case of a power loss; a diaphragm pump (10- to 100-gpm capability) with auto-start capability for use on glycol systems or an engine pump (100- to 1,000-gpm capability); or a backup generator.

FIGURE 16.9 | Engine pump (8 HP, 120 gpm) [3]

FIGURE 16.10 | Backup generator (40 HP, 1,000 gpm) [3]

Water Treatment

The following areas make up an effective water-treatment plan:

- ❖ Scale control
  - Not usually a problem in closed systems without evaporation
  - Must be controlled in open-loop systems
- ❖ Corrosion control
  - Oxygen pitting
  - Galvanic reaction – mixed metallurgy

- ❖ Biological control
  - Water treatment should be performed by local companies familiar with local water conditions.

For tower, plate and frame systems, the following is true:

- ❖ Must protect both ferrous and nonferrous components
- ❖ Sodium nitrite-based inhibitors (these are the most common) are used
  - Supplied with colorant for leak detection and treatment monitoring
  - Not compatible with glycol systems
- ❖ Requires biocides
- ❖ Dilution and discharge after use

For evaporative systems using glycol, the following is true:

- ❖ Ethylene glycol is the most commonly used coolant
- ❖ Propylene glycol is more expensive and less efficient
- ❖ Requires corrosion inhibitors (usually sold that way)
- ❖ Concentrations depend on climate (typically 30-50% by volume).
- ❖ Cannot be discharged to sewer

FIGURE 16.11 | Examples of scale and corrosion in untreated water systems [3]

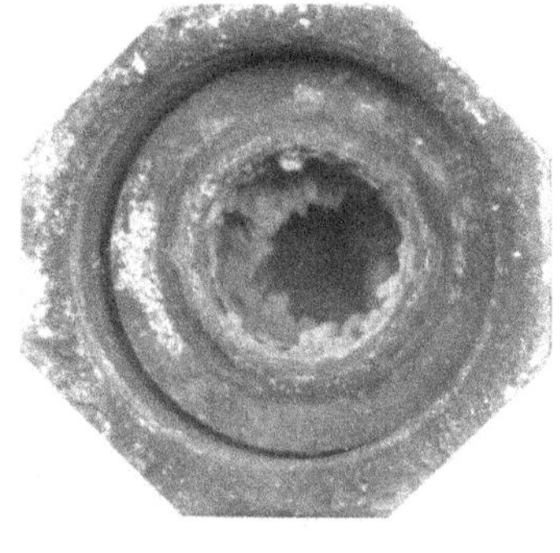

Maintenance

Depending on the type of water system employed, maintenance practices will vary. After the vacuum furnace has been in operation for some time, the water flow (collectively or in individual lines) and/or the exit water temperatures may decrease due to buildup of residue in the form of scale or rust. If the water temperature exceeds a predetermined maximum (e.g., 140°F) or if hot spots are observed on a particular area of the vacuum system, the water-flow path should be checked for restrictions. If the cooling-water flow decreases, the most common causes are:

- A gradual buildup of rust and/or scale in the vacuum vessel or cooling-water piping
- An increase in the amount of water bypassed from the discharge of the pump to the suction side of the pump
- A worn pump (or faulty pump)
- Electrical or thermal losses increasing in the cooling path

Chemical flushing is sometimes used to help correct the problem, but this requires a complete awareness of both the type and concentration of chemistry involved and an evaluation of the consequences of such treatments (good and bad). If such techniques are used, experts in this area should be consulted, as considerable damage has been reported by those without proper training and awareness.

Wintering is another practice that may be necessary given the climate, conditions and particulars of a system, especially to components located outdoors. Ethylene glycol antifreeze with low conductivity inhibitors is a popular choice as an addition to water systems given its compatibility with steel, copper, stainless steel and most other materials in the water-system piping and heat-exchange systems.

Finally, if a vacuum furnace is to be moved from one location to another, a careful inspection and close monitoring of the water system should be done in the months that follow the move. Dislodged scale can clog cooling paths and create hot spots. Corrosion effects can be accelerated, and the integrity of connections can be compromised. Older equipment that has not been on a treated water system of some type is especially vulnerable.

Summing Up

There are many choices of cooling systems available. A thorough analysis of whatever system is in use should be made in order to ensure that it is performing to expectation and that it will protect the investment and extend the life of the vacuum-furnace system.

REFERENCES

1. Brian Russell, Dry Coolers Inc., editorial review and private correspondence
2. *Vacuum Furnace Training Manual,* Abar/Ipsen "U"
3. "Vacuum Furnace Cooling Systems: Best Practices to Maximize Up-Time and Equipment Life While Saving on Maintenance and Energy Costs," ASM International Symposium on the Principles of Proper Vacuum Furnace Maintenance, 2008

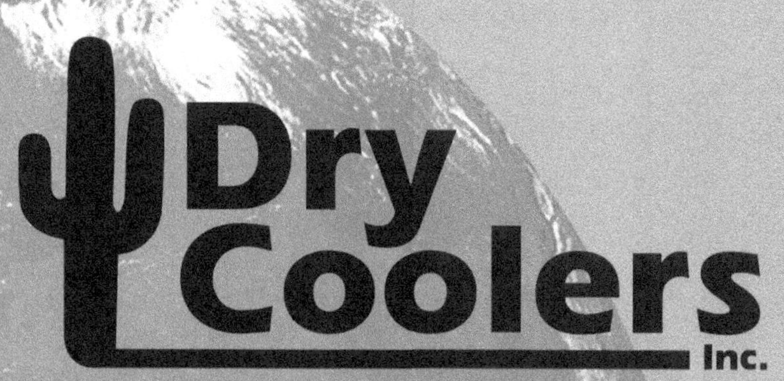

Dry Coolers Inc.
Industrial Process Cooling Systems

IHEA

MTI

CHAPTER 17

VACUUM PROCESS INSTRUMENTATION AND CONTROLS

Instrumentation and process controls used on vacuum furnaces in the heat-treatment industry (Fig. 17.1) are extremely diverse due in large part to the fact that the life of a vacuum furnace can range from 20 to 50 years. This survey of major equipment manufacturers and suppliers will allow the reader to better understand the products being offered today and provide a glimpse into the future with respect to state-of-the-art instrumentation and controls.

FIGURE 17.1 | Typical family of instrumentation (courtesy of Yokogawa Corp.)

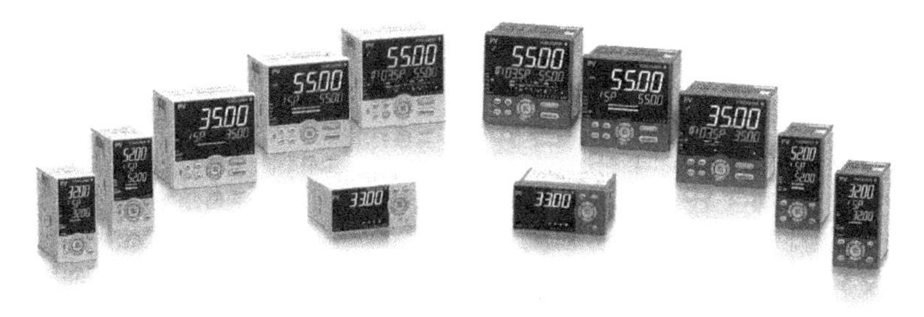

Introduction

"Adaptive" process-control systems are utilized on today's advanced vacuum furnaces. Depending on the machine or process, different variables exist that must be monitored, controlled and/or changed during the cycle to achieve maximum throughput, repeatable processes and stringent quality results. Sensors monitor a selected process or equipment parameter and send the gathered data back to a controller, which then compares it to a predetermined value or setpoint. Through calculations, a controller sends a signal back to the device to make the proper adjustments to obtain a "controlled" process.

An everyday example would be cruise control on an automobile where the speed setpoint must be maintained. The variables of speed, acceleration and resistance are monitored, and then adjustments are made to reach the desired end results. Control systems on vacuum furnaces function similarly by, for example, optimizing and regulating temperature and pressure to achieve the required process conditions and produce repeatable results. Programmable logic controllers (PLCs), sensors and computers make this all possible. In turn, data trending, real-time process monitoring and data collection for permanent retention is commonplace. By analyzing this data, new cycles containing modified variables can give better results in less time.

There are a number of hardware/software manufacturers that make this adaptive technology possible. Allen-Bradley, Siemens, Yokogawa, Honeywell and Eurotherm are a few of the recognized leaders in industrial controls, especially in the vacuum-furnace industry.

A Look at the Industry Today

Here's a look at what the various vacuum-furnace manufacturers use for instrumentation and controls throughout the industry. We begin by considering the following subjects:

1. Simple process-control systems
2. Temperature ramping requirements
3. Who makes what?
4. PLCs for cycle framework and single-chamber vacuum furnaces
5. Hybrid process-control systems
6. Utility monitoring and interlocks
7. Interlocks for vacuum pumps, heating systems and gas quenching
8. Control needs for a diffusion pump
9. Partial-pressure control and vacuum sensors
10. Multi-chamber vacuum furnaces and control of motions
11. Vacuum carburizing controls
12. Supervisory computer systems

SIMPLE PROCESS-CONTROL SYSTEMS

Vacuum-furnace process-control systems (Fig. 17.2) are somewhat similar to atmosphere-style batch-furnace control systems. However, they tend to be somewhat more sophisticated, especially from a temperature-control standpoint.

FIGURE 17.2 | Typical vacuum-furnace control cabinet showing ramp/soak programmer, high-limit instrument, digital chart recorder, vacuum controller, alarm/status indicating lamps, selector buttons and power amp meters (courtesy of Carter Manufacturing Co./Carter Bearings)

TEMPERATURE RAMPING REQUIREMENTS

Naturally, temperature control is the first requirement. Whereas most atmosphere furnaces tend to change setpoints only a couple of times per cycle and usually heat as fast as possible, vacuum-furnace cycles typically incorporate a heating ramp and quite often a reverse or backward ramp. Ramp rates are usually displayed in degrees per hour or degrees per minute. A traditional single-loop PID (proportional integral derivative) controller would not typically be used for temperature control alone.

Based on this need for more complicated temperature cycles, a control system with setpoint ramping, as well as multiple setpoints, is a must. It is not uncommon for vacuum-furnace recipes to have 10 or more segments in a given recipe.

As a minimum, the temperature controller would have a PID control function (as opposed to simple on/off control) coupled with a ramp/soak programmer. Some controllers will also have auto-tune capability along with anti-overshoot algorithms, which are often dubbed "fuzzy logic."

WHO MAKES WHAT?

The world of process control and programmable controller suppliers is vast, as is any electronic offering in the digital world. The process controllers listed in this chapter have been referenced due to their popularity within the general heat-treat industry and only a brief summary of their features given. If one were to investigate the purchase of one of these control systems, there are a great number of other worthy suppliers with comparable or even more sophisticated systems.

A typical entry-level controller commonly used for temperature control would be a Honeywell DCP 550 ramp/soak programmer (Fig. 17.3). Similar products from other suppliers are also available and comparable. Based on user preferences, one could also consider a Eurotherm model 2704, a Yokogawa model 8838 or an SSi model 9220 depending on features or experience with the product line.

FIGURE 17.3 | Honeywell DCP 550

We will also delve into the use of PLCs, which are used for logic control and interlocks and can be expanded to include the functions of the discrete ramp/soak programmers. A PLC will typically offer the ultimate in designer flexibility. The Eurotherm 2704 ramp/soak programmer with recipes and expandable digital I/O (Fig. 17.4) is one such device. This unit has up to 60 recipes or 600 total segments. This unit also has eight analog inputs, eight analog outputs and up to 43 digital I/O.

FIGURE 17.4 | Eurotherm 2704

Another typical PLC is the Yokogawa UP55 ramp/soak setpoint programmer (Fig. 17.5) with 30 recipes, each having 99 steps. This unit has nine digital inputs, 18 event outputs, eight PV events, 16 time events and eight alarms.

FIGURE 17.5 | Yokogawa UP55

PLCs FOR CYCLE FRAMEWORK AND SINGLE-CHAMBER VACUUM FURNACES

To work in conjunction with the ramp/soak programmer, it is common to use a small programmable controller. An entry-level unit for the PLC would be an Allen-Bradley MicroLogix PLC (Fig. 17.6). This unit, or equal, simplifies the control of the many event situations and cycle interlocks.

FIGURE 17.6 | Allen-Bradley MicroLogix PLC

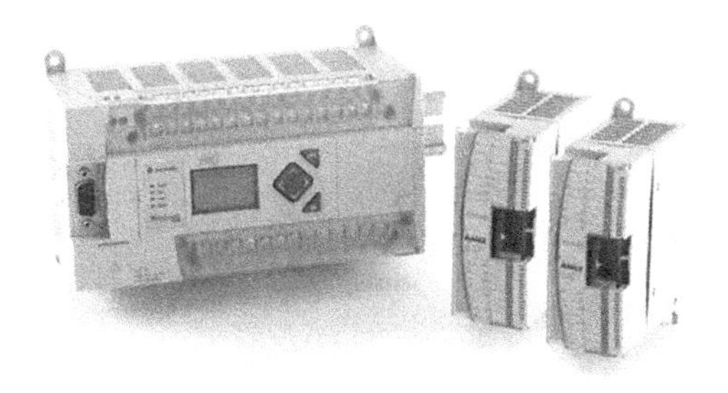

In a single-chamber furnace, the heating chamber starts cold or at room temperature. From that point, a load is placed in the heating chamber and the furnace door is closed. Once the door is closed, a clamping or locking ring is often used to ensure the door is safely closed should the furnace be equipped with positive-pressure gas quenching. The PLC can test that the door is fully closed and activate the evacuation valve to allow the vacuum pumping system to evacuate the heating chamber of air and other contaminants.

The logic of the PLC will also include alarming if the required utilities are not turned on. Hardware relays are used for safety interlock purposes, but PLCs lend themselves to displaying that an interlock condition is present.

Vacuum furnaces usually have a two-stage vacuum pump and often a diffusion pump for achieving higher vacuum levels. The mechanical vacuum pump (roughing pump) is controlled by a selector switch, motor starter with overloads and fuses. The vacuum blower, or booster, which compresses gas or atmosphere going to the mechanical pump, has the same components and usually a pressure switch to prevent the vacuum blower from turning on until the appropriate vacuum threshold is obtained. A typical threshold might be 30 torr. The pressure switch can be electronic, such as a Pirani gauge, or it might be a simple mechanical switch. Turning on the vacuum blower too early will overload the drive motor and cause overloads to trip or fuses to blow.

Additional interlocks exist that do not allow the heating system to be enabled until the furnace has pumped down to a low-vacuum setting, usually on the order of 0.05 torr (50 microns). Should the furnace be equipped with a diffusion pump, the heating system and temperature control are further held off until the furnace reaches sub-micron levels (for example, 5 x 10^{-3} torr). For simpler control schemes, these values are usually programmed into the vacuum-measuring instrument, or they could be part of a recipe on more exotic control systems.

HYBRID PROCESS-CONTROL SYSTEMS

An alternate process-control system often used is a hybrid system. The hybrid controller is a combination PID process controller with many PID loops available through software and an integral ramp/soak programmer. Instead of a dedicated loop controller with ramping capability, the system is programmable to have one or more PID loops available, with any of them coupled to ramp/soak programmer modules. The Honeywell HC900 hybrid controller (Fig. 17.7) with close-up of the operator interface (Fig. 17.8) is one such example.

FIGURE 17.7 | Honeywell HC900

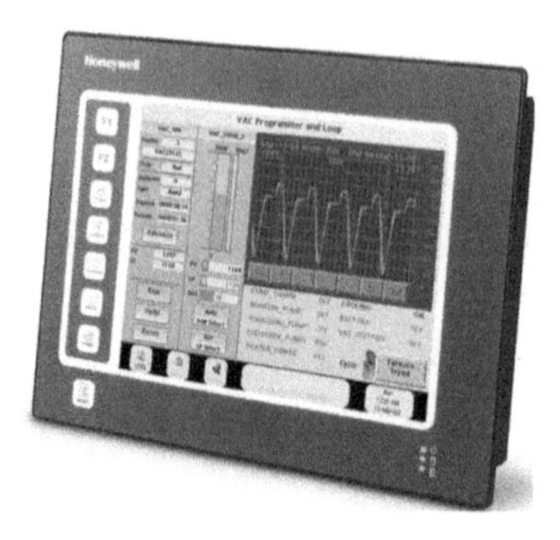

FIGURE 17.8 | Honeywell HC900 system

Note: The personal computer is not required after the system has been programmed, unless computer trending of the process data is desired.

The hybrid controller also has PLC logic capability, so this type of controller replaces the ramp/soak programmer and also the conventional PLC. Most systems have a color human machine interface (HMI) to allow the operator to take control of the entire furnace operation (Fig. 17.9). Most PLCs are programmed in ladder-style logic. Hybrid controllers, however, are generally programmed through the use of function blocks.

FIGURE 17.9 | Rockwell Automation panel HMI (courtesy of C.I. Hayes)

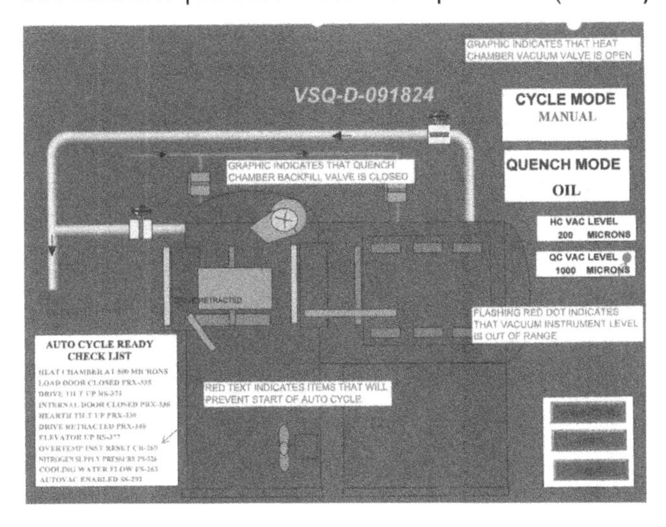

The advantage of the hybrid controller is that the logic and process control is performed in the same box. This approach saves money on inputs and outputs that are otherwise required to pass information from the PLC to the ramp/soak programmer and vice versa.

The hybrid controller typically has preformatted displays. Using preformatted displays on different furnaces makes it more user-friendly for the furnace operator. This also allows the design engineer to quickly configure the unit for the operator. The drawback to this approach is the dedicated layout does not always allow for custom displays. This limitation can be avoided by using higher-end HMIs from the supplier or third-party HMIs, which typically provide a blank-sheet-of-paper approach where displays can be laid out to meet any configuration desired.

Third-party HMIs using Wonderware development software (Fig. 17.10) are ideal when "canned" or programmed screens will not work well for an application.

FIGURE 17.10 | Wonderware software package (courtesy of C.I. Hayes)

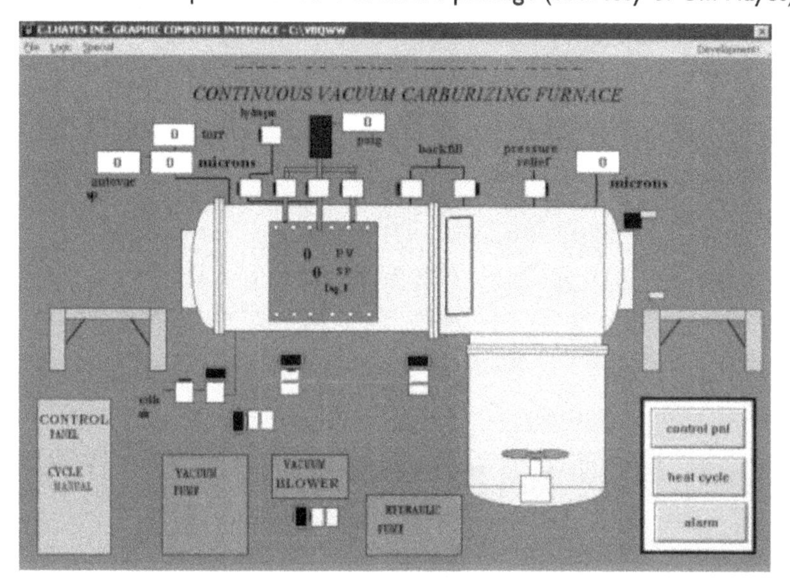

UTILITY MONITORING AND INTERLOCKS

Due to the very high operating temperatures possible with vacuum equipment, extreme care must be given to ensuring utilities are available at all times and, in the event of loss of utilities, that the furnace is automatically put into a safe condition.

Checks and precautions center around verifying, for example, water pressure and water flow to all water-cooled components. Without water flow to the heating-system transformer, casing, power feed-through, diffusion pump and other water-cooled members, severe damage could occur to the furnace as well as unsafe operating conditions. The alarm circuits are usually provided by hardwired relays, alarm contacts and interlocks.

Display on an HMI is often provided to alert operators as to the problem, but alarm action and intervention is always hardwired. Alarm circuits are usually provided by a separate source of electrical power as standard practice. Gas supplies for quenching (usually nitrogen or argon) are also tested by the use of pressure switches and must be "ON" before a typical vacuum-furnace cycle can start. Recommended practice is also for a furnace operator to visually verify the above situations are met as a further precaution. Under loss of power, the evacuation valve is always closed to prevent air from backstreaming through the vacuum pumps and into the furnace. Many customers maintain water flow with a natural gas or gasoline-powered backup water pump. Some systems go to city water during power failures.

Once the safety requirements (which include all utilities being present, water is flowing to all water-cooled components, no alarms are active and the furnace has been evacuated to the preset levels) are satisfied, the heating system is then allowed to turn on. On multi-chamber furnaces, similar precautions and interlocks are required. Once heated, however, the heating chamber will stay hot and evacuated.

CONTROL NEEDS FOR A DIFFUSION PUMP

Some vacuum furnaces are equipped with one or more diffusion pumps (Fig. 17.11) to allow the vacuum levels to be further reduced by a factor of 1,000 or more. While a typical vacuum furnace can operate at 0.05 torr (50 microns) while heating, diffusion pumps can further reduce vacuum pressure to 1×10^{-6} torr (0.001 micron) or even lower with larger pumps.

FIGURE 17.11 | Typical diffusion pump

Diffusion pumps have no real moving parts. They have an electric heater, however, much like the heater on a home electric range that heats the diffusion-pump oil to approximately 175°C (350°F) or higher. The electric heaters typically free run and have a mechanical high-temperature switch to turn off the heaters should water flow be inadequate. The diffusion pump has a large inlet valve, often called a right-angle valve, that is the same diameter as the diffusion pump. A downstream valve, called the foreline valve, is provided for connection of the diffusion-pump outlet to the mechanical vacuum pump and vacuum blower. The right-angle valve and the foreline valve are controlled by the PLC when the diffusion pump is to be activated. Prior to being able to use the diffusion pump, the diffusion-pump heaters must be on for at least 30-60 minutes to allow the pump oil to become vaporized. A vacuum sensor is also used downstream of the diffusion pump to ensure the gases from the diffusion pump are being sufficiently taken away. Should the downstream sensor exceed a preset level, the right-angle valve will immediately close until the micron level returns to a safe level.

The PLC is used to enable the heaters, both valves and, most importantly, "valve timing," which is critical to prevent any form of backstreaming from occurring. The diffusion pump also requires a small holding pump, which is another vacuum pump downstream of the diffusion pump. The holding pump is not typically controlled by the PLC. It is simply connected to a motor starter, overload and fuses.

PARTIAL-PRESSURE CONTROL AND VACUUM SENSORS

Most of the time, a vacuum furnace will operate attempting to remove all air (atmosphere) and contaminants from a hot zone, and the pumps are running full out. There are times, however, when too hard of a vacuum can vaporize materials being processed. A typical application is high-speed M-series steel, which is routinely processed at 1175°C (2150°F) or higher. This series of steel has chromium, among other alloys, that would deplete the chromium content through vaporization if exposed to high vacuum levels.

To alleviate this condition, partial-pressure control is activated where argon or nitrogen gas is bled into the furnace while the vacuum pump is trying to take it away. A typical partial-pressure setting for M-series steel would be 0.75 torr (750 microns). The partial-pressure circuit is usually activated above a threshold temperature such as 1010°C (1850°F). At temperatures below 1010°C (1850°F), the vacuum system runs full out and vacuum levels could easily be under 0.05 torr (50 microns). The partial-pressure circuit is activated (usually by the PLC from a recipe request) above 1010°C (1850°F), and a solenoid valve allows nitrogen or argon gas to flow into the vessel. Once the partial-pressure setpoint is reached – 0.75 torr (750 microns) in this example – the solenoid is closed. A dead band prevents the solenoid from cycling too quickly.

Another variation of partial-pressure control allows a high-flow sweep gas to pass through the furnace. In this situation, furnace pressure may be much higher, easily in the 2-10 torr range. This higher pressure is usually just a result of the higher gas flow going into the furnace. This high flow of gas is designed to move contaminants in the furnace quickly away through washing or sweeping. The vacuum pump typically runs full out in this situation, and the inlet gas flow rate is controlled via a fixed orifice or mass-flow controller.

Vacuum furnaces require at least one vacuum sensor, and some furnaces can easily have three or four. Vacuum sensors are made by a number of suppliers, including MKS, Varian, Televac and Leybold Heraeus to name a few.

The vacuum furnace generally uses a small sensor to measure the actual vacuum level, and a vacuum instrument digitally displays the vacuum level and generates a recorder or controller output voltage. The vacuum instrument also provides switch contacts or trips. Traditional output voltage ranges are 0-10 VDC or 4-20 ma. "Trip contacts" are important and are used as thresholds to turn on the heating systems, activate valves or turn on the vacuum booster.

The common sensors used to measure vacuum levels down to 1 x 10^{-3} torr (1 micron) are thermocouple gauge tubes, Pirani vacuum sensors and capacitance manometers.

Thermocouple gauge tubes work on the principle of thermal conductivity using a heater and thermocouple. The lower the vacuum level, the less heat is transferred from the heater to the thermocouple. Vacuum readings are detected from the thermocouple voltage.

Pirani sensors are the most common and are modestly priced. They must be standardized to gas present in the furnace. Therefore, if they are measuring air or nitrogen, they read normally. If the atmosphere is argon or hydrogen, they must be standardized to those gases.

Capacitance manometers do not need an instrument and are not sensitive to the gases in the furnace. They cost more than a Pirani sensor or thermocouple gauge tube. However, they can be exposed to any type of gas, including vacuum carburizing gases, and will read accurately. The above sensors are typically used down to vacuum levels of 0.001 mbar (1 micron).

Cold-cathode sensors are used to measure high vacuum levels when diffusion pumps are in use. They measure down to sub-micron levels and generally have a logarithmic output. Cold-cathode sensors can be easily attacked by furnace contaminants, so it is important to have a spare sensor available.

As part of vacuum-sensor calibration, a mercury manometer designed for vacuum service should be used for testing. Also, calibrate the vacuum sensor on a periodic basis. Accurately knowing the vacuum level is as important as accurately knowing the furnace's temperature.

MULTI-CHAMBER VACUUM FURNACES AND CONTROL OF MOTIONS

In addition to single-chamber vacuum furnaces, there are also multi-chamber vacuum furnaces (Fig. 17.12) that have dedicated chambers for heating and quenching. Quenching can be under vacuum in an oil-quench tank or under inert gas in a gas-pressure quench. Since there are multiple chambers for heating and quenching/cooling the workload, the furnace must also have material-handling apparatus to transfer the load from one chamber to another.

FIGURE 17.12 | Two-chamber gas-quench vacuum furnace (courtesy of Surface Combustion Inc.)

Multi-chamber vacuum furnaces are generally controlled by a PLC. Some, however, may be controlled by a hybrid control system. This furnace has an Allen-Bradley PLC. The PLC controls all valves, all motions, temperature control, partial-pressure control and everything except safety circuits, which are hardwired.

The control of a multi-chamber furnace is substantially more complicated than that of a single-chamber furnace. In traditional two- or three-chamber vacuum furnaces, control of chamber pressure must be done in tandem with the actual motion control. Therefore, before a door can be opened, pressure between the two chambers must be equalized by either venting a chamber under vacuum up to atmospheric pressure or by evacuating a chamber to a hard vacuum should the door be opening to another chamber under hard vacuum.

A PLC will typically be used for the control of motion as well as control of evacuation valves, backfill valves and vent valves. A typical vacuum-furnace HMI using Allen-Bradley PanelView Plus (Fig. 17.13) has screens to allow furnace operators to easily access controller settings and real-time process variables and know where the load(s) are at any given time. The furnace shown in Fig. 17.14 has a primary PLC for control of the system with remote I/O to individual chambers.

FIGURE 17.13 | HMI interface

CONFIGURATION FURNACE VACUUM - ACTIVE ALARM(S)

PUMP OFF DELAY (SECONDS): 10
BLOWER ON (TORR): 30.000
BLOWER DEADBAND (TORR): 2.000
VALVE OPEN DELAY (SECONDS): 10
HEATING ON (TORR): 0.050
HEATING DEADBAND (TORR): 0.900
PUMP DOWN LEVEL (TORR): 0.020
PUMP DOWN ALARM (MINUTES): 30

LEAK RATE TEST
TEST INTERVAL (CYCLES): 5
DURATION (MINUTES): 30
STABALIZE DELAY (SECONDS): 30
ALARM RATE (MICRON/HOUR): 50.000

RECIPE GUARANTEES
HIGH (TORR): 0.050
LOW (TORR): -0.050

DIFFUSION PUMP: YES
HEATERS ON (TORR): 0.100
HEATERS DEADBAND (TORR): 0.350
WARMUP (MINUTES): 30
HEAT ENABLE (TORR): 5.0 x 10 -4
HIGH VAC ON (TORR): 0.050
HIGH VAC DEADBAND (TORR): 0.200

SELECTOR SWITCHES | PREVIOUS <<< | NEXT >>> | SILENCE ALARM(S)
MAIN MENU | SUB MENU | ALARMS

FIGURE 17.14 | Multi-chamber vacuum carburizing system with four heating chambers, one oil-quench chamber and one high-pressure gas quench (courtesy of MMS Thermal Processing LLC)

VACUUM CARBURIZING CONTROLS

Carburizing in a vacuum furnace has become a popular process (c.f. Chapter 26). Vacuum carburizing furnaces require the most sophisticated control systems. In addition to a recipe for temperature, these systems require pressure-control loops to operate the furnaces in the 2-10 torr range if they are low-pressure carburizing systems. The pressure-control loops place a restricting valve between the vacuum pump and the carburizing chamber. As the valve closes, pressure in the chamber rises and will move toward lower torr levels as it opens. The control valve, along with a PID pressure controller, maintain vessel pressure at a given torr setpoint while the carburizing gases are flowing.

Vacuum carburizing operates in a series of carburizing "boosts" followed by a related "diffuse" segment. These times can be as short as a few seconds to as long as many hours. The control-system recipe must be capable of a high number of segments for these boost and diffuse segments. Cooling to a pre-quench temperature, as in atmosphere carburizing, is also common. Computer programs (Fig. 17.15) are used to determine the actual boost and diffuse times based on: desired case depth; carburizing temperature; base carbon; diffusion-slowing elements, such as nickel; and the material's carbide-forming elements, such as chromium, molybdenum and vanadium.

FIGURE 17.15 | Typical process recipe screen

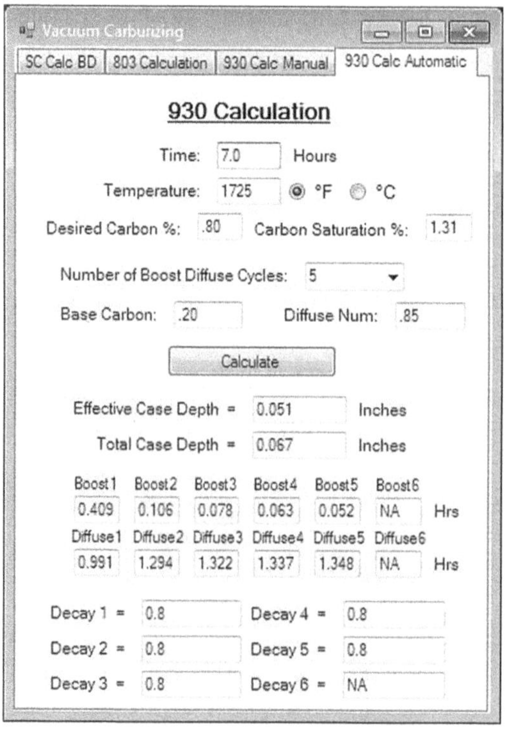

Vacuum Carburizing

SC Calc BD | 803 Calculation | 930 Calc Manual | 930 Calc Automatic

930 Calculation

Time: 7.0 Hours

Temperature: 1725 °F °C

Desired Carbon %: .80 Carbon Saturation %: 1.31

Number of Boost Diffuse Cycles: 5

Base Carbon: .20 Diffuse Num: .85

Calculate

Effective Case Depth = 0.051 Inches

Total Case Depth = 0.067 Inches

Boost1	Boost2	Boost3	Boost4	Boost5	Boost6	
0.409	0.106	0.078	0.063	0.052	NA	Hrs
Diffuse1	Diffuse2	Diffuse3	Diffuse4	Diffuse5	Diffuse6	
0.991	1.294	1.322	1.337	1.348	NA	Hrs

Decay 1 = 0.8 Decay 4 = 0.8

Decay 2 = 0.8 Decay 5 = 0.8

Decay 3 = 0.8 Decay 6 = NA

Vacuum carburizing recipes also require setpoints for carburizing-gas flows. These setpoints are typically channeled to mass-flow controllers for traditional gases such as methane, propane, acetylene and often hydrogen. The "units of measure" are often "liters per minute" for these gases. For liquid vacuum carburizing media, such as cyclohexane, an injector pulse width in milliseconds is also provided as a setpoint.

SUPERVISORY COMPUTER SYSTEMS

A customer often desires further control either for the furnace operator or for enhanced documentation involving critical processes. This desire can be accomplished by adding either a personal computer or an industrial PC that would reside by the furnace.

There are many directions one can pursue, and the actual selection should be based on what features are needed to be fulfilled along with other issues like compatibility with other furnaces or other production equipment. Some companies offer "canned programs" where the end user or the OEM simply fills in the blanks and the programming is complete. Other companies offer programs that require a great deal of configuration and can easily require several months of programming. The latter approach provides the end user with a product that fits seamlessly into their operation. The canned program is quick to implement and usually simple to operate. Should the end user need a feature that is not provided with the program, they may be stuck at that point in time.

The first requirement of the supervisory system is to make sure the furnace controllers are capable of communicating with it. The equipment described in this chapter has models that are available with computer communication capability. This is usually Ethernet or Modbus, but other protocols exist and will work.

What should the supervisory system (Fig. 17.16) do? There are many functions that a supervisory system can perform, including some of the following:

- Recipe cycle upload and download, which simplifies entering recipes, especially if more than one furnace exists
- Controller configuration settings download, which allows rapid setup if a controller is replaced
- Operation using part numbers and assurance the recipe run agrees with the part number
- Load scheduling helps manage which furnace should process a given load or batch of parts.
- Loading instructions, photos, fixture requirements, part orientation, etc.
- Bar-code operation to eliminate mistyped recipe or part numbers

- ❖ Alarm tracking, which documents (to the hard drive) furnace alarm issues that may occur and when they are corrected
- ❖ Maintenance and routine calibration reminders as well as documenting when they are performed
- ❖ Equipment checks such as temperature surveys or leak-up rates
- ❖ Process-variable trending, along with setpoints and controller outputs (can also include load thermocouples, vacuum level, mass-flow controller levels, backfill pressure, etc.)
- ❖ Real-time remote viewing of the equipment in the office or remote site, which often helps in equipment troubleshooting and allows experts at a remote location to analyze equipment or process troubles
- ❖ Lab testing requirements and integration with lab metallurgical findings
- ❖ Online operating instructions, which can also include start-up or shutdown procedures
- ❖ Integration into an enterprise system

FIGURE 17.16 | Computerized data storage for a vacuum carburizing furnace

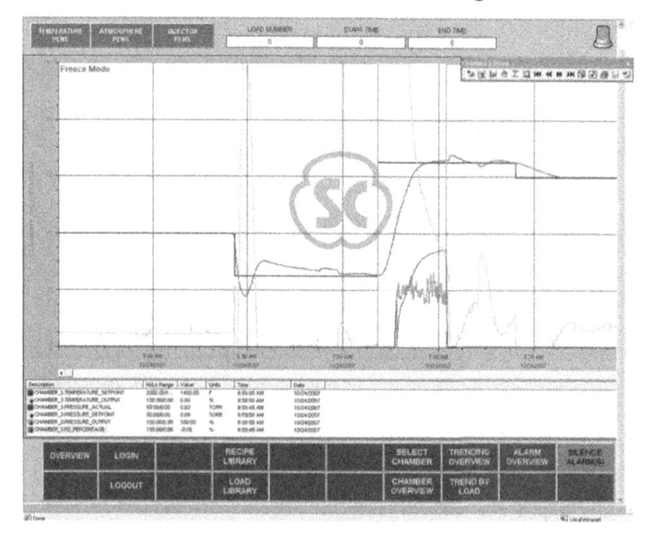

Temperature control in a vacuum heat-treating environment can be difficult because of the changing heat-transfer characteristics of the furnace as it moves from convection to radiation and conduction. The rapid heating rate of a vacuum furnace demands precise control, including setpoint program control with soak-guarantee inputs.

Vacuum furnaces (Fig. 17.17) are often used for a variety of products and processes by the heat treater, making recipe management an important function. Overshoot of temperature setpoints is usually not tolerated for metal-treating applications. Setpoint program control is often applied to the temperature, vacuum level and gas pressure with extensive interaction between these programs and also with the logic control.

FIGURE 17.17 | Modern vacuum-furnace controls

Vacuum furnaces are also used in the metal-treating industry for applications such as hardening, case hardening, brazing, melting and thin-film deposition. They are used to bring materials to high temperature with a minimum of surface interaction reaction (e.g., oxidation). In addition, surface and internal contaminants on the metal surface are volatized and removed.

Vacuum furnaces for heat treating and brazing are typically single-chamber furnaces operating batch cycles. The batch cycles vary between processes but commonly require regulation of temperature, vacuum and sequence logic. The temperature and vacuum interact extensively with the logic.

A typical heat-treating cycle starts after the product is loaded into the furnace and the door is clamped shut. Some users secure the furnace and perform a leak test before proceeding. A roughing vacuum pump lowers the pressure to about 0.05 torr (50 microns). An optional diffusion pump can lower the pressure to below 1 x 10^{-3} torr (1 micron). Some processes require an inert gas such as argon to be fed into the furnace at a low flow rate, allowing the pressure to rise to about 0.50 torr (500 microns). This is called partial-pressure control.

The pressure increases as the temperature rises and contaminants volatilize. Control of the vacuum is maintained at about 0.50 torr (500 microns) in partial-pressure processes or below 10^{-4} torr in high-vacuum processes. If the vacuum deviates from the setpoint by more than a specific value, the temperature program is held until the condition is corrected. The temperature program goes through a series of ramps and soaks. After a high-temperature soak, the quench process activates and the temperature is allowed to drop. An increased flow of inert gas and circulation of cooling water in the furnace walls and heat exchanger cools the work. The cooling lowers furnace pressure, requiring additional pressure control. During cooling, the pressure is typically controlled between 0.85-10 bar depending on process type. A light or horn usually activates as an indication to the operator that the cycle is complete. The operator then brings the furnace back to atmospheric pressure manually and unloads the product.

Control Implementation

During the last 40 years of vacuum-furnace manufacturing, different hardware platforms for controls have been used. A common platform is based on the Honeywell HC900 hybrid controller together with Experion Vista SCADA software.

The HC900 hybrid controller combined with Experion Vista interface (Fig. 17.18) meets all of the requirements for safe and productive process operation with maximum operator convenience, including:

- Program control of sequencing and variables vs. time
- Proportional (PID) modulating loop control
- Logic functions for equipment and process status
- Alarm detection, annunciation and logging
- Data acquisition and data logging
- Recipe configuration, local storage and download capability
- Easily programmable by operators in engineering units
- Sixteen programmable events for integration with sequence-control functions
- Alarms and events can be programmed to send an e-mail message
- Modbus/TCP protocol allows interfacing to HMI, data acquisition and OPC server software
- Ethernet port supports direct PC connection or external modem connection for configuration upload, download and maintenance
- Isolated, universal analog inputs allow mix of analog input types on same card, saving I/O cost

- Auto tuning and fuzzy overshoot protection for quick start-up and proper control operation
- Storage of up to 1,000 recipes for fast, error-free product selection
- Storage of up to 1,000 time/temperature profiles, each of which can be part of a recipe
- Any HC900 can support up to eight peer controllers for exchange of analog or digital data over Ethernet

FIGURE 17.18 | Control-panel assembly for HC900 Experion Vista process controller

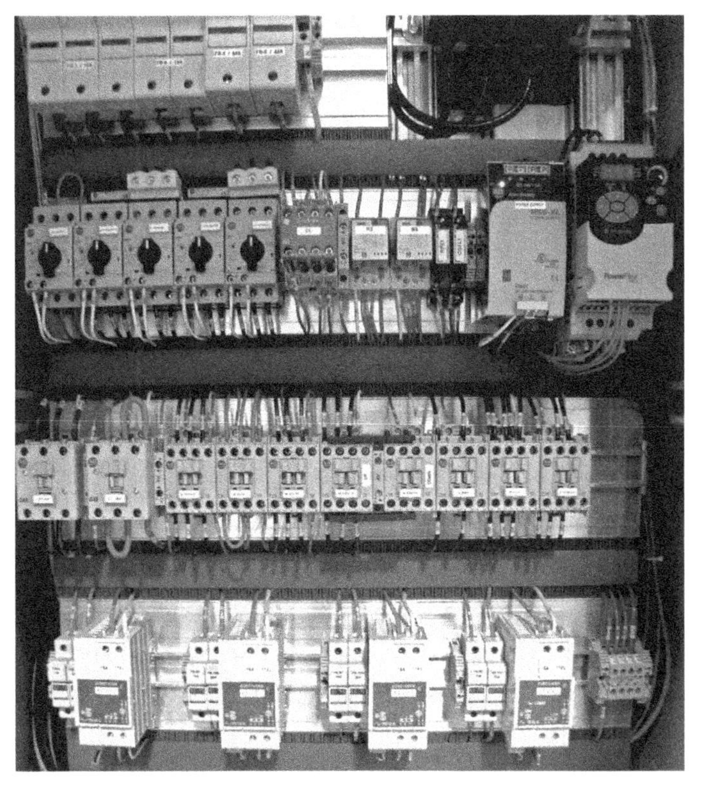

VACUUM FURNACE EQUIPPED WITH A HONEYWELL HC900 SYSTEM

Control of temperature is executed with a powerful algorithm set that satisfies most application requirements. Multiple tuning constants may be used to tailor the control response to the dynamic characteristics of the furnace.

Approach limits allow maximum heating rates without overshoot, reducing cycle time and optimizing efficiency. The HC900 integrates the setpoint programmer, loop and logic functions within a single device. The HC900's setpoint program capability is used to control up to 1,000 different temperature profiles appropriate for a wide range of products that can be created and stored for use when these products are processed.

A typical vacuum-brazing cycle profile (Fig. 17.19) uses load guarantee-soak function (as event 7) to control critical soak temperature. The cycle profile also contains other events used to control diverse functions required by a heat-treatment cycle (high and low vacuum level, partial pressure, quench, etc.).

FIGURE 17.19 | Typical vacuum-brazing cycle profile

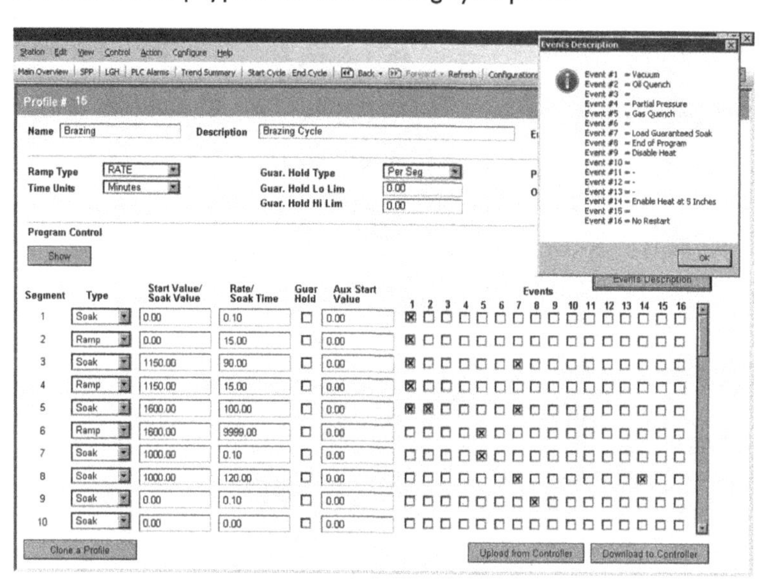

A single configurable database integrates both the loop (proportional, modulating) functions and the logic (discrete, Boolean) functions required by the process. User-friendly operator displays provide the operator with dynamic information about the status of each run as it progresses. Alarms are announced in color on dedicated displays and can be acknowledged directly from the operator interface.

The data acquisition and control capability of the HC900 permits ongoing process analysis to define and implement the various control strategies.

IMPLEMENTATION

The HC900 is a panel-mounted controller connected to a computer-based operator interface (Fig. 17.20). All field signals terminate at the controller. The controller has universal analog inputs, analog outputs, and a wide variety of digital input and output types. This controller will provide all the vacuum-furnace control functions.

CONFIGURATION

The Hybrid Control designer tool (Fig. 17.21) provides advanced configuration techniques that allow a variety of strategies to be easily implemented. The run-mode configuration monitoring and editing capability allows these strategies to be tested and refined as process knowledge is gained.

FIGURE 17.20 | Operator interface main overview screen

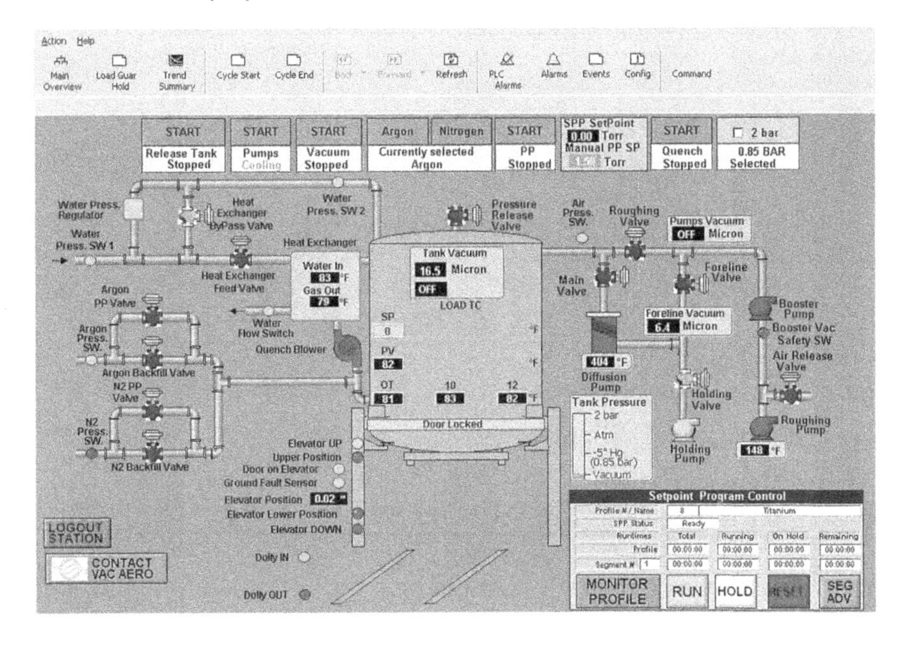

MONITORING

The complete operation can be monitored and controlled from the display screens. Standard and customized displays make it simple for operators to learn and use the system (Figs. 17.22-17.27).

FIGURE 17.21 | Hybrid Control designer tool (HC900 configuration screen)

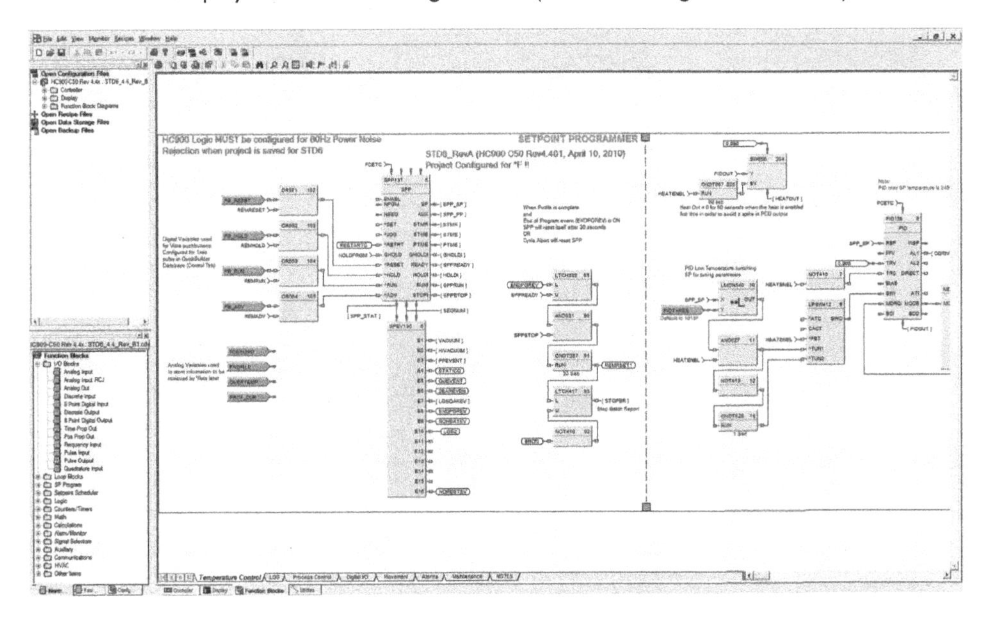

FIGURE 17.22 | Heat power adjustment screen

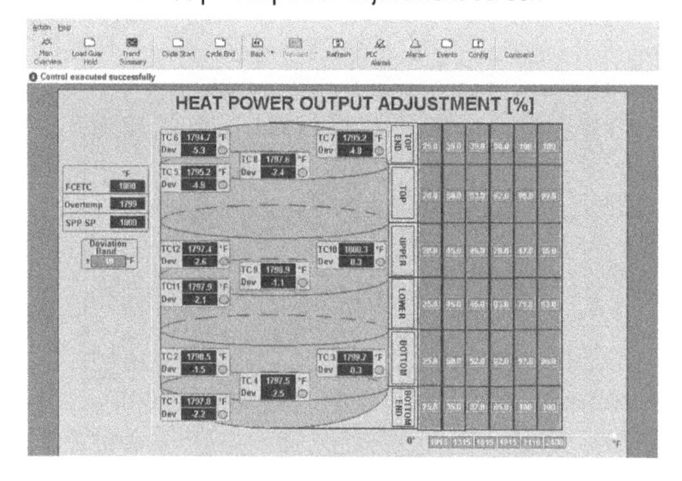

FIGURE 17.23 | Load guaranteed-hold configuration screen

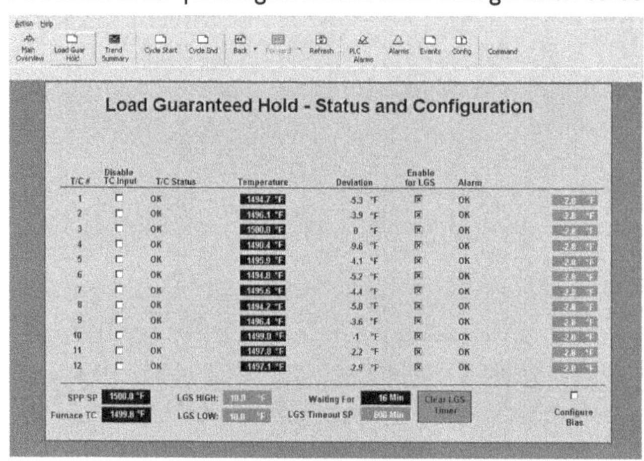

FIGURE 17.24 | Trend screen example

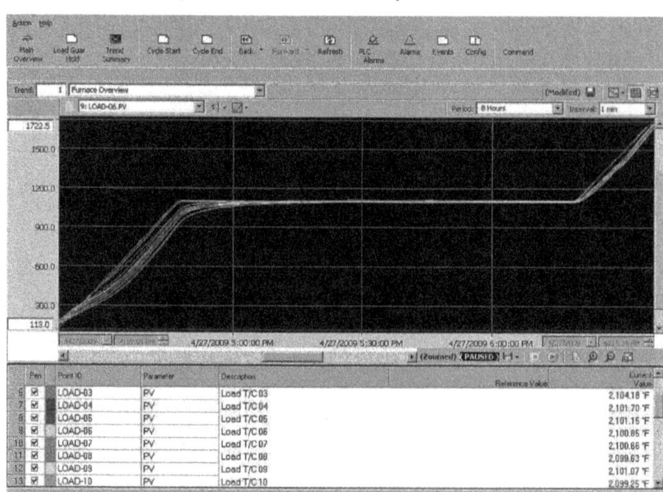

FIGURE 17.25 | Event log screen

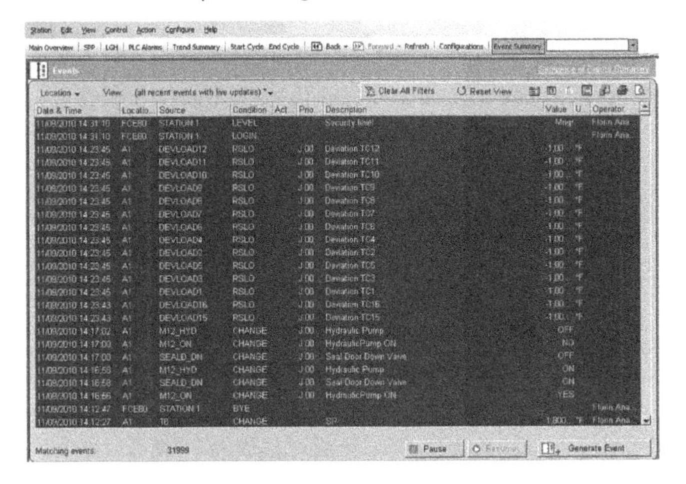

FIGURE 17.26 | Profile configuration screen

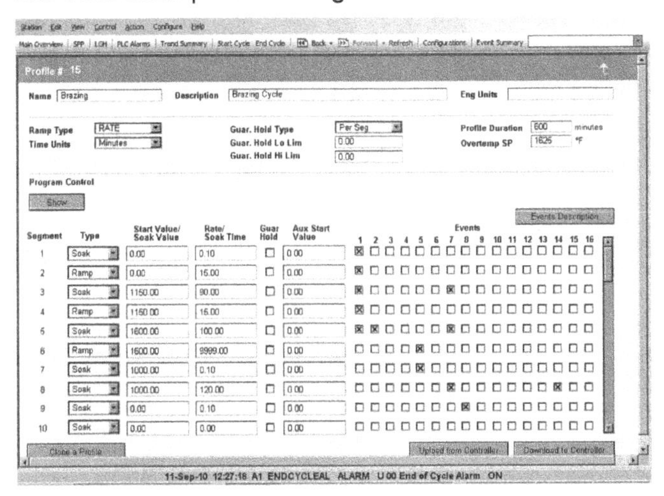

FIGURE 17.27 | Maintenance timers screen

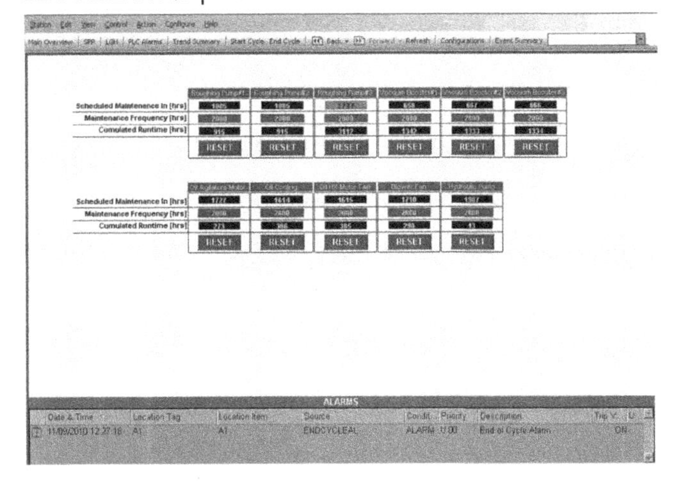

DATA COLLECTION AND STORAGE

The supervisory system provides many built-in reporting functions. Standard alarm and report functions include:

- Alarm/Event Log reports all alarms and events in a specified time period.
- Alarm Duration Log reports the time of occurrence and elapsed time before return-to-normal for specific alarms in a specified time period.
- Alarm Pager (optional) – setpoint alarms may be sent to an alarm paging or messaging system.
- Integrated Excel reporting provides the ability to launch a report built using Microsoft Excel.
- Batch reports collect history for points and events that occurred during a process production run. Static batch data such as batch number, customer name and lot size may also be added to the report.
- Bar-coded data functionality may be used to enter batch information.
- Reports may be generated periodically or on an event-driven or demand basis. The report output may be directed to screen, printer, file or directly to another computer for analysis or electronic viewing.
- History collection is available over a wide range of frequencies in both average and snapshot/production formats. A large amount of history can be retained online, with automatic archiving allowing retention of and access to unlimited quantities of historical data.
- Flexible trend configuration allows trends to be configured. Real-time and historical data are presented together on the same trend.
- The data-storage feature can be used to log process information during the cycle to an internal hard-drive disk or to a plant network storage device for a permanent record.

CONTROL IMPLEMENTATION SUMMARY

The Honeywell HC900 control system is capable of being used in an industrial shop environment, and numerous systems are operating in the field. The software has been optimized to anticipate all normal operating and alarm conditions. The software also provides supervisory control and data acquisition (SCADA) using a touch-screen LCD for operator interface for features such as:

- Compatibility with plant-wide SCADA and network integration
- Process cycle validation
- Extensive alarm and event management and reporting
- Temperature control using advanced algorithms, auto tuning and multiple-zone digital trimming
- Operator sign-on/sign-off security to limit operator control of individual functions

- Enhanced maintenance and troubleshooting management
- Extensive set of advanced algorithms for maximum process performance
- Open Ethernet connectivity via Modbus/TCP protocol that provides plant-wide process access and data acquisition
- Extensive equipment diagnostic and monitoring to maximize process availability

Other Types of Systems

Allen-Bradley's PanelView operator terminal offers electronic interface solutions in a variety of sizes and configurations. Each system is capable of providing process information over a variety of communication protocols by Ethernet, ControlNet, DeviceNet, DH+, DH 485 and RS-232-C. Most are offered in touch screen or keypad, and they include tools such as alarming, quality imaging and data trending.

Siemens' technology, which is more widely used in the European market, continues to grow in popularity in the United States. For example, S7 PLC products offer very fast scanning rates and networking capabilities with RS-232, RS-485, Profibus-DP and MPI protocols. Siemens also provides a range of touch screens and push-button interfaces for machine operation. There is a strong force overseas that guides American sister companies to use Siemens controllers. However, acceptance of these controls is an issue for engineers and maintenance personnel more familiar with other products. Nonetheless, Siemens continues to expand into the heat-treat industry, lessening these concerns.

Honeywell also has dedicated product lines specifically for temperature and process control. The company's UDC temperature controllers have a proven track record. The PLC merely sends the setpoint to the controller, and the UDC takes over control from there. Honeywell also has its own modular controller (UMC 800 – Universal Multi-loop Controller) that addresses the analog and digital control requirements of small-unit processes. Using its strong algorithmic background for PID control, it combines PLC function block programming for machine functionality. This is a good solution for small furnaces with limited input/output (I/O) needs. The UMC 800 provides integrated-loop and logic control.

For simpler vacuum furnaces that do not require the power of a PLC, a Digital Control Processor (DCP) like Honeywell's DCP550 can run a furnace program. Given 99 programs with 99 segments, ramp rate, soak setpoint, soak time and events are all parameters that can be entered and run within this controller. From there, its PID loops maintain certainty between the temperature setpoint and the furnace control thermocouple.

The method of controlling basic vacuum-furnace process parameters has changed over the years. PLC controls tied to versatile HMIs are doing a job that previously required several devices.

External temperature controllers are frequently used for adjusting the heat on vacuum furnaces, even though the control is offered within the PLC. Properly tuning PID control within the PLC can be very troublesome. With better PID algorithms being developed and proven by PLC manufacturers, coupled with the option to eliminate the external temperature controller, greater PID functionality will be realized within the PLC in the near future.

Cooling for a vacuum furnace is another important variable that must be addressed. Two types of control typically exist: uncontrolled and controlled. Uncontrolled cooling simply depends on the heat retained in the parts. Then it's just a matter of time until either the furnace or parts (work thermocouples) are brought to a satisfactory temperature under the correct pressure to continue or end the process. Controlled cooling allows the parts to be cooled at a specified rate in order to obtain desired end results. By means of a variable-frequency drive (VFD) to control the speed of the cooling fan or a variable (damper) valve, the amount of cooling gas introduced into the furnace chamber can be monitored and controlled through precise adjustments made by the PLC controller.

Isothermal hold is an option normally used when heat treating large parts such as dies while controlling distortion and avoiding cracks (Fig. 17.28). Coupled with controlled cooling, it allows more accurate control of the quenching process. By taking a thermocouple's reading on the inside of the part and comparing it to a thermocouple on the outside of the part, heat treaters are able to monitor the difference between the two thermocouples. With this calculation, cooling can be controlled so that the difference between the two thermocouples does not exceed a bandwidth requested by the operator. Thus, a more uniform, controlled quench is achieved.

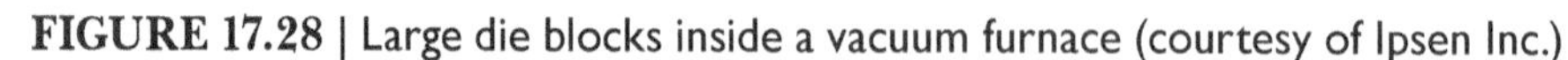

FIGURE 17.28 | Large die blocks inside a vacuum furnace (courtesy of Ipsen Inc.)

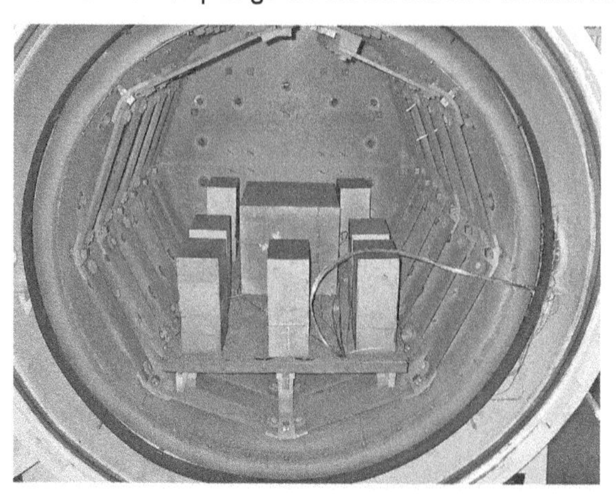

Multiple thermocouples placed in a tray or basket in different areas with the parts allow monitoring of the temperature spread or uniformity throughout the load. Each thermocouple will give a temperature reading (Fig. 17.29). Cooler or hotter areas within the furnace can be visible at different locations.

Convection heating is a desirable choice for heat treaters that have dense or irregularly shaped loads. Heat will find all surface areas of a part when the heat is circulated by a fan in the furnace through a positive-pressure medium. Convection also reduces the heating time and may reduce distortion.

FIGURE 17.29 | Thermocouple survey screen

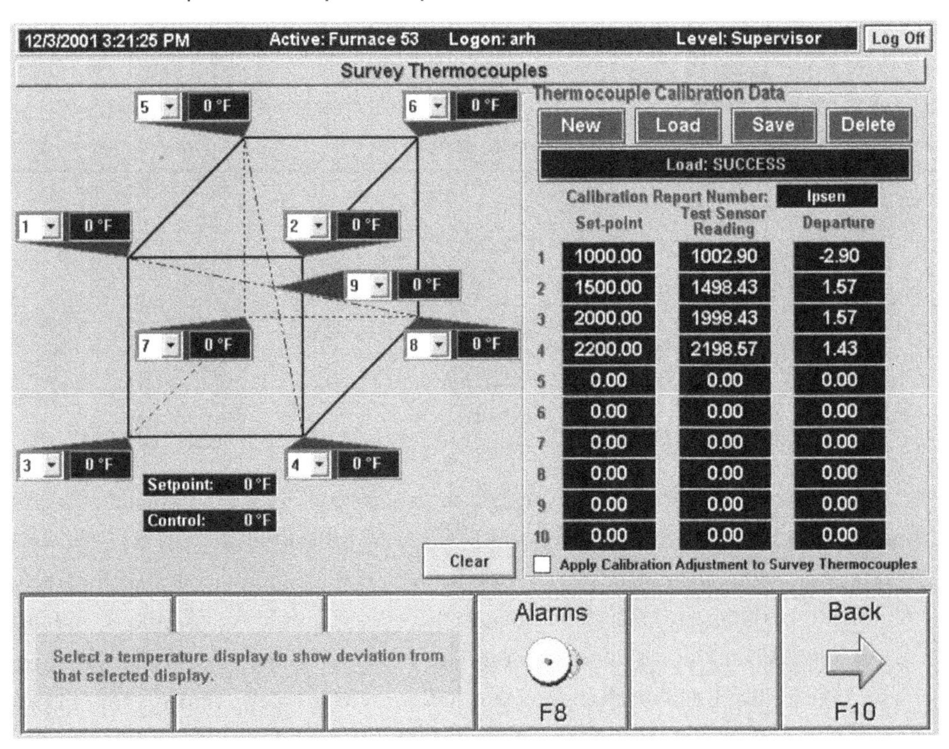

Cycle time for a particular load is simply based on the ramp rate or soak time that is specified in the recipe at a certain temperature. The soak time starts when a temperature setpoint has been reached or when temperature is within a bandwidth specified by the operator. Timers maintain the temperature within said bandwidth for the amount of soak time specified.

Another variable that can be controlled is the pressure inside the vessel. Using appropriate gauging for the vacuum levels required, valves can be opened and closed as needed in conjunction with a pumping system to maintain an accurate and acceptable level. Switches and transducers are used for monitoring these levels for a positive or negative pressure. Transducers offer greater flexibility for running a process at several different ranges of pressure.

Figure 17.30 | Graphical overview of a vacuum furnace

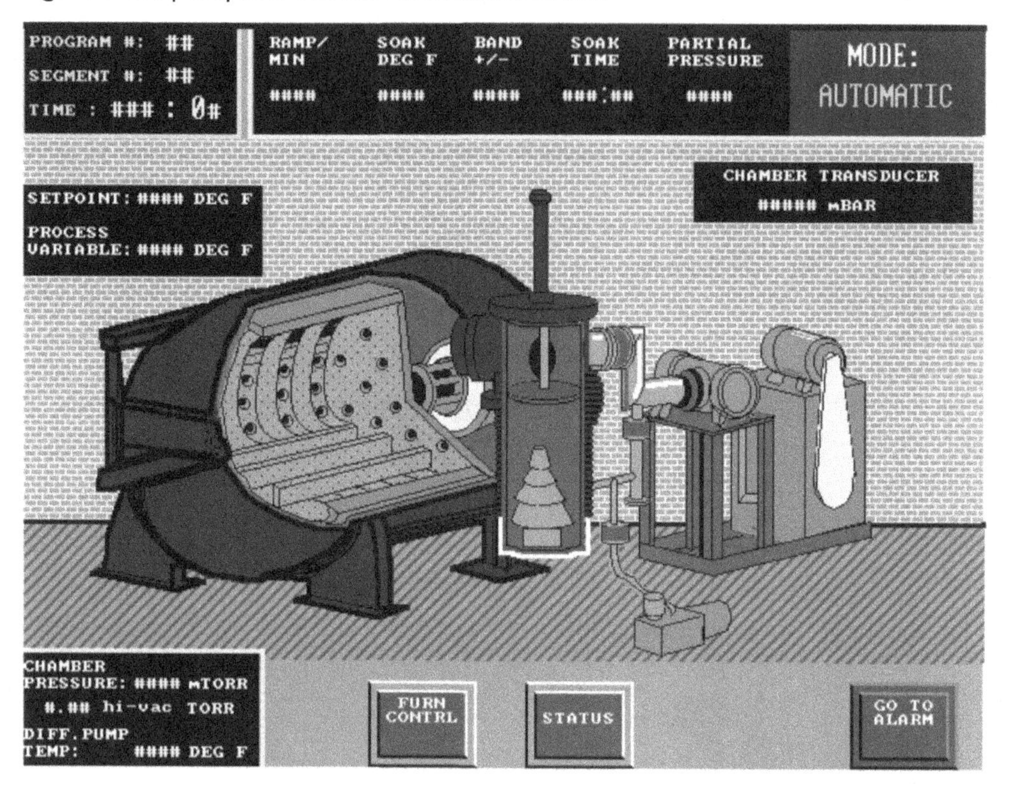

Graphical interfaces at the furnace or at a remote location provide easy-to-understand information on what the furnace is doing (Fig. 17.30). CTC Parker Automation, Allen-Bradley's PanelView interfaces and Siemens OP terminals are just a few of the many HMIs used to monitor, record and store machine variables as needed. Visual representations of the status of valves, motors, pumps and fans can be displayed. Color change or simple animation of the main components on a furnace can provide information not readily seen on the furnace exterior. Having detailed information in graphical format along with other important readings from the furnace can be very informative at a quick glance.

Other screens are also helpful by way of displaying specific data, such as I/O status of the PLC. Real-time trending and historical trending offer a comparative tool to help improve processes or determine machine repeatability. It is a tool that allows us to see the differences and effects that specific loads or programs have on a cycle. Maintenance screens give information on hourly usage of motors, pumps or other systems so that a maintenance schedule can be created and followed, which extends the life of each component (Fig. 17.31). Operator screens can also give us detailed alarms at the time of occurrence.

FIGURE 17.31 | Sample preventive-maintenance screen

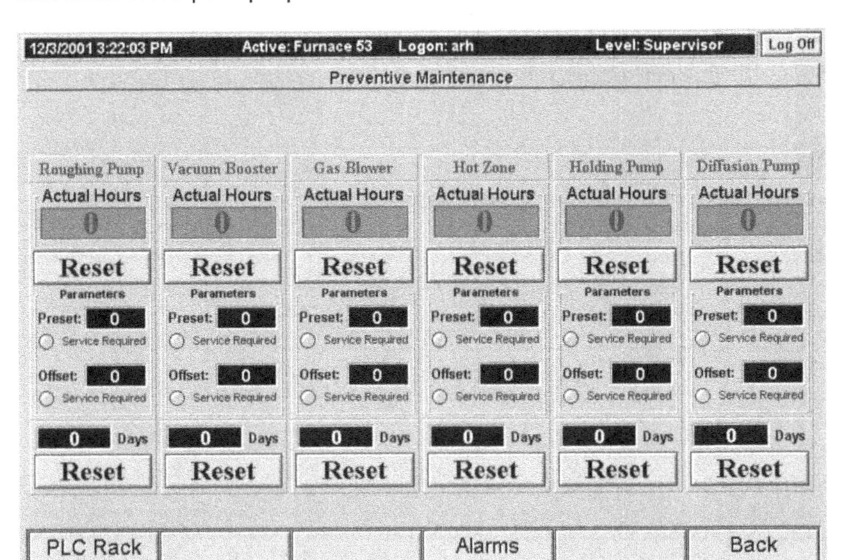

Recipe creation provides the operator or supervisor with the means to enter values, ramp rate, soak temperature, soak time and associated events for a furnace to run its cycle automatically. Once the recipe is created, cycle start is the only other button needed to run the furnace. After that, all functions are automatic. When the recipe has finished, data can be interpreted from graphs or trends to verify furnace control followed the instructions entered within the recipe.

Supervisory monitoring/limited control is becoming increasingly popular. Remote systems are ideal for data collection and real-time information purposes. Wonderware's InTouch package is a prime example. Recipes for loads can be stored and created remotely; graphics can show machine systems running to individuals in a different location; and information can be saved to various databases or plant networks for backup. These systems give heat treaters increased versatility and expandability. Several machines or entire lines throughout a plant can be connected to one supervisory PC that monitors and collects all machine variables. These systems can also manage and optimize workflow throughout a plant with proper part-tracking tools. Utilizing the capabilities of a networked system keeps heat treatment on the leading edge of this communication revolution.

Predictive software that calculates or simulates process cycles is available (Fig. 17.32). Entering the required hardness level, the material to be treated and the required case depth (in the case of atmosphere or vacuum carburizing) allows the software to generate the required recipe to achieve the requested results. This is very beneficial for heat treaters that process several types of parts. Once the recipe is generated, this information can be sent to the furnace's control system for execution. Material results can then be compared with the cycle run, giving a metallurgist the ability to improve or adjust parameters if needed.

FIGURE 17.32 | Simulation software

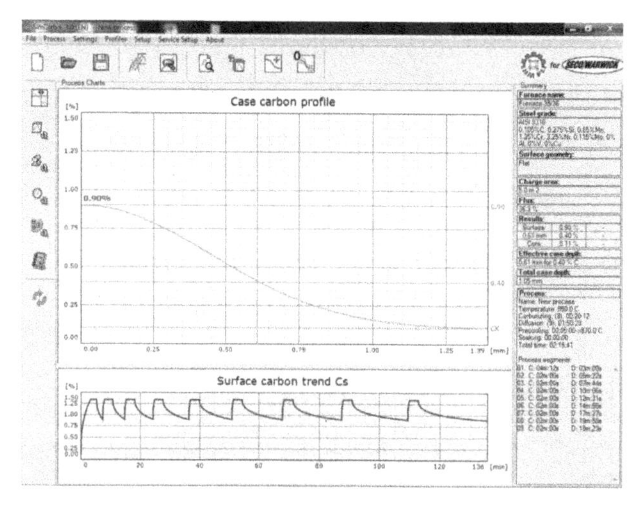

More on Control Systems

Control systems that are both intuitive and easy to operate are at the heart of industrial heating equipment. SCADA-based control systems are standard on many vacuum furnaces with PLCs, industrial PCs with touch screen and visualization software, making up the human-machine interface.

Control packages based on HMI systems such as Wonderware's InTouch offer software tools to simplify daily use, including: production management and process programming; collecting and safe storage of batch data; process reporting; reminders of basic maintenance operations; and troubleshooting capability.

It is important for a vacuum-furnace manufacturer to have a dedicated programming team intent on continuously improving the control systems. This group should be capable of answering questions in areas such as switching between different engineering units and operator languages; interpreting system-help screens; report generation; and verification of process recipes to name a few. Help should also be available via online tips, text-messaging systems, online recipe editing and more.

Using Ethernet network to communicate PLC and HMI, these furnaces can be integrated into an office network. This creates extensive possibilities in areas of data acquisition, production management and supervision of processes and equipment. It allows integrating furnaces with a company ERP system in order to gather all information about products from different stages of the production process in one safe place.

Industry regulations and aerospace specifications software should incorporate operations required by AMS 2750 into the furnace control system (Fig. 17.33). Advanced recorders (Fig. 17.34) simplify the task. TUS and system-accuracy test (SAT) support with multi-point correction factors and offsets for all temperature sensors and thermocouple-life counters are important features to include.

FIGURE 17.33 | Support for AMS 2750 pyrometry requirements

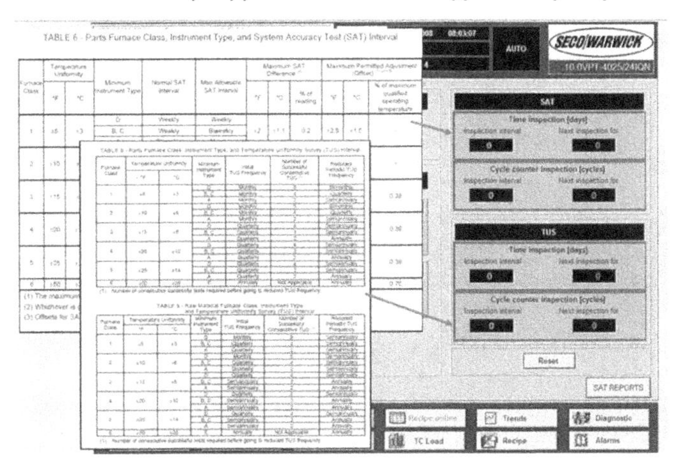

FIGURE 17.34 | TUS recorders (courtesy of Yokogawa Corp.)

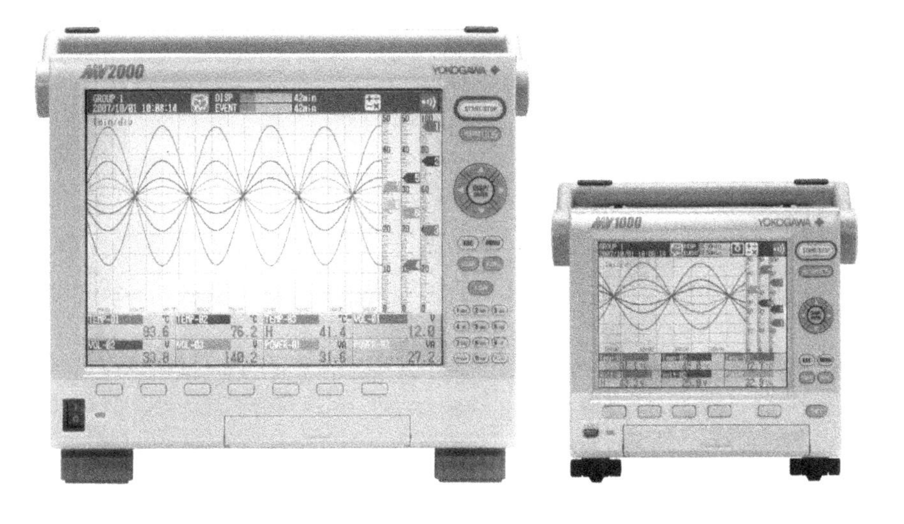

SATs became more important with newer revisions of AMS 2750. It is important that suppliers offer upgrades to software packages for vacuum furnaces to support these changes.

Smart control systems can remind users about a scheduled test; collect data from a furnace and test instrument via industrial communication interfaces; provide maintenance reminders; and even create and store complete SAT reports that meet requirements of Nadcap documentation. After completion of testing, recorded data can be compared and results can be calculated. A complete report can be printed or stored in an internal database. It is easy to organize, and it is quickly available during audits (Fig. 17.35).

Control systems allow for entering temperature offsets at several stages for each control thermocouple according to SAT results. Those offsets are independent from correction factors taken from the sensor's certificate of calibration.

FIGURE 17.35 | SAT report preview

System Accuracy Test (SAT)
Calibration Report

Report Date: 2008-19 12:39
Furnace: VTR-5050/216HV
Specification: AMS 2750
Test Instrument: Fluke S/N 25346346

Test Sensor: 31244

		Indicator	Control Zone #1 Recorder	Load Sensor
Date/Time		2008-08-02 15:32	2008-08-02 15:32	
Set Temperature		920°F	920°F	
Instrument Indicated or Recorded Temp.	A	921.4°F	921.4°F	
Correction	B	None	None	
A+B = C	C	921.4°F	919.°F	
Test instrument reading	D	919.8°F	919.8°F	
Test T/C correction factor	E	- 1.6°F	- 1.6°F	
Test instrument correction factor	F	- 0.6°F	- 0.6°F	
True temperature D+E+F=G	G	917.6°F	917.6°F	
SAT difference C - G		3.8°F	3.8°F	
Result		Fail (1)	Pass	
Technician		RK	RK	

Notes: (1) Test results do not meet requirements. SAT difference exceeds maximum allowable 3°F

SAT reporting can provide the following information:

- Test-sensor identification
- Test-device identification
- Checked-sensor identification
- Date/hour of test
- Monitoring-device reading
- Test-device reading
- Adjusted test-device reading
- Computed variation of system accuracy
- Identification of the test-performing technician
- Indication of acceptance or rejection of the test
- Report notes

TUS's should be performed monthly for most furnaces (e.g., Class 1 and 2). Systems should be designed to support customers in this endeavor with a SAT, TUS module (Fig. 17.36) that also has reminder features, data acquisition and reporting. TUS reporting can include the following information:

- Device identification

- Device class
- Temperature range
- Acceptable temperature variation
- Dimensions and volume of the operation space
- Specification used
- Heating method
- Date and time of start and end of each test
- Measuring-element type
- Measurement-device type
- Temperature setpoint
- Thermocouple layout map
- Minimum, maximum and average value for each of the measurement elements
- Confirmation of reset occurrence
- Test result
- File path to the Excel file with all the measurements
- Report notes

Figure 17.36 | TUS report preview

TUS Summary

Channel (°F)	1	2	3	4	5	6	7	8	9	10
High	912.3	912.4	912.2	912.6	912.8	912.4	917.2	912.5	912.7	912.8
Low	911.8	912.4	912.2	912.6	912.8	912.4	912.2	912.5	912.7	912.8
Average	912.5	912.3	912.6	912.2	912.8	912.1	912.4	912.5	912.3	912.6

Channel (°F)	11	12	13	14	15	16	17	18	19	20
High	912.3	912.4	912.2	912.6	912.8					

Temperature Uniformity Survey (TUS)
Calibration Report

Report Date: 2008-08-19 12:39
Furnace: VTR-5050/216HV

Furnace Information

Furnace Class	2	Temperature Range:	850°F - 1150°F
Furnace Instrumentation class	C	Temperature Tolerance:	+/- 10°F
Work Zone Dimensions	11 ft x 7 ft x 5 ft	Work Zone Volume:	385 cubic ft
Controlling Specifications	AMS 2750D	Heating Method:	Electric

Schedule

Test Date	2008-08-19	Next Due Date:	2008-10-19

TUS Test Load Data

Test T/C Identification	310-1/-15		
Test T/C Wire Type	"K", special limit	Temperature Setpoint:	920°F
Test Instrument	2352356/36	Time Stabilized:	12:55
Time in Furnace	12:30		

Note: Correctiion factors applied to the test instrument prior to TUS

Test Instrument C/F:	+ 0.4°F
Thermocouple C/F:	- 1.7°F

Door
C
H
L

TUS report capability should allow for collecting and storing data from all test and furnace sensors. Test temperatures can be received from most test recorders via industrial communication interface or can be attached as a data file. Correction factors for each test sensor can be applied according to certificate of calibration. After TUS completion, recorded temperatures are verified and the system indicates if any point has temperature variation behind allowed limits. A TUS test report can also be generated and stored on a computer's hard drive.

All furnace users in the aerospace industry have to track the life of base-metal temperature sensors and keep those records, then replace thermocouples when they are supposed to be replaced. Thermocouple usage counters (Fig. 17.37) are another outstanding feature of these control systems. These lifetime counters keep tracking temperatures of each thermocouple and compute them as required by AMS 2750. Each time the sensor is exposed to a certain temperature, a count is taken along with the time. When the thermocouple is used a certain number of times at a certain temperature or when its maximum usage time has elapsed, the operator will get a warning and alarm message with detailed information of what happened and what has to be done. This function is very useful because it eliminates the human factor and reduces the number of operations.

FIGURE 17.37 | Thermocouple usage counters

Thermocouple Usage

Load TC #1			
SEL	Time & Temp.		Reset
			Used
2300 °F	-	----->	0 / 1
2200 °F	-	2299 °F	3 / 10
1801 °F	-	2199 °F	7 / 90
1200 °F	-	1800 °F	20 / 180
----->	-	1200 °F	20 / 270
PLC Date			2010-09-13
Next inspection - warning			2010-10-08
Next inspection - alarm			2010-10-13

Load TC #2			
SEL	Time & Temp.		Reset
			Used
2300 °F	-	----->	0 / 1
2200 °F	-	2299 °F	3 / 10
1801 °F	-	2199 °F	7 / 90
1200 °F	-	1800 °F	20 / 180
----->	-	1200 °F	20 / 270
PLC Date			2010-09-13
Next inspection - warning			2010-10-08
Next inspection - alarm			2010-10-13

Low-Pressure Carburizing Simulation Software

New technologies for low-pressure carburizing include simulation tools and the resources to support them. Customized software determines the carburizing process parameters according to specified carbon case parameters.

The application allows the user to define the steel grade and part geometry and select the material or process specifications. This information is then used for further calculations of cooling speeds based on the physical properties of treated elements as well as the mass and geometry of the charge. As the result of simulation, the user gets the proposed heat-treat recipe along with graphic representation of carbon profile.

Results of simulations can be printed and attached to the batch report, and a created recipe can be easily transferred to SCADA systems such as SecoVac. This software has a built-in database of steel grades according to different international standards. The customer can also define different steel grades if needed. A database of basic shapes of parts (Fig. 17.38) allows the software to estimate the surface area of the batch. Active surface area is an important parameter for vacuum carburizing processes as far as amount of carbon gases supplied to the chamber.

New software packages can now predict the hardness profile of carburized and gas-quenched parts (Figs. 17.39). Determining the cooling speed at a particular distance from the surface is necessary to calculate the hardness profile in the carburized layer of a particular geometry. This module also takes into consideration different cooling abilities of individual furnace types at different pressures of quenching gas. After simulation completion, the user receives the predicted hardness profile with suggested heat-treat recipe.

FIGURE 17.38 | Database of basic shapes built into simulation software

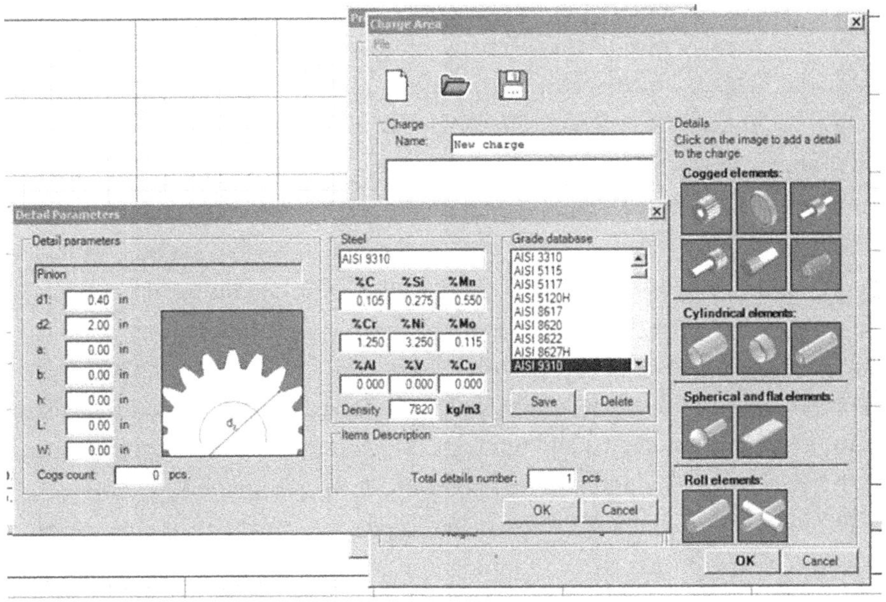

FIGURE 17.39 | Hardness simulator to determine hardness profile

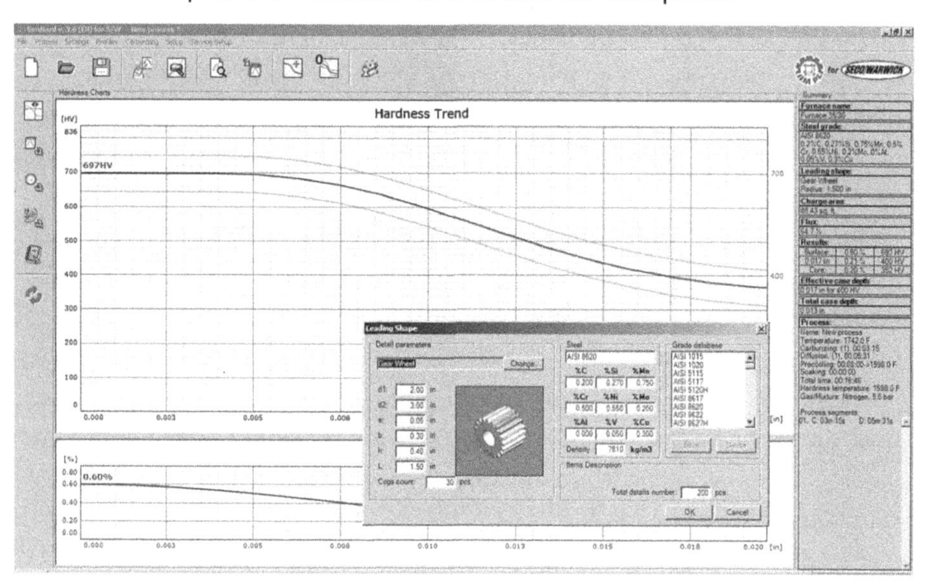

The Future of Instrumentation and Controls

Over the past 10 years, instrumentation for vacuum furnaces has experienced a major change, from stand-alone analog to integrated microprocessor-based instrumentation at an ever-increasing rate. For process control, relay logic has long since been replaced with microprocessor-based PLC. Newer control panels feature touch-screen interactive controls to operate vacuum pumps, control valves, hot zone power, partial-pressure control, gas backfill, quench blower motor, alarms and more. Full temperature and vacuum programs can be installed on these panels and monitored or altered via the operator. Most importantly, this instrumentation can be monitored over the Internet remotely, via computer, and the process can be upgraded or changed with proper password.

Vacuum gauges have seen a similar transition away from meters to digital readout and microprocessor-based controls, allowing vacuum scales stretched from atmosphere to high vacuum over hundreds of decades of pressure range unheard of years ago. Newer vacuum gauges are corrected for residual atmosphere and now operate on an absolute-pressure basis. "Smart vacuum-gauge heads" are already available, eliminating a vacuum-gauge controller reading out directly to a panel view and logging directly to data collection.

The most useful change is probably in recorders. Gone are strip-chart recorders and with them the task of retaining a mountain of charts. Data collection is now digital and stored on central computers, available for quality control or transmission to the customer over the Internet.

The future will see additional changes hard to visualize as microprocessors and computers advance further. This means smaller, more easily assessable, portable, probably voice-actuated equipment at lower cost with more uses and capacity. However, instrumentation will become outdated faster. Manufacturers will not support older models, forcing heat-treating operations to invest in the future. A major trend is and will be increasing the dependence on electronic control and instrumentation to take control away from human operators in an effort to avoid costly mistakes with workloads that cost thousands if not tens of thousands of dollars or more.

Final Thoughts

It is natural to wonder how vacuum-furnace control systems will evolve in the future. What we know for sure it that customers are emphasizing the need for accurate data collection and archiving services. Information produced by a machine must be available at any time (and in real time if at all possible). Furnaces must have the ability to be networked, yet operate independently.

Whatever type of control is used, the data must be accessible through a remote or networked system. By analyzing this information from virtually anywhere, including hand-held devices, heat treaters will have the ability to reduce energy and run shorter cycles with the same or better results in real time. Utilization of more interactive simulators with improved feedback circuitry to predict an accurate furnace program to give repeatable cycles will become a reality. In the future, more automation – such as automatic loading and unloading, part-tracking systems, and "lights-out" operations – will be a reality. The communication revolution is evolving, and the vacuum-furnace industry will be at the forefront of using these innovations.

REFERENCES

1. Ralph Poor, Surface Combustion, content contributions
2. Bill Jones, Solar Manufacturing, content contributions
3. Hagler, Alex, "State-of-the-Art Controls for Vacuum Furnaces," Ipsen Inc., white paper
4. Alan Charky, VAC AERO International, content contributions
5. Rafal Walczak, SECO/WARWICK, content contributions

CHAPTER 18

VACUUM MAINTENANCE PRACTICES, PROCEDURES AND TIPS

To ensure reliability and repeatability of operation as well as uncompromising safety, maintenance practices need to be well defined, understood by all and implemented in a prudent and well-thought-out manner. Only trained personnel experienced in vacuum technology should be allowed to service vacuum-furnace systems.

The frequency of maintenance (i.e. interval between routine repairs) is highly dependent on factors such as:

- Type and number of heat-treating processes performed
- Skill level of the operators
- Equipment design
- Quality of prior maintenance and type of spare parts used
- Quality of the water system, gas system, etc.

When performing maintenance, it is important to have a written plan defining the specific task to be performed and a reason why a particular task is necessary (i.e. purpose of the task). A work order should be issued and the work signed off upon completion, which includes testing to ensure that the repair was successful.

The following conditions should be met before any repairs are undertaken:

1. Power should be switched off for any repairs not directly involved with the electrical systems, controls or instrumentation. Lockout/Tagout procedures should be in place (Fig. 18.1).
2. The furnace should be cool, less than 50°C (120°F).
3. The furnace door(s) should be in the open position and secured so that they cannot be closed.
4. Disconnect all utilities including gases, water and air. Lockout/Tagout procedures should be followed.
5. Check that the furnace environment is safe and that adequate ventilation is in place and functioning properly.
6. Wear protective clothing, including safety glasses and safety shoes.
7. Be sure that all confined-entry procedures are thoroughly understood and followed without exception.
8. Use the buddy system.
9. Before entering the vacuum furnace, confirm that the oxygen level is safe for human exposure.

FIGURE 18.1 | Lockout/Tagout
(courtesy of Carter Manufacturing Co./Carter Bearings)

FIGURE 18.2 | Typical horizontal vacuum-furnace maintenance areas (courtesy of SECO/WARWICK Corp.) [2]

Key:

1. Chamber door seal care (after each run)
2. Water systems and hoses (weekly visual inspection)
3. Pumping system (daily/weekly checks)
4. Hot zone inspection (daily checks)
5. Instrument calibration (per AMS 2750)
6. Hearth inspection (after each run)

FIGURE 18.3 | Typical vertical vacuum-furnace maintenance areas (courtesy of VAC AERO International)

Key:

1. O-ring seal care (after each run)
2. Water systems and hoses (regular visual inspection)
3. Cooling and blower motor (inspect every six months)
4. Heat exchanger (annual inspection)
5. Vacuum gauge calibration (biannual inspection)
6. Pumping system; hidden behind panel (daily/weekly checks)
7. Hot zone inspection (daily checks)
8. Hearth inspection (after each run)

General Maintenance

Safety is **mandatory** and cannot be compromised during any maintenance activity. Safety interlocks must never be bypassed, and verification that all potentially hazardous energy sources have been isolated and disabled is a necessary first step in the maintenance process (reference NFPA 86 and NFPA 70).

Lockout/Tagout procedures are required to disable machines or equipment during maintenance to prevent injury and are part of OSHA code (Regulation 1910.147). A vacuum furnace may have different places in which electrical power must be disconnected – a single main electrical disconnect (i.e. circuit breaker) for the entire furnace or power supplied from several electrical sources, each with a separate disconnect device. The electrical drawings for the specific unit in question should be reviewed and physical inspections should be conducted in the event that undocumented changes have taken place.

In addition, a vacuum furnace may also have pneumatic or hydraulic systems, including sources of compressed air, inert gases and process (reactive) gases that must be isolated from the system. Confined-entry space restrictions also apply.

All safety interlocks present in normal equipment operation should be tested on a regular basis to ensure proper operation. These include:

- ❖ Over-temperature instrumentation (test monthly)
- ❖ Process interlocks (test semiannually)
- ❖ Water interlocks (test semiannually)
- ❖ Air interlocks (test semiannually)

Several additional points regarding items that impact vacuum maintenance can be summarized as follows.

- ❖ Workload cleanliness is essential.
- ❖ Proper part placement and location (in baskets or on grids) is critical.
- ❖ Loading and unloading the equipment must be done with care.
- ❖ Backfill piping must be leak free, and backfill gas must be high quality.
- ❖ Small parts in loads must be adequately constrained (Fig. 18.4).
- ❖ Maintenance must be done with extreme care to avoid damaging adjacent components or systems.

FIGURE 18.4 | Small parts embedded in the heat exchanger system (courtesy of SECO/WARWICK Corp.)

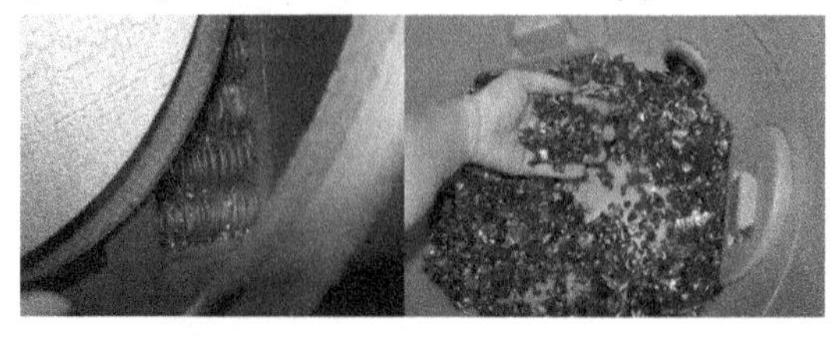

Vacuum-Vessel Maintenance

The most common maintenance practice performed on the vacuum vessel is leak detection (c.f. Chapter 12).

For most furnaces, the front-door O-ring or door seal is the single most important area of focus. Inspect and wipe this seal clean each time the door is opened after a furnace cycle. Look for cracks, lack of elasticity, flat spots, dirt and metallic fines on O-ring seals. Wipe the door O-ring and flange with a clean, lint-free rag before closing the door. Reapply a thin coating of vacuum grease as necessary. Once a week, wipe the door O-ring and mating flange with a clean rag soaked in denatured alcohol. Reapply just enough vacuum grease to produce a sheen on the O-ring surface, and check that a rubber-gloved finger glides freely along it. [6]

Vacuum Pumping-System Maintenance

In most cases, the vacuum pumping system requires the most maintenance of any area on the vacuum furnace.

MECHANICAL PUMP

Mechanical (wet) pumps rely on oil for proper operation, and the correct oil type for the pump in question should be used. The oil level and the condition of the oil must be checked daily, usually at the beginning of the first shift of the day. The correct oil level for most pumps is at the center of the site glass or just below center when the pump is operating at high vacuum. Oil should be added when the pump is stopped. It can be added in some instances, but not if the vacuum level is below 1 torr. Overfilling will cause a loss of efficiency as well as create the potential for backstreaming of oil into the hot zone.

The oil condition should be checked daily. Good oil is translucent and clean. Cloudy or milky oil indicates the presence of moisture. If this is observed, the pump should be ballasted off-line in accordance with the manufacturer's instructions. Dark or discolored oil indicates the presence of dirt, carbon (often in the form of soot) or other contaminants. If this is observed, the oil should be changed as needed or after 300 operating hours (whichever comes first). The oil reservoir should be drained and cleaned with denatured alcohol and clean, lint-free rags every six months or when the oil is excessively dirty. At the same time, the exhaust valve springs (poppet valves) and all discs should be replaced.

Observe and check the oil temperature under normal operating conditions. Normal operating temperature is between 60-71°C (140-160°F) as indicated by the gauge on the side of the vacuum pump.

Check the drive belts monthly for wear and adjust for proper tension. The pump manufacturer's instructions should be followed regarding proper belt tension settings. Here's a useful tip: With a new belt, at the belt midpoint (between the drive and motor pulleys) apply pressure (typically 5-7 pounds) to the belts. Record the resulting deflection, and use this value for future adjustments. If the belt tension is too tight, damage can occur to shaft bearings. If the belts are too loose, slippage will occur, causing excessive wear.

The gas ballast valve and spring should be checked and replaced, if necessary, every three months.

Vacuum-Valve Maintenance

Vacuum valves require the least amount of routine maintenance of any component on a vacuum furnace due in large part to their design. In most cases, no lubrication or adjustments are required. However, maintenance is not only necessary but critical at certain intervals in a valve's operating life.

For example, butterfly-type roughing and foreline valves should be removed from the vacuum line every two years to inspect their rubber seats for cracking and dryness. When the valve is replaced between flanges in a vacuum line, the disc should be rotated to the open position before the flange bolts are tightened.

Poppet-valve pistons and shafts should be lubricated monthly with vacuum grease through the fitting on the cylinder mounting block. The valve disc O-ring should be cleaned and lightly lubricated semiannually, and it should be replaced annually.

BOOSTER PUMP

The booster pump does not normally require a great deal of maintenance. However, the oil level in all reservoirs should be checked weekly and drive belts checked monthly for wear and adjusted for proper tension, similar to the mechanical-pump belts. The oil in the bearing and gear oil reservoirs must be changed every 2,000 hours of operation.

DIFFUSION PUMP

The diffusion pump should be taken down, jet assembly pulled and thoroughly cleaned, as well as the inside of the pump, and recharged with oil annually.

The oil level on the diffusion pump should be checked weekly. Oil must only be added to a cold (<130°F oil temperature) pump and only to the cold mark on the site glass. Never open the drain plug when the pump is hot. (Caution: There is a risk of explosion.) An oil change is recommended every six months or immediately if discoloration or contamination is observed. One very common mistake found is to not use the correct diffusion-pump oil.

Cooling water to the pump must be checked daily for adequate flow. If proper cooling is not provided to the diffusion pump, the oil can fractionate (i.e. break down and form solid carbon deposits) and damage the pump, which may need to be factory repaired. The heating elements should be checked for tightness and proper operation at least once a year. A common cause of diffusion-pump failure is heater burnout, so the diffusion-pump heater current should be checked regularly.

HOLDING PUMP

The oil level and condition of the oil are critical for efficient operation on holding pumps. The correct oil level is at the center of the site glass or just below center with the pump operating at high vacuum. Oil should be added only when the pump is stopped. The oil condition should be checked weekly, and (similar to the mechanical pump) the holding pump should be ballasted or replaced if the oil is not translucent and clean.

Hot Zone Maintenance

After ensuring proper ventilation and following all safety guidelines with respect to asphyxiation and confined-space entry, the interior of the hot zone should be inspected after every load. The bottom of the hot zone should be cleaned of all debris and foreign matter, and the heating elements and heating-element connections inspected for damage and tightness.

Graphite heating elements can be patched in some instances, and the damaged section can be replaced with a new element section. Molybdenum heating elements can be repaired, although no more than three repairs are recommended per element band. Special procedures are required since molybdenum is brittle and molybdenum dioxide fumes should not be inhaled. Once a month, heating-element resistance to ground should be checked with a volt/ohm meter. A good reading is between 90-100 ohms for most furnaces. As molybdenum elements age or the element standoff metallizes, their resistance to ground drops. A failed reading would be 10 ohms or less.

Low resistance to ground is an indication of metallized ceramic insulators and also an indicator for the need of replacement or cleaning. Otherwise, heating elements will arc and fail. In some cases, the ceramic insulators can be removed and baked out in air to remove contaminants.

A hot zone bake-out cycle should be run every 200 hours or when deemed necessary by the performance of the equipment.

Several additional points regarding hot zone maintenance can be summarized as follows:

- Check for insulation degradation
- Maintain proper tension on electrical connections (e.g., heating elements, power feed-throughs)

- Inspect resistance to ground
- Inspect heating elements for wear and/or oxidation (e.g., thinning or a "sugar cube" appearance, indicating attack by oxygen)
- Check that thermocouples and controls are functioning properly

Water-System Maintenance

Most vacuum furnaces are cold-wall designs with an annular spacing between an inner and outer shell. As such, proper conditioning of the water is important for effective cooling. Water should be treated for pH, hardness, bacteriological agents and (if appropriate) have rust inhibitors present to help minimize sediment and scale buildup, particularly in the bottom portions of the shell. A blockage of the vessel wall will result in a hot spot, so the shell should be periodically checked since most blockages occur slowly over time.

Several additional points regarding water-system maintenance can be summarized as follows:

- Checking water quality
- Cleaning/maintenance of the heat exchanger(s)
- Establishing corrosion protection
- Maintaining coolant levels to various subsystems (e.g., pumps power feed-throughs, vessel)

Record Keeping

Accurate record keeping is an often overlooked aspect of a successful maintenance program. It is at the heart of any efficient and effective plan. Record keeping should start on equipment installation and document any and all changes to the equipment over its lifetime. After the unit is put into service, it is essential to create and maintain a performance log containing information such as:

- Blank-off pressures (particularly mechanical pumps)
- Pump-down time to a given pressure
- Ultimate vacuum and the time required to achieve it
- Leak-up rate when the chamber is blanked off
- Heating rate (empty, fully loaded) to processing temperature

This type of data is invaluable when evaluating a future problem or when trying to determine if the vacuum system has deteriorated.

Training

The value of training should never be underestimated. Over the years, the majority of vacuum furnace failures can be traced to the following causes:

- Inadequate training of operators or maintenance personnel

- Lack of proper maintenance
- Improper use of equipment
- Improper record keeping

All operating personnel, supervisors, maintenance and quality-control individuals should have a good understanding of what heat treating is, how vacuum technology differs from other types of heat treatment, and how the equipment should be operated and maintained to ensure safety, efficiency and proper results. It is further recommended that annual retraining be conducted to maintain a high level of proficiency and effectiveness.

Preventive-Maintenance Checks

Setting up a planned preventive-maintenance program will minimize equipment downtime, ensure that proper spares are on hand for repairs and simplify the overall maintenance effort. As a minimum, the following checks should be performed.

EACH RUN

The following activities should be performed before each run.

1. Inspect the front door O-ring for cleanliness and damage. Clean and regrease as necessary.
2. Inspect hot zone insulation and heating elements for signs of damage and deterioration. Ensure connections are snug and secure.
3. Inspect the load thermocouple(s) for damage (if appropriate).

DAILY

The following activities should be performed daily.

1. Inspect the exterior and interior of the vacuum furnace for indications of damage, discoloration, dripping fluids and the presence of foreign material (e.g., dirt, grease, oil).
2. Check the water flow and temperature from each drain line.
3. Check the oil level on all pumps.
4. Ballast the vacuum pump (15-20 minutes minimum) before processing the first workload of the day.
5. Inspect for hot spots, leaking fluids, excessive noise and/or vibration during operation.

WEEKLY

The following activities should be performed weekly.

1. Perform a leak (rate-of-rise) test on the main vacuum vessel and pumping system. The furnace should be clean, dry, empty and outgassed before testing.
2. Check mechanical-pump oil for contamination (e.g., dirt, particulates, water).
3. Check instruments for functionality.
4. Inspect the pumping system (pumps, valves, piping).
5. Visually inspect control and over-temperature thermocouples for damage.

MONTHLY

The following activities should be performed monthly.

1. Check for hot zone deterioration (insulation and heating elements), including doors.
2. Check calibration of vacuum instruments.
3. Check all thermocouples (e.g., control, over-temperature, load).
4. Change vacuum-pump oil.
5. Check belts for proper tension (e.g., mechanical pump).
6. Change all filter elements.

SEMIANNUALLY

The following activities should be performed semiannually.

1. Replace or recalibrate all thermocouples.
2. Flush all cooling lines and clean all in-line filters, strainers, etc.
3. Clean and replace diffusion-pump oil (if applicable).
4. Inspect all vacuum gauges.
5. Replace door gasket or O-ring seals.
6. Remove, clean and reinstall thermocouple vacuum gauges.
7. Test the pressure-relief valve in accordance with manufacturer's instructions.
8. Make all necessary repairs to hot zone components (including power feed-throughs).
9. Clean all mating flanges.

ANNUALLY

The following activities should be performed annually.

1. Drain and inspect the cooling-water system (including temperature sensors).
2. Check all electrical connections. (Caution: Only a licensed electrician trained in the procedure should perform this activity).
3. Service all motors.
4. Clean the furnace heat exchanger (if applicable).
5. Drain and filter the quench oil (if applicable).
6. Check the convection fan and/or oil agitators for proper operation (if applicable).
7. Check, remove and inspect vacuum valves for proper operation, sealing and wear (if applicable).

Summing Up

Maintenance should be performed in such a manner as to return the equipment to full operational service. Never compromise – a job worth doing is worth doing right. This will ensure years of productive service from your vacuum furnace.

REFERENCES

1. *Vacuum Furnace Training Manual,* Abar/Ipsen "U"
2. Kowalewski, Janusz, "Maintenance Program Essential For Safe, Consistent Heat Treating," *Modern Application News,* March 2000
3. Kowalewski, Janusz, "Maintaining Vacuum Furnaces," *Advanced Materials and Processes,* April 2000
4. Craig, Roger A., "Vacuum Furnace Maintenance: Essential for Reliable Operation," Conference Proceedings, Society of Manufacturing Engineers, February 1999
5. DeWeese, Ted, "Furnace Maintenance," Furnaces & Atmospheres for Today's Technology, Conference Proceedings, June 2010
6. Herring, D.H., "The Ubiquitous O-Ring," *Industrial Heating,* November 2009

CHAPTER 19

GAS QUENCHING

What is pressure quenching?

The description most often used to define high-pressure gas quenching (HPGQ) is "accelerating the rate (speed) of quenching by densification and cooling of gas."[2] One of the many reasons for the intense interest in this quenching technique is related to improved part distortion with full hardness. A critical concern in using this technology is to avoid sacrifice of metallurgical, mechanical or physical properties (i.e. retain the ability to transform a material to a microstructure that is similar, identical or superior to that of a known quenching medium such as oil or salt).

Pressure Levels

Selecting the optimum gas pressure for quenching is highly dependent on a number of factors, including material, component geometry, loading, net-to-gross load ratio, gas parameters, equipment design, etc. (Fig. 19.1). Pressure ranges (Table 19.1) are typically classified as sub-atmospheric, low, medium, high and ultrahigh pressure, irrespective of the type of gas used.

FIGURE 19.1 | Variables in high-pressure gas-quenching performance [8]

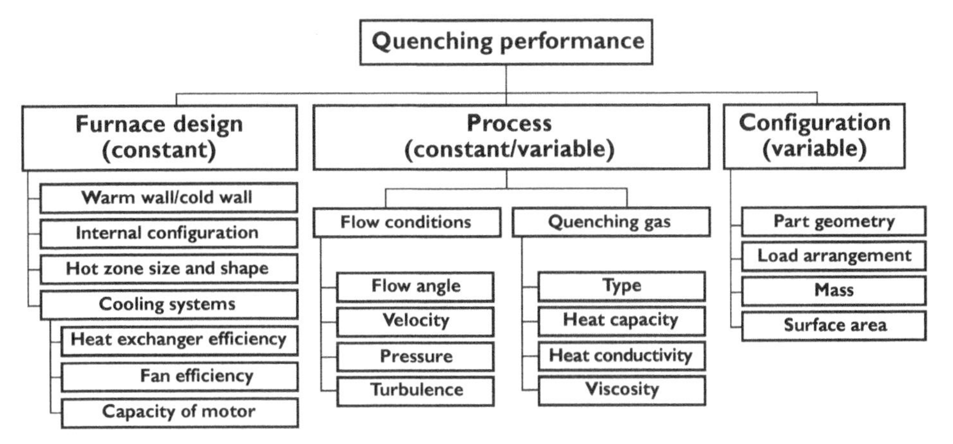

TABLE 19.1 | Classification of gas-quenching pressure ranges

Classification	Pressure range, bar
Vacuum cooling	< -0.67
Sub-atmospheric pressure	-0.67 to -0.17 [a]
Low pressure	2-4
Medium (mid-range) pressure	5-9
High pressure	10-20
Ultrahigh pressure	>20

Notes:

[a] Pressure is normally limited to -0.17 bar (-5 inches Hg) to prevent the furnace from reaching atmospheric pressure and an outer door from opening.

A Theoretical Understanding of the Heat-Transfer Coefficient

A key difference between quenching in a liquid and quenching in a gas lies in the different mechanisms involved in their heat-transfer characteristics. Most liquids (Fig. 19.2), such as water, polymer or oil, have distinct boiling points and, therefore, different heat-transfer mechanisms (and rates) at various temperature stages. For example, oil has three distinct heat-transfer phases, namely: vapor blanket or "film" boiling stage; nucleate or "bubble" boiling; and convection. For gaseous media (Fig. 19.3), heat transfer takes place by convection only.

FIGURE 19.2 | Heat transfer in liquid quenching (courtesy of ALD Vacuum Technologies GmbH)

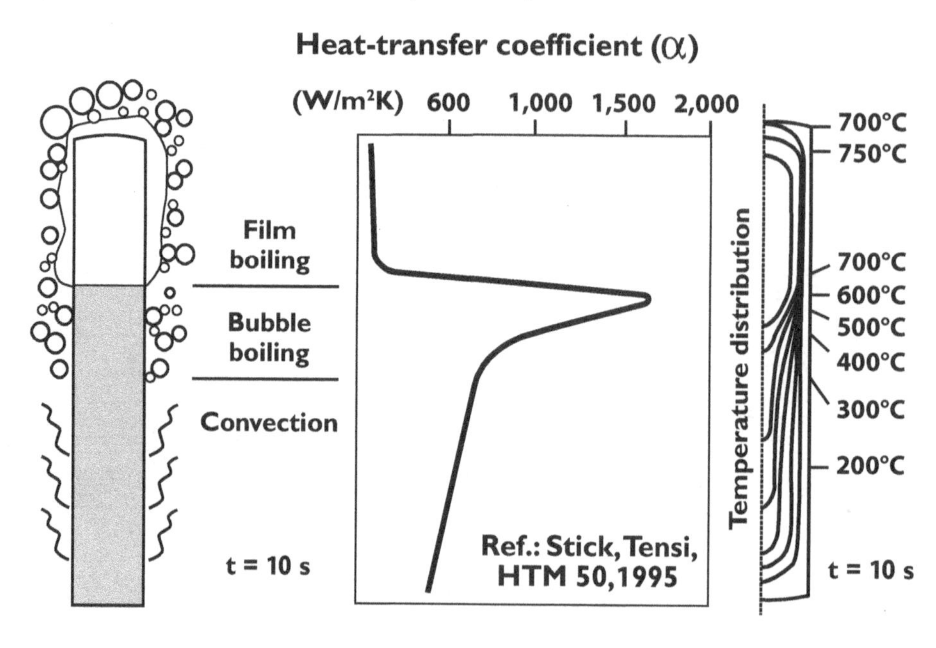

FIGURE 19.3 | Heat transfer in gas quenching (courtesy of ALD Vacuum Technologies GmbH)

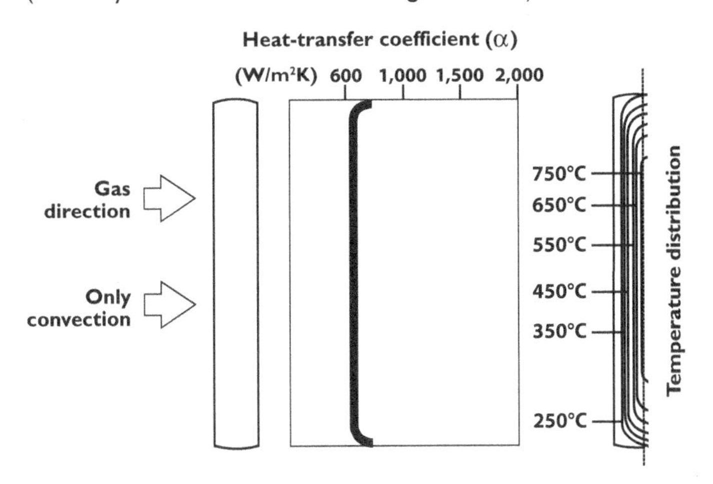

There is a relationship (Equation 19.1) that describes the convective heat transfer (q') in terms of the heat flux (or heat flow) through the surface of a part, which is proportional to the difference in temperature of the part surface (T_s) and the temperature of the quench gas (T_g) at any given moment in time.[3] This proportionality value, α (alpha), is the heat-transfer coefficient.

19.1) $q' = \alpha\ (T_s - T_g)$

In convective heat transfer, a complex formula (Equation 19.2) describes the heat-transfer coefficient (α) as dependent on the gas type as described by its heat capacity (c_p), its heat conductivity (λ) and the dynamic viscosity (η) of the gas and on a geometric factor (d) that represents the area of free passage of the gas for a specific load arrangement. The physical laws that illustrate how velocity (v) and pressure (ρ) influence the heat-transfer coefficient are given by:

19.2) $\alpha = C\ v^{0.7}\ \rho^{0.7}\ d^{-0.3}\ \eta^{-0.39}\ c_p^{\ 0.31}\ \lambda^{0.69}$

where C is a constant[4]

Therefore, the greatest influence on the gas cooling rate is the magnitude of the convective heat-transfer coefficient, α, which in gas quenching is dependent on velocity and pressure as well as gas temperature. In other words, for a given geometrical arrangement (e.g., transverse gas flow over a cylindrical body) and a given gas type (e.g., nitrogen or helium), the heat-transfer coefficient is directly proportional to the product of gas velocity and pressure raised to an overall (system-dependent fluid-dynamics constant) power, which is generally between 0.5-0.8.[5]

A Practical Understanding of the Heat-Transfer Coefficient

In simple terms, gas quenching of a component is required to improve its mechanical properties (e.g., strength, hardness, corrosion resistance, fatigue life) and better manage distortion. The ideal situation is to control the cooling process by changing the heat-transfer rates to keep distortion to a minimum while satisfying hardness and residual-stress requirements.

Distortion as a result of the quenching process is predominantly due to the thermal gradient and phase transformations within the component part. These can be controlled during gas quenching by adjusting the gas pressure for a given gas type and flow speed for a fixed quench-nozzle configuration.

Liquid quenchants such as water, polymer or oil have the characteristic that extremely high cooling rates (Fig. 19.4) result in very high instantaneous heat-transfer coefficients at the onset of the nucleate-boiling phase. This is a distinct advantage in the temperature range where pearlitic transformation occurs and one not possessed by gas quenching. With the breakdown of the vapor phase at the onset of boiling, however, the so-called Leidenfrost phenomenon occurs. The result is a totally non-uniform heat-transfer rate on different surfaces of the different parts that is dependent on a variety of variables and factors. This uneven transitory step creates huge temperature differentials and is the major factor in distortion when quenching in these media.

FIGURE 19.4 | Increase in cooling rate in quenchants at onset of nucleate boiling

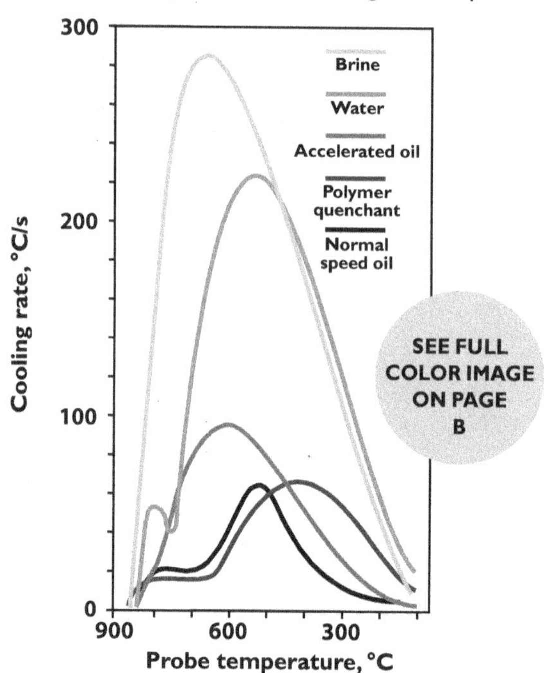

Although the maximum quenching "power" may be described by the instantaneous value of the heat-transfer coefficient, the average heat-transfer coefficient (Table 19.2) provides a better relative comparison of the various quenching media since it represents the value of the heat-transfer coefficient over the entire range of cooling (from the start to the end of quenching).

TABLE 19.2 | Average heat-transfer coefficient values

Quench media	Average α-value (W/m²-°K)	Instantaneous α-value (W/m²-°K)	Remarks
Brine or caustic	3,500-4,500	>15,000	>15,000
Water	3,000-3,900	>12,000	15-20°C
Oil, highly agitated	2,000-2,500	4,000-6,000	20-80°C
Polymer	1,500-2,000	3,000-4,500	20-50°C
Oil, agitated	1,500-1,750	3,000-4,000	20-80°C
Oil, still	1,000-1,550		20-80°C
Gas, ultrahigh	1,000-3,000	3,000-4,500	20-40 bar He/H_2
Gas, high pressure	300-1,750	2,500	5-20 bar N_2/He
Gas, low pressure	100-500		<5 bar N_2
Salt	400-500		500-600°C
Air	100-300		20°C

Gas quenching avoids the Leidenfrost phenomena and, therefore, has an inherent capability to produce smaller temperature differences in a part during quenching and less dimensional variation. As can be seen, however, its relative ability to successfully transform the microstructure of a part may be limited (Fig. 19.5).

FIGURE 19.5 | Microstructural variation in gas quenching

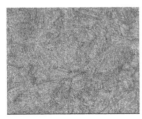

Microstructure is martensite. Core hardness = 44 HRC

Microstructure is primarily martensite with transformation products (bainite and ferrite) present. Core hardness = 28 HRC

Transmission gear
12 kg (26 pounds)/piece
AISI 8822H

Factors That Influence Quenching Performance

The calculation of the heat-transfer coefficient helps us to determine how quickly a workload will cool. As we have previously seen, however, there are other factors that significantly influence the cooling rate and overall time involved. These are generally classified as "external" factors and are sometimes included in the formula for the heat-transfer coefficient (Equation 19.2) as the multiplier f^a. The exponential value (a) for this factor is often equipment-dependent. In the case of single-chamber vacuum furnaces, it has been found to be in the range of 0.25-1.55.[7]

An understanding of the influence of these factors might be seen in a comparison of cooling rates between a single-chamber vacuum furnace (hot chamber) and a cold ("black body") chamber (Fig. 19.6). It should be noted that this test involved an ISO-9950 test probe. A small test specimen (12.5 mm diameter x 60 mm long) that was developed for laboratory testing of quench oils was used.

FIGURE 19.6 | Comparison of cooling rates in different quench chambers [10]

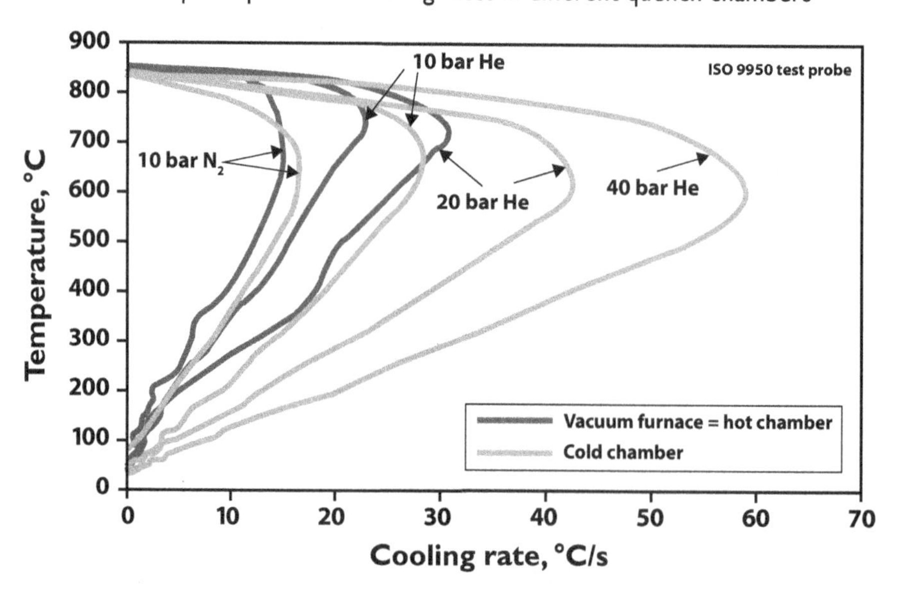

There are several concerns when using this small test sample in gas-quenching applications, namely:

1. A small mass cools much quicker than a typical part and, as such, fails to represent the entire quenching process for most components.
2. The small volume, having the thermocouple at its geometric center, fails to register a difference in cooling relevant to its position (vertical or horizontal) or to the gas-flow direction (top to bottom, bottom to top or alternating).

Embedded thermocouples in specific workloads or probes designed to measure the heat-transfer coefficient within the workload are used during gas quenching. These are available from a number of manufacturers.

Advantages and Disadvantages of HPGQ

When properly applied, HPGQ has several recognized advantages, which include:[2]

- Safety
- Overall economics
- Reduction of secondary manufacturing operations
- Optimization of dimensional variation
- Controllable cooling rates
- Part cleanliness
- Overall environmental impact

Of course, there are disadvantages that must be factored into any consideration to use this technology.[2] These include:

- Cooling-rate limitations (i.e. quench severity)
- Reversed application of heat-transfer rates (i.e. slow cooling rates in the pearlitic transformation range and high cooling rates in the martensitic transformation range)
- Regulations and (pressure vessel) codes
- Noise levels

Typical Applications

TOOL AND DIE INDUSTRY

The tool and die industry requires the processing of a variety of materials, geometries, cross-sectional areas and load sizes. Heat-treatment specifications can be extremely diverse. The principal application areas include die processing (Table 19.3), hot/warm-working steels (H11, H13), cold-working steels (e.g., ASP 23, M4, T15, CPM 9V, CPM 10V, CPM 15V) and high-speed steels.

HIGH-SPEED STEELS

Cutting tools in the form of hobs, shapers, shavers, milling cutters, broaches (flat and circular), form tools (dovetail and circular), end mills, reamers, taps and drills benefit from gas quenching. These tools are typically made from M-series (e.g., M1, M2, M3 Type 1 and 2, M7 and M42), T-series (e.g., T1, T15) and various powder-metal grades (Table 19.4). Gas quenching at pressures from sub-atmospheric to 5-12 bar are common for various small (≤50 mm; 2 inches) and medium-sized (≤150 mm; 6 inches) cross sections.

TABLE 19.3 | Tool-steel applications [18]

Tool type	Material	Typical service hardness requirements – as tempered (HRC)
Molds/holder blocks	4135, 4150	25-30
	P20	28-38
	J13, S7	42-55
	NAK 55 (Ni + Al PH)	38-40
	420, 420 modified	48-52
	17-4, 15-5	30-42
	Maraging steels	48-56
	A2, D2, 440C	56-60
	CPM 9V, CPM 10V, 440V, M390	54-62
Injection molding	Alloy and nitriding steels	28-32+
	H13, S7	42-55
	A2, D2, 440C	56-60
	17-4	30-42
	CPM 9V, CPM 10V, CPM 15V, MPL-1	54-62
Pelletizing/ granulating knives	Hard-faced alloy steels	58+
	D2, 440C	58+
	17-4	40-42
	M4, CPM 9V, CPM 10V, 440V	54-62

TABLE 19.4 | Powder-metal tool-steel applications [18]

Grade	%C	%W	%Mo	%V	%Co
M2HC/ASP 23	1.3	6.2	5	3	
M4HC	1.4	5.5	5.25	4	
Rex 45/ASP 30	1.3	6.25	5	3	8
Rex 20	1.3	6.25	10.5	2	
T15	1.6	12		5	5
Rex 76	1.5	6.5	7	6.5	10
ASP60	2.3	6.5	7	6.5	10

AEROSPACE/LAND-BASED TURBINES

Many aerospace heat-treating specifications now permit vacuum hardening of a defined section thickness of certain high-strength steels by gas quenching (Table 19.5). Pressures of 2 bar or higher are typical. The process drivers include cost, pure convective heat transfer, distortion, variable-outcome control (i.e. controllable cooling rates) and environmental factors.

TABLE 19.5 | Aerospace/land-based turbine applications [18]

Material type	Examples	Process
Stainless steel		
300 series	304, 316, 321	Solution annealing
400 series	403, 410, 416, 431	Hardening
Precipitation hardening	17-4, 17-7, 15-5, 13-8	Solution heat treating
Maraging steel	Vascomax C-250	Solution annealing
Iron-based superalloys		
Solid-solution strengthening	Incoloy 800, MA 956, RA330	Solution annealing
Precipitation hardening	A286, Incoloy 901	Solution heat treating
Nickel-based superalloys		
Solid-solution strengthening	Hastaloy B, C, X, RA330, Inconel 600, 625, MA 754, 718, X750	Solution annealing
Precipitation hardening	Nimonic 80A, 90, Rene 95, Waspaloy, Inconel MA 6000	Solution heat treating
High-strength alloys	300M, 4340	Hardening

The development of HPGQ was based on the principle that the denser the cooling medium, the more heat will be extracted from the load. By pressurizing, the quench gas becomes denser and has improved heat-transfer properties. In addition to the pressure of the quench gas, the cooling efficiency of the system is influenced by factors such as the type of quench gas used, the velocity of the gas and the design factors related to the size and shape of the chamber in which the quenching is taking place.

In terms of distortion control, it is widely assumed that gas quenching should produce less quench-related distortion because it creates less thermal shock than oil quenching.

Recent Developments

A study[16] of gas-quenching flow and heat transfer in quench chambers (including the interior components) and for external cooling flows (including multi-jet impingement) on different load geometries highlights the fact that flow non-uniformity in quenching is caused primarily by the chamber design. Hence, uniformity, as well as pressure drop, can be controlled by proper design of the flow passages. This can be most effectively done by Computational Fluid Dynamics (CFD) and Finite Element Methods (FEM) simulation and modeling (Fig. 19.7) followed by experimental testing in the field. When designing ducts with flow-area changes, low length-to-inlet diameter ratios and high exit cross-section aspect ratios increase both distortion of the exit velocity field and the overall pressure drop.

FIGURE 19.7 | Gas flow simulation [16]

SEE FULL COLOR IMAGE ON PAGE C

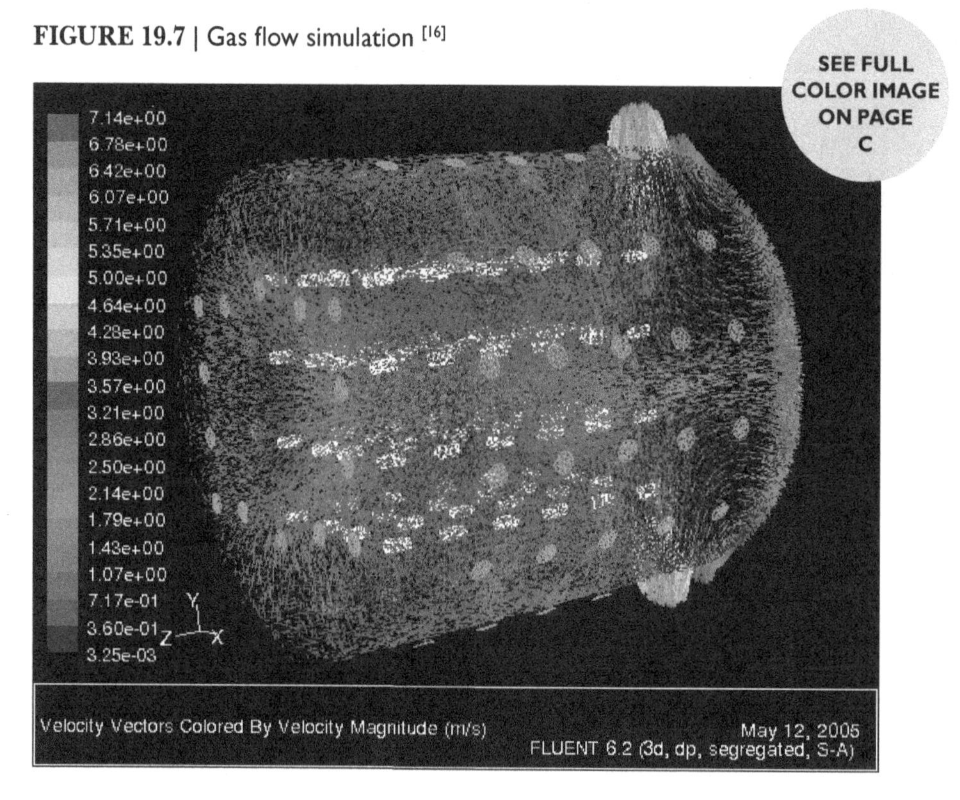

It has been reported[17] that variations in the heat-transfer coefficient (of up to about threefold) occur along the cylinder surface in unidirectional cross- or axial-flow cooling of round or square cross-section specimens. These non-uniformities are somewhat diminished by increasing the intensity of the upstream turbulence.

Knowledge of the heat-transfer coefficient distribution along the quenched part surface allows, in conjunction with solid mechanics and phase transition modeling of the quenched part, the prediction of items such as the extent of phase transformation, mechanical properties and dimensional changes (i.e. part distortion). Distortion and property non-uniformity are somehow proportional to the non-uniformity in the surface heat-transfer coefficient. In a workload composed of a number of cylindrical parts, the upstream cylinder cooling differs significantly from that of those downstream, which experiences much closer cooling rates on average. The cooling generally depends on the specific location of the part.

High and uniform heat-transfer coefficients can be obtained in multi-jet impingement cooling, and this process can be optimized by the design of the impingement system and its adaptation to the specific parts that are being quenched.

Work is also under way to predict and control phase transformations and limit dimensional changes in order to better understand the distribution of residual stresses and their relationship to mechanical properties. Mathematical modeling and simulation software are being combined with experimental field trials to refine the models.

Work is being done by a number of researchers to characterize steel hardenability. In HPGQ applications, the ability to change the quench behavior of gases by varying individual quench parameters (e.g., gas type and composition, pressure, circulation pattern and velocity) plays an important role in achieving improved hardenability. One example is research into the development of a gas-quench equivalent of the Jominy test.[11]

There are also a number of probes and other devices designed to measure the heat-transfer coefficient within the workload during the gas quenching.[12] The measurement and recording of the quenching intensity (and heat-extraction dynamics) in HPGQ applications have been solved by using these types of probes.

These types of sensors have thermocouples at precisely defined positions, enabling the calculation of the flow of thermal energy and the thermal heat-transfer coefficient from the temperature-measurement data during gas quenching. They are usually instrumented with three thermocouples – two of them near the surface of a cylindrical body, which enable the probe to distinguish different heat-transfer characteristics in both the vertical and horizontal positions, and one in the center. Thus, the influence of the "external" factors talked about above can be quantified. The measured quenching characteristics can be compared and data fed into a computer-based system for the prediction of hardenability as well as determining the required quenching parameters (gas pressure, fan speed, gas type, water flow to the heat exchanger, etc.) for a specific workload during the entire quenching process. This enables the quench intensity of the gas to be precisely determined.

Another technique for adjusting the end result is to adapt the quench intensity according to the specific hardenability of the quenched material via correlation diagrams (Fig. 19.8). The hardenability of a material can be expressed by its lambda (λ) value, a relationship between achieved hardness and cooling time in the range of 800-500°C (1475-925°F). Usually, this expression is given in the standard Jominy test or in a gas end-quench test (Fig.19.8, top left). Knowing the cooling time in a given furnace configuration (Fig. 19.8, top right), a correlation can be drawn to set the optimal quenching parameters to reach a minimum required hardness for materials with different hardenabilities (Fig. 19.8, bottom right). In addition, the diagram may be set up for other quenching parameters such as type of gas, position of the part in the load or the particular characteristics of the quenching-chamber flow conditions.

FIGURE 19.8 | Example of correlation diagram for selecting proper process parameters [21]

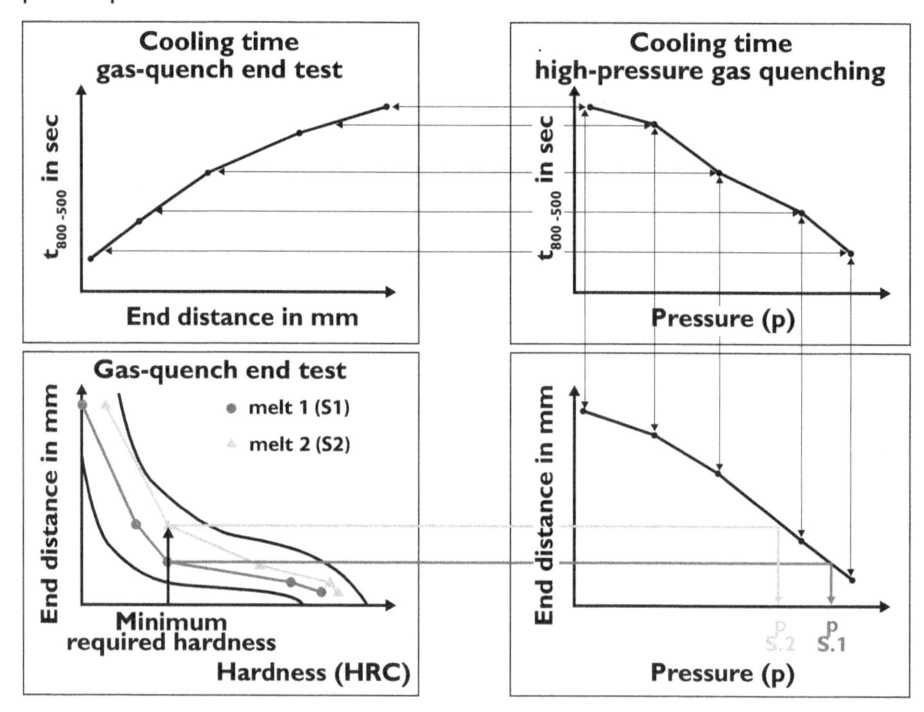

Another emerging area is that of controlled heat-extraction (CHE) technology.[13] CHE explores the possibility of automatically following a predetermined heat-extraction (temperature-time) cycle during HPGQ. The reason for interest in CHE is because:

1. The HPGQ process lasts much longer than quenching in liquid quenchants (assuming comparable dimensions and masses of the workpiece).
2. Influential quenchant parameters (e.g., medium temperature, pressure and agitation rate) cannot be changed during the quenching process when using liquid quenchants, but they can be automatically changed during HPGQ. This versatility offers new possibilities to influence the heat-extraction dynamics during the quenching process itself. Furthermore, with the controlled addition of sprayed cryogenic liquid (e.g., nitrogen), an instantaneous drop of the cooling-gas temperature occurs with a substantial increase in the quenching intensity. This can have a drastic influence on hardness distribution throughout the part cross section as well as on residual stresses and distortion after quenching.

The influence of heat-extraction dynamics on hardness distribution after quenching is also being actively studied.

Summary

Over the last 20 years or so there have been a large number of published papers providing theoretical calculations and empirical data on the subject of gas quenching (a partial list is presented in the references that follow). The intense interest in this technology is due to that fact that high-pressure gas quenching offers a method of limiting dimensional variation within a part and within a workload.

The key to understanding the factors that influence gas quenching is to achieve a balance between the speed of quenching and the uniformity of quenching. The former requires holding constant as many of the process variables as possible, while the latter involves uniform heat extraction. In this way, both repeatability (quality) and performance (productivity) will be achieved with an optimized microstructure.

Gas-quenching technology is readily available to the heat treater. It should be applied in those applications where its advantages outweigh its disadvantages. As with all technologies, having as complete an understanding as possible of the performance requirements of the product helps to evaluate which quenching choice is best for the end-use application.

REFERENCES

1. Herring. D.H., "A Review of Gas Quenching from the Perspective of the Heat Transfer Coefficient," *Industrial Heating*, February 2006
2. Herring, D.H., "Pressure Quench, Furnace Design Extend Range of Applications," *Heat Treating*, September 1985
3. Edenhofer, B., F. Bless and J.W. Bouwman, "The Evolution of Gas Quenching in Today's Heat Treating Industry," Conference Proceedings, 11th IFHT Congress on Heat Treatment, October 1998, Florence, Italy
4. Preißer, F., "The Physics of Gas Quenching," Conference Proceedings, Workshop on High Pressure Gas Quenching, ASM International, 1996, Indianapolis, Ind.
5. Herring, D.H., "A Review of High Pressure Gas Quenching in Semi-Continuous Vacuum Furnaces, Proceedings," ASM Heat Treat Conference, 1997
6. Hick, A. J., "The Wolfson Test for Assessing the Cooling Characteristics of Quenching Media," *Heat Treatment of Metals*, 1983.3, p. 69
7. Carter, George C., "Optimizing Gas Quenching," *Advanced Materials & Processes*, February 1996
8. Thomas Wingens, Ipsen Inc., private correspondence
9. Prof. Bozidar Liscic, Faculty for Mechanical Engineering, Croatia, private correspondence

10. Segerberg, S. and E. Troell, "High-pressure Gas Quenching using a Cold Chamber to Increase Cooling Capacity," *Heat Treatment of Metals*, 1997, p. 21-24
11. Lohrmann, M., F. Hoffmann and P. Mayr, "Characterization of the Quenching Behavior of Gases," Conference Proceedings, Heat Treating Equipment and Processes, ASM International, 1994, Schaumburg, Ill.
12. Herring, D.H., "Applying Intelligent Sensor Technology to Problems Related to Distortion," SME Conference Proceedings, Quenching and Distortion Control, 1998
13. Liscic, B., "Critical Heat-Flux Densities, Quenching Intensity and Heat Extraction Dynamics During Quenching in Vaporizable Liquids," Conference Proceedings, 2003 Heat Treat Conference & Exposition, ASM International, Indianapolis, Ind.
14. Pritchard, Jeff and Scott Rush, "Vacuum Hardening High Strength Steels: Oil versus Gas Quenching," *Heat Treating Progress*, May/June 2007
15. Lior, N., "The Cooling Process in Gas Quenching," *Journal of Materials Processing Technology*, Elsevier, 155-156 (2004) 1,881-1,888
16. Center for Heat Treating Excellence (CHTE), Worcester Polytechnic Institute
17. Cheng, He-ming, Hong-gang Wang and Jian-bin Xie, "Calculation of Coupled Problem Between Temperature and Phase Transformation During Gas Quenching in High Pressure," *Applied Mathematics and Mechanics* (English Edition), 2006, 27(3): 305-311
18. Dixon, R.B., "Meeting Industry Demands for Rapid Gas Quenching Systems for Vacuum Heat-Treating Equipment," *Heat Treatment of Metals*, 1997, 37-42
19. Li, Zhichao and Raman V. Grandhi, "Multidisciplinary Optimization of Gas-Quenching Process," *Journal of Materials Engineering*, February 2004, Vol. 14 (1), 136-143
20. Lubben, Th., F. Hoffman, P. Mayr and G. Laumen, "Scattering of Heat Transfer Coefficient in High Pressure Gas Quenching," Conference on Heat Treating Process and Equipment, 1994, Schaumberg, Ill.
21. Fritsching, U. and R. Schmidt, "Gas Flow Control in Batch Mode High Pressure Gas Quenching," Innovation in Heat Treatment for Industrial Competitiveness, Verona, May 2008

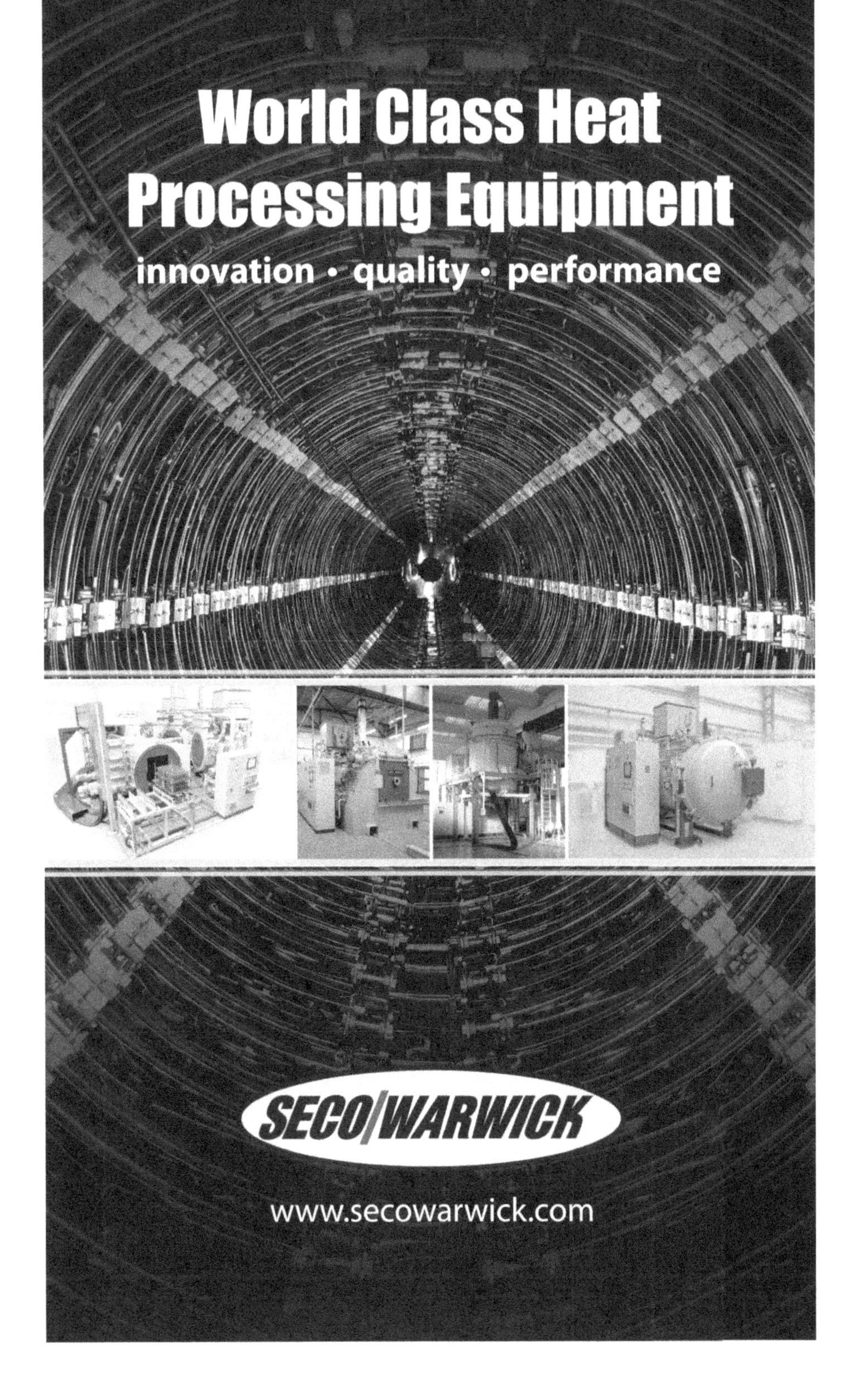
World Class Heat
Processing Equipment
innovation • quality • performance
SECO/WARWICK
www.secowarwick.com

COLOR IMAGE SECTION

FIGURE 8.6 | Debris buildup in a hot zone

FIGURE 13.1 | Typical interior of a contaminated vacuum-furnace hot zone

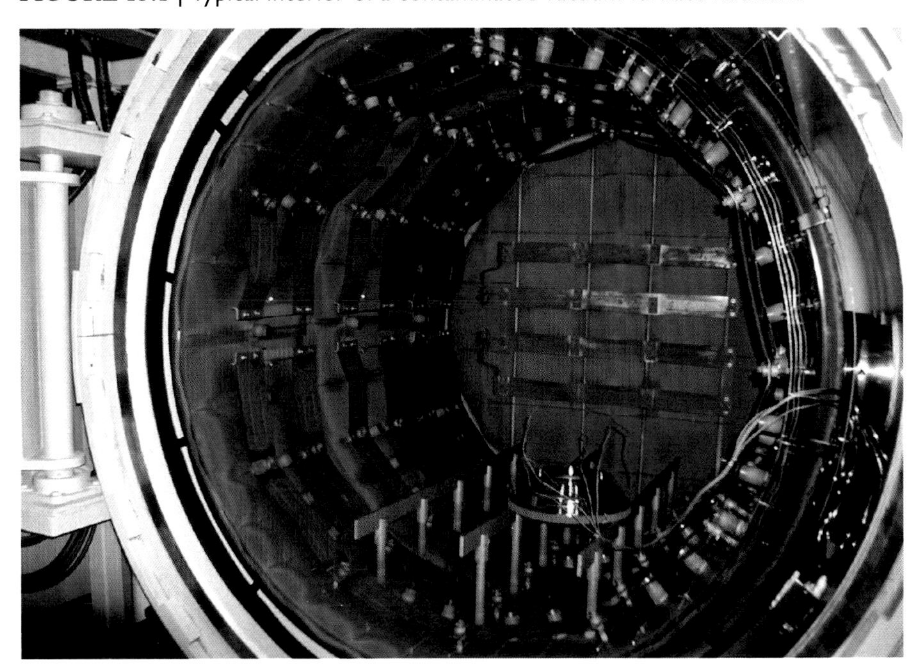

FIGURE 15.5 | Gas circulation pattern for an internal heat exchanger system

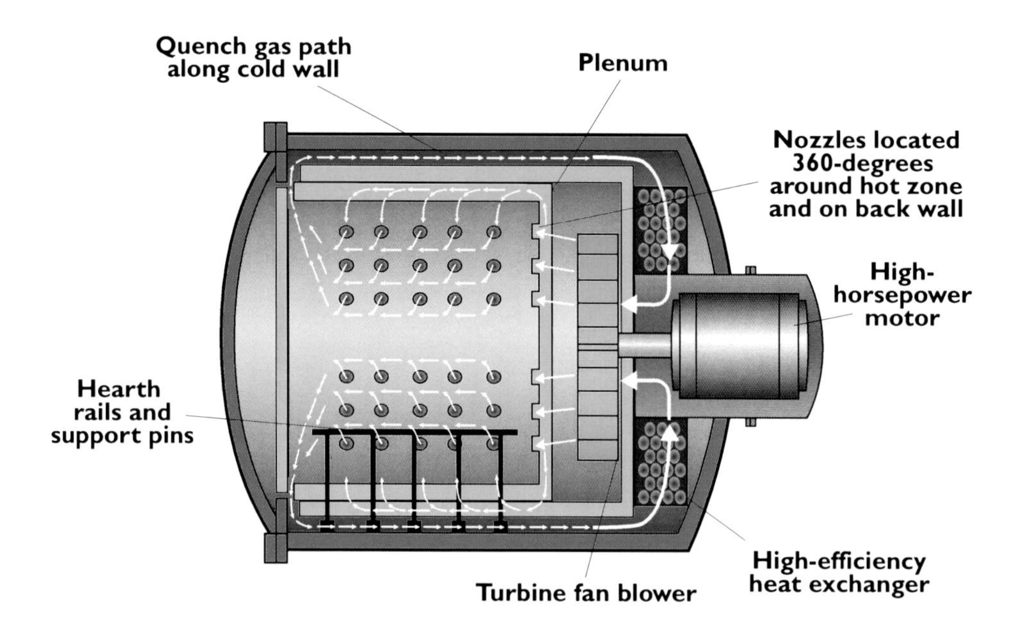

FIGURE 19.4 | Increase in cooling rate in quenchants at onset of nucleate boiling

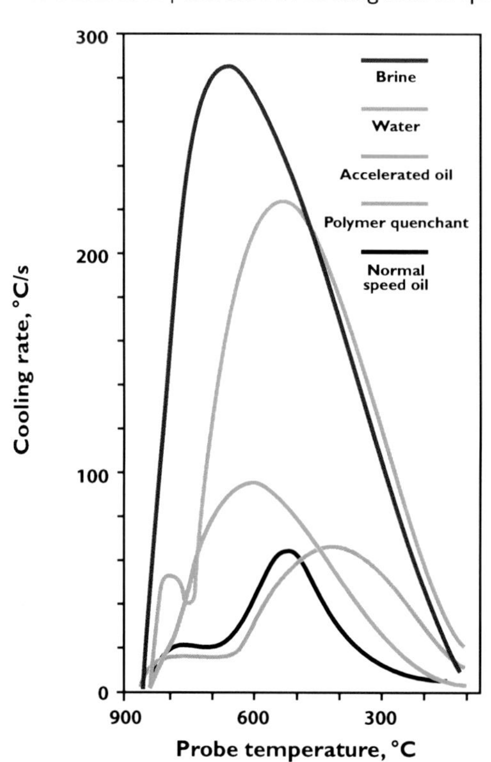

FIGURE 19.7 | Gas flow simulation

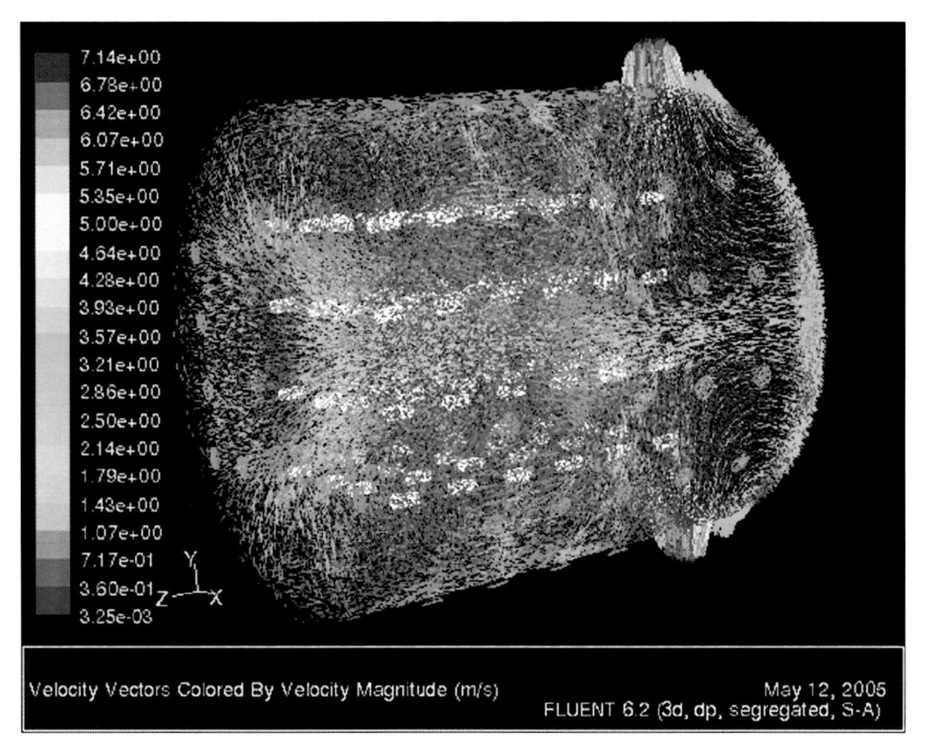

FIGURE 22.2 | Helium gas-recovery system

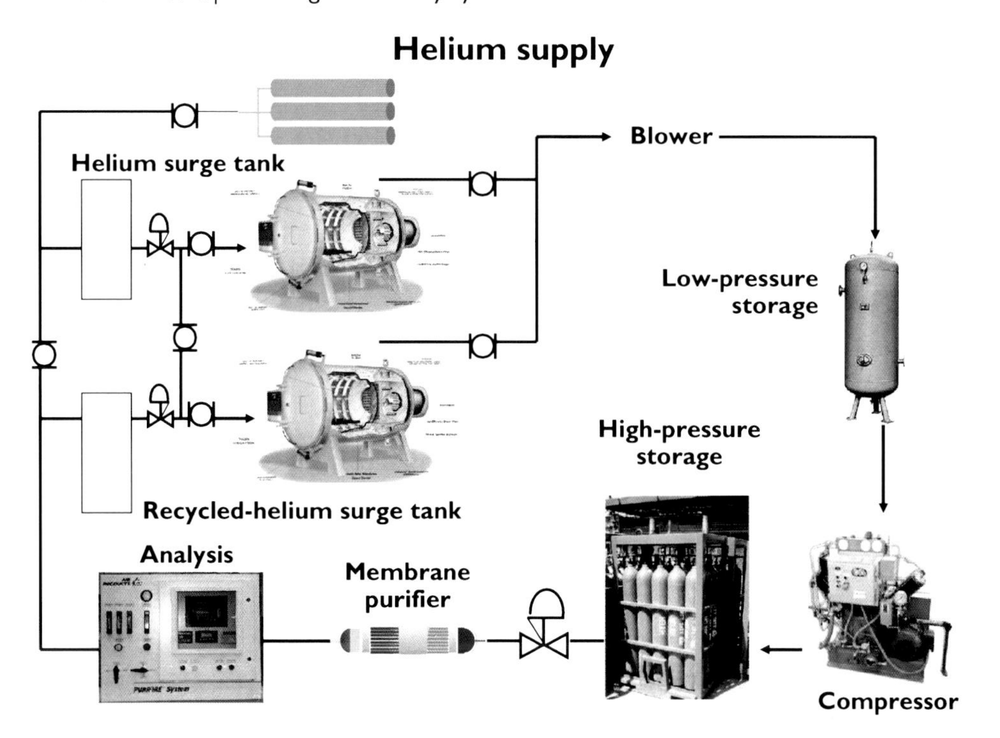

FIGURE 26.1 | First commercial heat-treat load, February 1969: gears carburized at 930°C (1700°F), 13 mbar (10 torr), methane (CH_4)

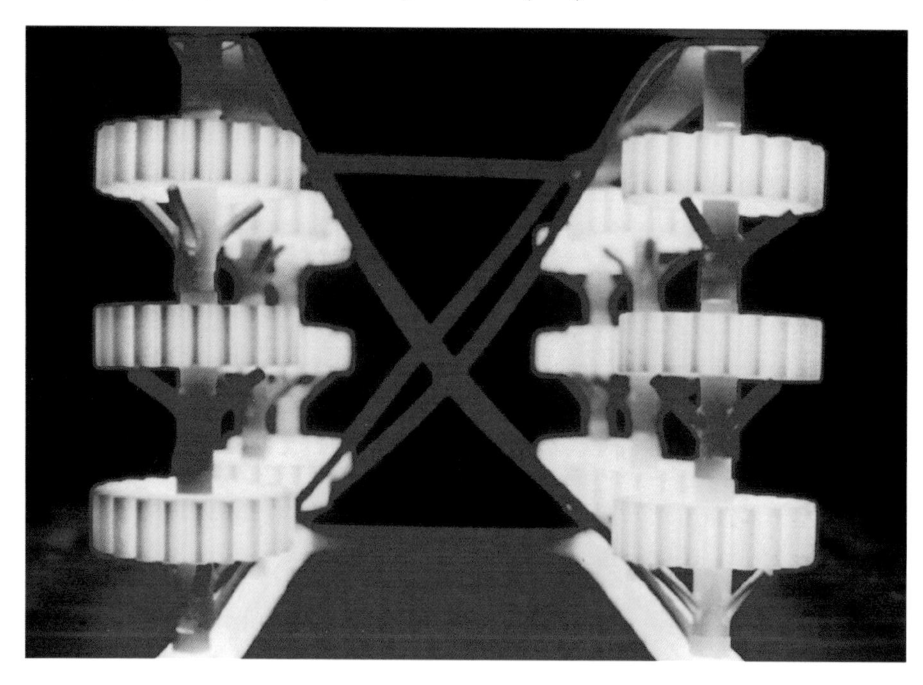

FIGURE 26.8 | Plasma-carburized workload

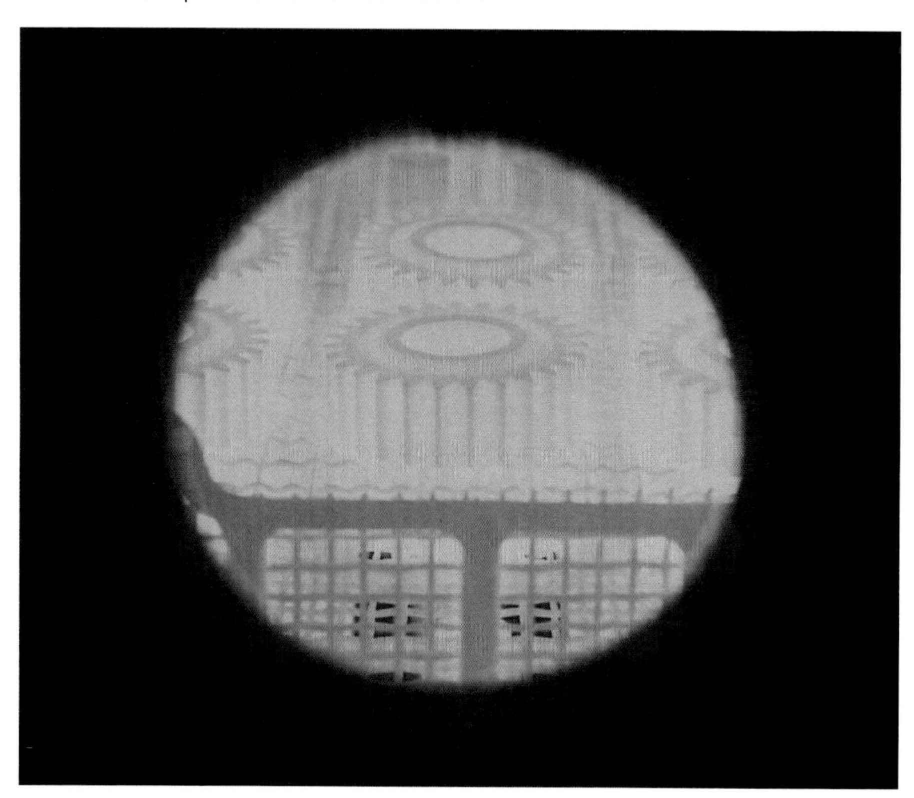

FIGURE 26.10 | Ion nitriding of automotive components

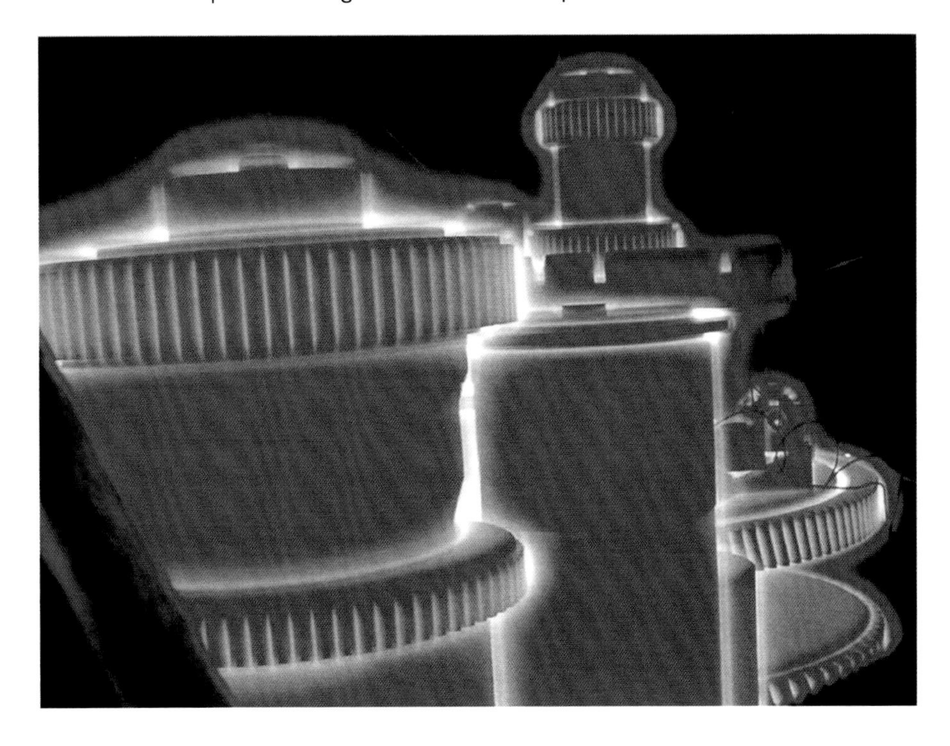

FIGURE 26.25 | Vacuum-carburized gearbox actuator gears

FIGURE 27.1 | Bright annealing of copper blanks for diode heat sinks

FIGURE 27.13 | a.) Hydrogen molecules adsorb onto the metal lattice; b.) absorption and chemisorption lead to an expanded metal lattice

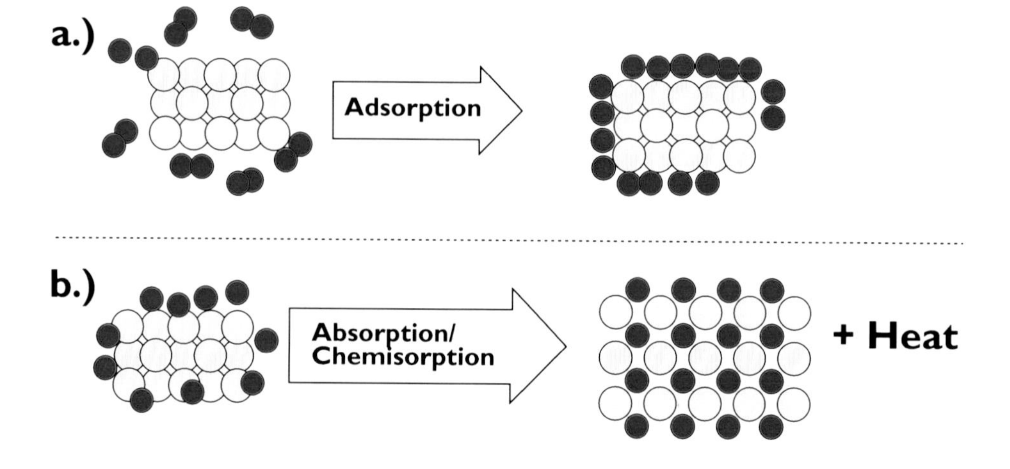

FIGURES 28.1-28.5 | Vacuum market share in the Americas

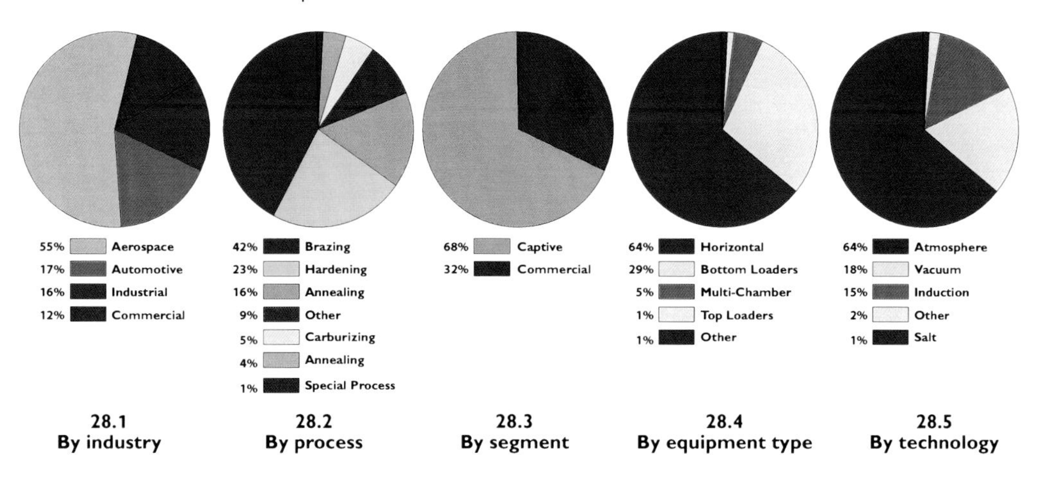

FIGURE 28.11 | Copper brazing of 304 SS housings

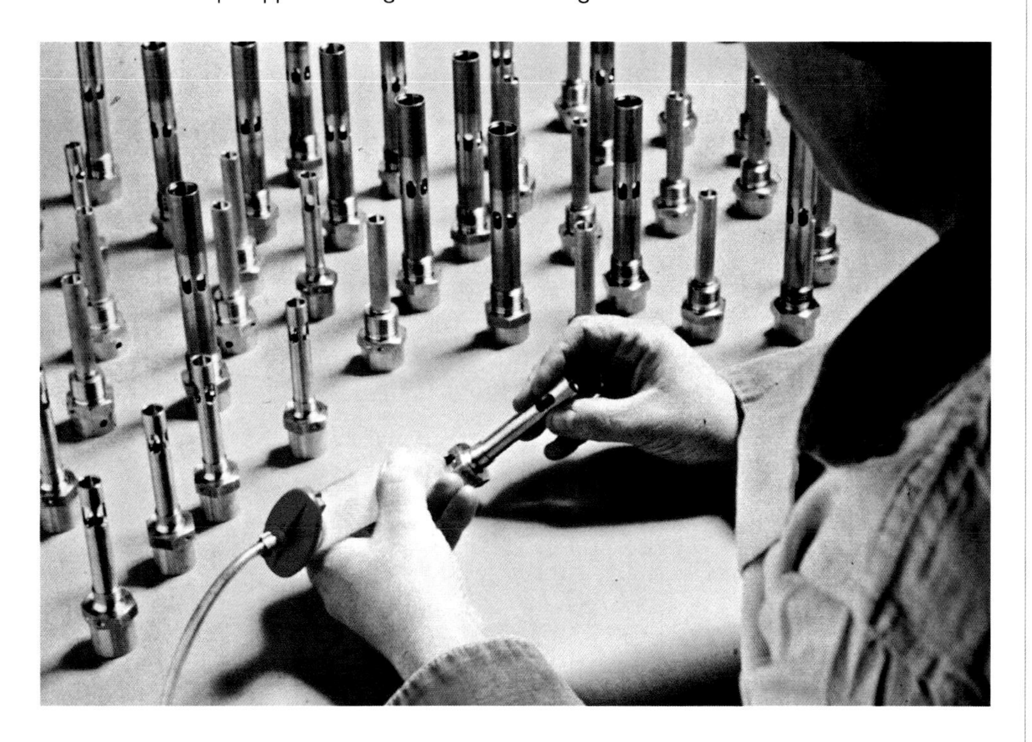

FIGURE 28.20 | Statistical process-analysis measurement over balls 0.036578 mm (0.0015 inches); two locations

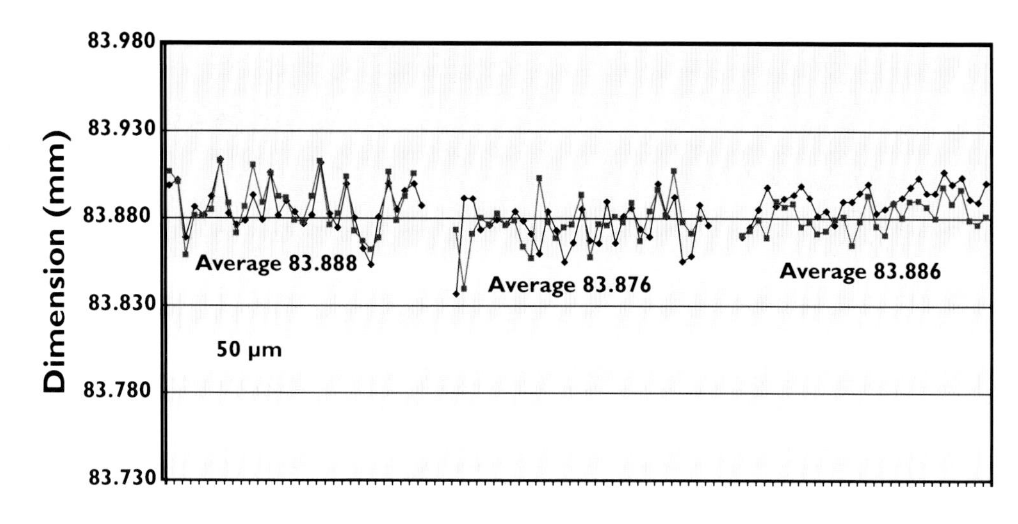

FIGURE 28.28 | Typical forged components

CHAPTER 20

OIL QUENCHING

Why oil quenching?

Many components use oil quenching to achieve consistent and repeatable mechanical and metallurgical properties as well as predictable distortion patterns. Oil quenching is so popular due to its excellent performance results and stability over a broad range of operating conditions. Oil quenching facilitates hardening of steel by controlling heat transfer during quenching, and it enhances wetting of steel during quenching to minimize the formation of undesirable thermal and transformational gradients which may lead to increased distortion and cracking. For many, the choice of oil is the result of an evaluation of a number of factors, including:[1]

- Economics/cost (initial investment, maintenance, upkeep, life)
- Performance (cooling rate/quench severity)
- Minimization of distortion (quench system)
- Variability (controllable cooling rates)
- Environmental concerns (recycling, waste disposal, etc.)

Oil Quenching in Vacuum

Oil-quench vacuum systems offer an attractive alternative to other technologies because of their ability to control the various quench variables. All oil-quench vacuum furnaces must meet NFPA 86 standards.

One difference is that oil-quenching furnaces can vary and control the pressure over the oil. This technique can extend the range of part cross sections and materials that can be processed. In addition, the use of a vacuum oil quenching has been found to reduce distortion in a wide variety of components such as gears, shafts and ball bearings.

Distortion-minimization techniques using pressure variation have proved effective.[2, 3] Altering pressure over the quench oil allows for a change in the boiling point of the quenchant. The position of the boiling point (i.e. characteristic temperature) determines where and for how long the various stages of oil cooling take place. The lower pressure allows for longer vapor-blanket stages and a somewhat long "vapor transfer" stage due to the reduced boiling point of the oil. This has been found to reduce distortion in some part geometries and provide the sought-after hardness, provided the material's hardenability is suitable.

Distortion-minimization methods have been used in combination with changes to flow characteristics (e.g., some manufacturers pull oil down through the workload as opposed to pushing it upward) and oil compositions specially blended for use in vacuum having low vapor pressure so that they are easily degassed (Table 20.1).

TABLE 20.1 | Vacuum quench oils (courtesy of C.I. Hayes) [4]

Characteristic	H-1	H-2	H-3	No. 420[a]	H-4	Accelerator
Maximum operating temperature, °C (°F)	60 (140)	82 (180)	82 (180)	120 (250)	230 (450)	
Normal operating range, °C (°F)	32-60 (90-140)	50-65 (120-150)	50-65 (120-150)	82-105 (180-220)	105-175 (225-275)	
GM Quench-O-Meter	9-11 @ 60°C (140°F)	13-15 @ 60°C (140°F)	12-14 @ 60°C (140°F)		27-28 @ 150°C (300°F)	n/a
Hot wire, amps	34.0 @ 60°C (140°F)	32.5 @ 60°C (140°F)	34.0 @ 60°C (140°F)	31.0 @ 60°C (140°F)	n/a	n/a
Specific gravity	0.87 @ 16°C (60°F)	0.86 @ 25°C (77°F)	0.86 @ 25°C (77°F)		0.90 @ 16°C (60°F)	0.92 @ 16°C (60°F)
Viscosity (SUS)						
@21°C (70°F)					6,500	
@38°C (100°F)	93	115	130	260	1,800	3,100
@55°C (130°F)	58					
@82°C (180°F)	42					
@99°C (210°F)				73	120	
Flash point (COC), °C (°F)	170 (340)	193 (380)	196 (385)	221 (430)	300 (575)	170 (340)
Free fatty acid	Nil	Nil	Nil	Nil	Nil	Nil
Conradson carbon residue	0.01	0.01	0.01	0.01	0.40	

Color (ASTM), Hellige	1.5	1.5	1.5	1.5		
Ash content (ASTM)	Trace	Trace	Trace	Trace		Trace
Pour point, °C (°F)	-7 (20)	-18 (0)	-18 (0)		-18 (0)	-4 (25)
Vapor pressure, mm Hg						
@ 38°C (100°F)	0.0001	0.0001	0.0001			
@ 93°C (200°F)	0.001	0.0103	0.0103			
@ 150°C (300°F)	0.50	0.45	0.45		0.25[b]	
@ 205°C (400°F)					4.0[c]	
Active sulfur	Nil	Nil	Nil	Nil	Nil	Nil
Halogens	Nil	Nil	Nil	Nil	Nil	Nil
% Additive	10.0	7.2	10.7		0.0	

Notes:

[a] Park Chemical oil product

[b] Reduced by 25%, distilled off to 0.0012

[c] Reduced by 25%, distilled off to 0.050

The design of an integral vacuum oil-quench system requires considerations beyond those of atmosphere oil quenching. For example, the boiling point and vapor pressure of the base oil as well as the accelerant additive's characteristics must be taken into consideration along with the quench-oil temperature, agitation, cleanliness, pH and viscosity. Also, the vapor pressure of the quench oil must be compatible with the selected operating vacuum level.

Finally, vacuum systems do not permit the build up of water in the quench tanks. In a vacuum-furnace system where vacuum is used to process the work or purge the quench environment, moisture will be removed as the system is evacuated and the oil circulated. The circulated oil brings any moisture to the surface, where it is vaporized and removed from the oil by the pumping system. There must be an interlock to prevent quenching if circulation (agitation) is lost.

Cooling-Rate Characterization

Measuring the efficiency, or speed, of an oil quench can be done one of two ways: by measuring the oil's hardening power (i.e. its ability to harden a steel) or by measuring the cooling ability of the liquid. Because cooling ability is independent of steel selection (composition and grain size), this method is popular since it provides information about the oil itself independent of its end-use application.

The preferred test method is cooling-curve analysis (ISO 9950), which involves a laboratory test using a nickel-alloy probe for the determination of the cooling characteristics of industrial quenching oils. The test is conducted in non-agitated oils and is therefore able to rank the cooling characteristics of the different oils under standard conditions and provide information on the cooling pathway, which must be known if the ability of quench oil to harden steel is to be determined. Older methods such as the GM Quench-O-Meter (ASTM D3520) or the hot-wire test are still in common use. The GM Quench-O-Meter, for example, measures the overall time to cool a 22-mm (7/8-inch) nickel ball from 885°C to 355°C (1625°F to 670°F), while the hot-wire test is influenced by the heat-extraction rate of the oil at temperatures close to the melting point of Nichrome, which is approximately 1510°C (2750°F).

Oils are generally classified by their ability to transfer heat as fast, medium or slow "speed" oils (Table 20.2). Fast (8-10 seconds) oils are used for low-hardenability alloys, carburized and carbonitrided parts, and large cross sections that require high cooling rates to produce maximum properties. Medium (11-14 seconds) oils are typically used to quench medium- to high-hardenability steels. Slow (15-20 seconds) oils are used where hardenability of a steel is high enough to compensate for the slow cooling aspects of this medium.[2]

TABLE 20.2 | Classification of quench oils [5]

Type	GM Quench-O-Meter rating, seconds [a]
Fast oil	7-10
Medium oil	11-14
Slow oil	15-20
Marquench oil	18-25

Note:

[a] A cooling-curve test better represents the condition of the oil. The GM Quench-O-Meter test is no longer an ASTM standard test method.

Mechanisms of Heat Removal During Quenching

The mechanism of cooling a gear in liquid is largely dependent on geometry for a given material that dictates the requirements of the quench system.

Traditionally, we talk about three distinct stages of cooling (Fig. 20.1). Stage A is called the vapor-blanket (or film-boiling) stage. It is characterized by the Leidenfrost phenomenon, which is the formation of an unbroken vapor blanket that surrounds and insulates the workpiece. It forms when the supply of heat from the surface of the part exceeds the amount of heat that can be carried away by the cooling medium. The stability of the vapor layer, and thus the ability of the oil to harden steel, is dependent on the metal's surface irregularities; oxides present; surface-wetting additives that accelerate the wetting process and destabilize the vapor blanket; and the quench oil's molecular composition, including the presence of more volatile oil degradation by-products.[6] In this stage, the cooling rate is relatively slow in that the vapor envelope acts as an insulator, and cooling is a function of conduction through the vapor envelope.

Stage B is the second stage of cooling known as the vapor-transport (nucleate or bubble-boiling) stage. It is during this portion of the cooling cycle that the highest heat-transfer rates are produced. The point at which this transition occurs and the rate of heat transfer in this region depend on the oil's overall molecular composition.[6] It begins when the surface temperature of the part has cooled enough so that the vapor envelope formed in Stage A collapses. Violent boiling of the quenching liquid results, and heat is removed from the metal at a very rapid rate, largely due to heat of vaporization. The boiling point of the quenchant determines the conclusion of this stage. Size and shape of the vapor bubbles are important in controlling the duration of this stage. The majority of gear distortion occurs during this stage.

Stage C is the third stage of cooling called the liquid (or convection) cooling stage. The cooling rate during this stage is slower than that developed in the second stage and is exponentially dependent on the oil's viscosity, which will vary with the degree of oil decomposition. Increasing oil decomposition will initially result in a reduction of oil viscosity followed by increasing viscosity as the degradation process increases. Heat-transfer rates increase with lower viscosities and decrease with increasing viscosity.[6] This final stage begins when the temperature of the metal surface is reduced to the boiling point (or boiling range) of the quenching liquid. Below this temperature, boiling stops and slow cooling takes place by conduction and convection. The difference in temperature between the boiling point of the liquid and the bath temperature is a major factor influencing the rate of heat transfer in liquid quenching. Viscosity of the quenchant plays a major role in the cooling rate in this stage.

FIGURE 20.1 | Three stages of liquid quenching [7]

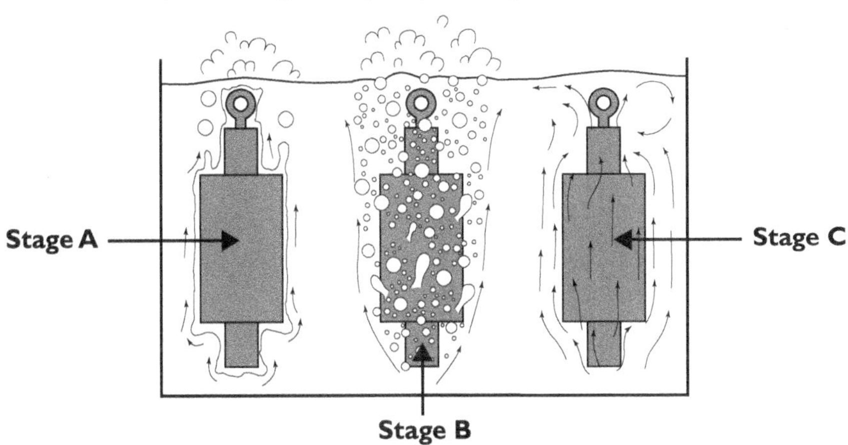

FIGURE 20.2 | Typical cooling curves and cooling-rate curves for new oils [8]

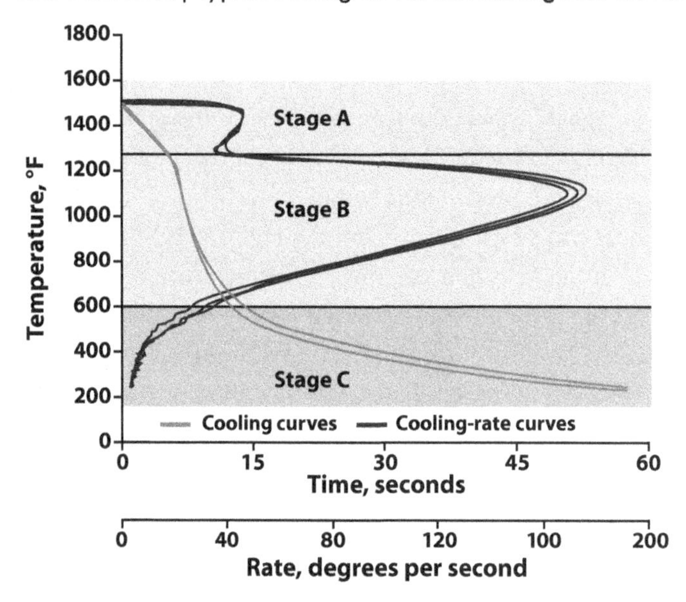

According to an investigation by N.I. Kobasko,[9] there are four modes of heat transfer around a part being quenched – namely shock-film boiling, full-film boiling, nucleate boiling and convection (Fig. 20.3a). At quenching, there are two critical heat-flux densities, qcr_1 and qcr_2 (Fig. 20.3b), for any vaporizable liquid. The first critical heat-flux density, qcr_1, is the heat-transfer rate necessary to form the compact vapor film around the part.

Due to the large temperature difference between the very hot surface of the part and the quench-bath temperature, the liquid layer in contact with the hot part's surface heats up to boiling temperature in about one-tenth of a second. First, small vapor bubbles occur. Then larger bubbles grow in size and number until they detach from the part surface forming the vapor blanket. This is the shock-film boiling stage. The second critical heat-flux density, qcr_2, is the minimal heat flux at which the transition from full-film boiling to nucleate boiling occurs. According to Tolubinski,[9] the critical heat-flux density qcr_1 is given as Equation 20.1.

20.1) $qcr_1 = 7r\ (af\rho'\rho'')^{0.5}$

where:

r = heat of vaporization

a = λ/ρc (thermal diffusivity)

f = frequency of vapor bubble detachment

ρ' = density of the liquid

ρ'' = density of the liquid's vapor

FIGURE 20.3 | a.) Four modes of cooling during quenching; b.) critical heat-flux densities [9]

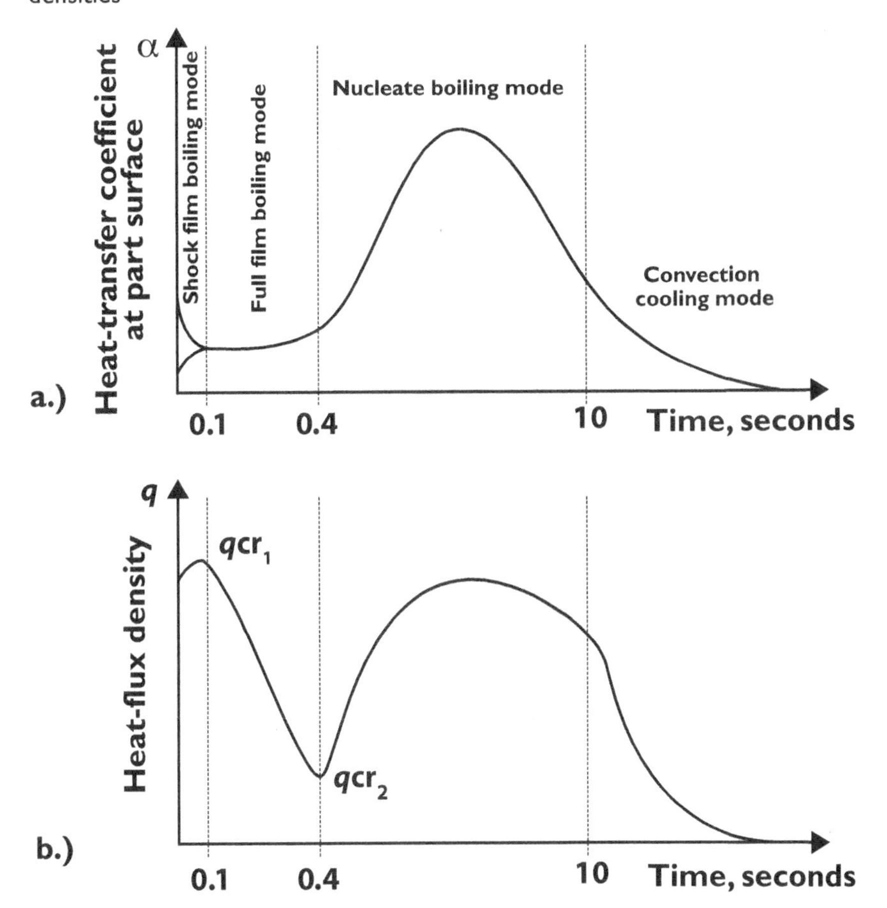

The only variable in Equation 20.1 on which qcr_1 depends is f (i.e. the frequency of vapor bubbles detaching from the surface). In other words, a compact vapor film around the part will be formed only if the critical heat-flux density qcr_1 is attained, which depends on the number of vapor bubbles and the frequency with which they detach from the part's surface.

In addition, these stages of cooling may not occur at all points on a part at the same time (Fig. 20.4). As the internal heat moves to the surface, differences in heat rejection may vary based on the surface configuration. Consequently, the need for a uniform and controllable agitation of liquid over the part surface is imperative. Controlled movement of the quenching liquid is vital, because it causes an earlier mechanical disruption of the vapor blanket in the first stage and produces smaller, more frequently detached vapor bubbles during the vapor-transport cooling stage. Agitation constantly provides a cooler liquid to the part surface, which provides a greater temperature difference that allows for improved heat rejection.

FIGURE 20.4 | Influence of part geometry on cooling [10]

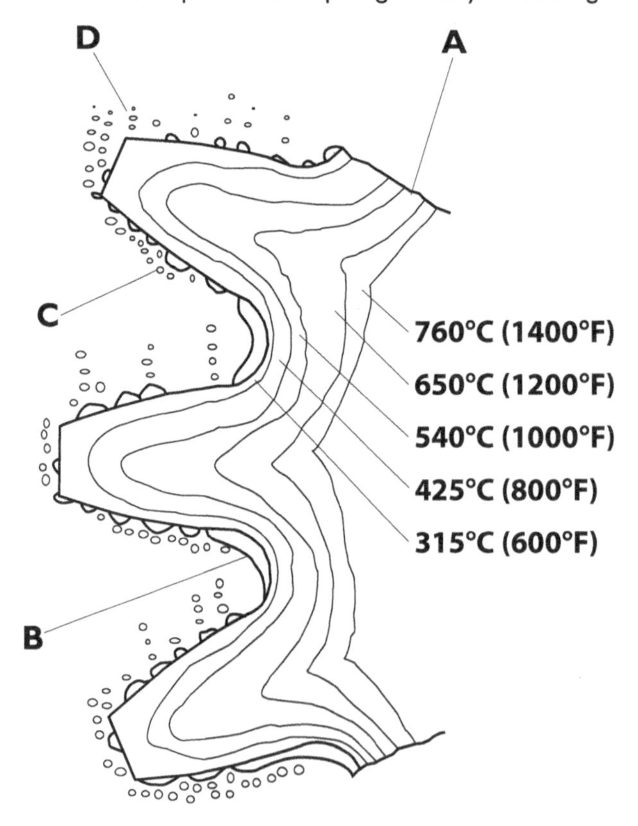

Properties of an Ideal Quenching Medium

The ideal quenching medium is one that would exhibit high initial quenching speed in the critical hardening range (through Stage A and B) and a slow final quenching speed through the lower temperature range (Stage C). Therefore, the ideal quenchant is one that exhibits little or no vapor stage, a rapid nucleated boiling stage and a slow rate during convective cooling. The high initial cooling rates allow for the development of full hardness by getting the steel past the nose of the isothermal transformation diagram (quenching faster than the so-called critical transformation rate) and then cooling at a slower rate beginning at the time the steel is forming martensite. This allows stress equalization, and distortion and cracking are reduced. The first criterion that any quenchant must meet is its ability to approach this ideal quenching mechanism.

When conventional quenching oils are used, the duration of Stage A is longer, the cooling rate in Stage B is considerably slower and the duration of Stage C is shorter. As such, the "quenching power" of oil is far less drastic than that of water. Water and water solutions exhibit high initial cooling rates. Unfortunately, because of water's low boiling point, this fast cooling persists until the steel is cooled to below 150°C (300°F). As most steels have formed or are forming martensite by this point, stresses are given little time to equalize. Therefore, water is typically limited to simple shapes or low-hardenability materials.

Oil has a major advantage over water due to its higher boiling range. A typical oil has a boiling range between 230-480°C (450-900°F). This causes the slower convective cooling stage to start sooner, enabling the release of transformation stresses. Therefore, oil is able to quench intricate shapes and high-hardenability alloys successfully.

As it is heated, oil has a proportional drop in viscosity. This allows the quenchant to move more freely, increasing (in general) the tendency to break the vapor-blanket layer. The nucleate-boiling stage is not drastically altered by changes in bath temperature. The cooling rate in the convection stage of an oil quench will slow as the bath temperature increases. This is advantageous for obtaining a slower rate of cooling through the austenite-to-martensite transformation range. In general, as the temperature of a quenching oil increases, the overall quenching rate increases.

Practical heat-transfer coefficient (α) values in the 1,000-2,500 $W/m^2 \cdot K$ range can be achieved based on oil characteristics and degree of agitation. Peak values of α in the cooling range of oil are 4,000-6,000 $W/m^2 \cdot K$ or a cooling rate greater than 100°C/second (180°F/second).

Effect of Increasing Bath Temperature

The temperature of the quenchant has dramatic influence on the rate of part cooling. Increasing the bath temperature from 21-120°C (70-250°F) produces slightly faster cooling in Stage A because the viscosity of the oil decreases. In Stage B, cooling is only slightly increased. In Stage C, cooling decreases near the end of the quench because the temperature differential between the bath and the steel is decreased.

Quenching oils have various characteristics that allow for variable cooling rates as a function of not only their boiling point but also their temperature. This has a direct bearing on properties such as viscosity, conductivity and heat rejection (based on the log mean temperature differential) of the heat-exchange system. Bubble size as well as the conductivity of the vapor barrier in all three stages of cooling are, to a degree, selectable by oil choice.

For most quench oils other than marquench oils, the optimum rates of cooling are normally obtained when the bath temperature is between 50-65°C (120-150°F). In this temperature range, properly refined mineral oils are indefinitely stable, and the effect of viscosity is drastically reduced. Various manufacturers usually have an optimum temperature range for their product. The instantaneous rate of rise of the entire quench bath is also important. This is normally dependent on design, but it usually averages between 6-22°C (10-40°F).

Effect of Increasing Degree of Agitation

Even with a properly selected oil and correct quench-environment design, the stages of cooling described above may not occur at all points on a part configuration at the same time. Part cooling is a function of the quality of the oil as well as the part's geometry, fixturing and loading techniques. As the part's internal heat moves to the surface, differences in heat rejection will vary based on the surface configuration. Consequently, the need for a uniform and controllable agitation of clean, properly controlled liquid flow over the part surface is imperative. Controlled movement of the quenching liquid is vital, as it causes an earlier mechanical disruption of the vapor blanket in the first stage and produces smaller, more frequently detached vapor bubbles during the vapor transport cooling stage. Agitation constantly provides a cooler liquid to the part surface, which provides a greater temperature difference that allows for improved heat rejection.

Cleanliness of the oil is an important issue (i.e. the oil must be free of particulate materials such as carbon, sludge and water). Carbon is formed after evaporation and fractionation under conditions of insufficient oxygen, or it is introduced by processes such as carburization. Oil breakdown on the part surface may occur if sufficient quenchant agitation is not provided.

Important considerations with respect to agitation are as follows: type and design of agitators (mixers) or pumps, and draft-tube design. For example, we often give little consideration to an internal component such as a draft tube, but we should. Draft tubes are important in the overall performance of the system and should have the following general characteristics:[11]

- A down-pumping flow path (to take advantage of the tank bottom)
- An angle of 30 degrees on the entrance flare (to minimize head loss and establish a uniform velocity profile at the inlet)
- Liquid coverage over the top of the draft tube of at least one-half the tube diameter (to avoid flow restriction and disruption of the inlet velocity profile)
- Anti-cavitation or internal-flow straightening vanes (used to prevent fluid swirl)
- Proper impellor positioning (both insertion depth into the draft tube – a distance equal to at least one-half the tube diameter as dictated by the required inlet velocity profile – and diameter), fitting tight enough to prevent fluid flow along the sides of the draft tube
- Anti-deflection capability to compensate for occasional high deflection

Effect of Quench-Tank Design

Draft tubes are just one component that highlights an often-overlooked aspect of quenching. The limitations imposed by the design of the quench tank can have a significant (negative) effect on the ability of the quench oil, or any quench medium, to perform properly. The volume of oil, the localized instantaneous temperature rise of the bath, the ability to circulate the quench medium through the load (measured in meters/second or feet/second), the capacity of the heat exchanger system, and the overall maintenance of the tank all influence quenching.

The volume of oil contained in a quench tank is important for controlling the overall rate of temperature rise after quenching. The type (Figs. 20.5, 20.6) and design of the equipment play an important role in the ability of the quench system to perform properly. The common rule of thumb for oil-quench tanks is 0.12 kg/l or 1 gallon per pound of steel. Quench tanks that utilize less than this ratio must be designed with highly effective agitation systems. Of course, having a large volume of oil is no guarantee of success if the quench-tank design is inadequate for the job.

FIGURE 20.5 | Typical dual-chamber horizontal vacuum oil-quench furnace (courtesy of SECO/WARWICK Corp.)

FIGURE 20.6 | Typical vertical vacuum oil-quench furnace (courtesy of VAC AERO International)

Distortion and Cracking

Another important advantage of oil quenching is that it minimizes the tendency to cause distortion and cracking. While better than some mediums (such as water or brine), other mediums (such as salt or high-pressure gas quenching) tend to produce less overall distortion. A key consideration, however, is the uniformity/repeatability of the distortion profile, and oil quenching has the ability to produce a very consistent profile.

Fast oils that are highly agitated tend to produce the highest rates of distortion, while slow (marquench) oils tend to minimize distortion. Quenching in still (non-agitated) oil is often used as a means of distortion control on critical parts.

Water in Quench Oil

One of the concerns regarding oil quenching in atmosphere furnaces is the presence of water in the quench oil. It is dangerous because it will form steam on quenching, which results in a volume expansion. As the steam bubble rises out of the quench tank, its surface is coated with oil, and as it exits from the furnace (usually under extremely high pressure), it is readily ignited. Some manufacturers believe that as little as 0.1% of water in oil may cause dramatic changes in quenching and part surface contamination.

Relationship of Physical Properties of Quench Oils to Performance

Oil is often analyzed to determine its performance characteristics, and the testing laboratory issues a report that contains information about the physical property characteristics of the oil. Below is a listing of various test procedures and insights into the meaning of the results obtained.[12]

- **Viscosity:** As discussed earlier, quenching performance is dependent on the viscosity of the oil. Due to degradation (the formation of sludge and varnish), oil viscosity changes with time. Samples should be taken and analyzed for contaminants, and a historical record of viscosity variation should be kept and plotted against a process-control parameter such as part hardness.
- **Water content:** Water, from oil contamination or degradation, may cause soft spots, uneven hardness, staining and (perhaps worst of all) fires. When water-contaminated oil is heated, a crackling sound may be heard. This is the basis of a qualitative field test for the presence of water in quench oil. The most common laboratory tests for water contamination are either Karl Fisher analysis (ASTM D 1744) or distillation.
- **Flash point:** The flash point is the temperature where the oil in equilibrium with its vapor produces a gas that is ignitable but does not continue to burn when exposed to a spark or flame source. There are two types of flash-point values that may be determined: closed cup or open cup. In the closed-cup measurement, the liquid and vapor are heated in a closed system. Traces of low-boiling contaminants may concentrate in the vapor phase, resulting in a relatively low value. When conducting the open-cup flash point, the relatively low boiling by-products are lost during heating and have less impact on the final value. The most common open-cup flash-point procedure is the "Cleveland Open Cup" procedure described in ASTM D 92. The minimum flash point of an oil should be 90°C (160°F) above the oil temperature being used.
- **Neutralization number:** As an oil degrades, it forms acidic by-products. The amount of these by-products may be determined by chemical analysis. The most common method is the neutralization number, which is determined by establishing the net acidity against a known standard base such as potassium hydroxide (KOH). This is known as the acid number and is reported as milligrams of KOH per gram of sample (mg/g).
- **Oxidation:** This variable may also be monitored and is especially important in tanks running marquenching oil or oils being run above their recommended operating range. Oxidation is detected by infrared spectroscopy. Nitrogen blanketing of the oil is one way to reduce both oil oxidation and sludge formation when the vacuum furnace is open.

- **Precipitation number:** Sludge is one of the biggest problems encountered with quench oils. Although other analyses may indicate that a quench oil is performing within specification, the presence of sludge may still be sufficient to cause non-uniform heat transfer, increased thermal gradients, and increased cracking and distortion. Sludge may also plug filters and foul heat exchanger surfaces. The loss of heat exchanger efficiency may cause overheating, excessive foaming and possible fires.
- **Sludge formation:** This is caused by oxidation of the quench oil and by localized overheating (frying) of the quench oil. The relative amount of sludge present in a quench oil may be quantified and reported as a precipitation number, which is determined using ASTM D 91. The relative propensity of sludge formation of a new and used oil may be compared to provide an estimate of remaining life.
- **Accelerator performance:** Induction coupled plasma (ICP) spectroscopy is one of the most common methods for the analysis of quench-oil additives. When additives (such as metal salts) are used as quench-rate accelerators, their effectiveness can be lost over time by both drag-out and degradation. By performing ICP spectroscopy, which is a direct analysis method for metal ions, compensating measures (such as the addition of a specific percentage of new accelerator) can be taken.

Application Examples

Truck transmission shafts (Fig. 20.7) of SAE 8620 material are vacuum carburized followed by a drop temperature to the hardening temperature of 845°C (1550°F) prior to oil quenching to develop the required properties. The effective case depth (measured at 50 HRC) achieved was 1.2 mm (0.05 inches). A core hardness of 25 HRC at mid-radius is achieved by quenching in 90°C (195 °F) oil. Load weight was 450 kg (1,000 pounds).

FIGURE 20.7 | Load of truck transmission shafts (courtesy of ECM USA) [13]

Die-cutting punches (Fig. 20.8) of SAE 1018 material are oil quenched after vacuum carbonitriding to develop the required properties. A core hardness of 25 HRC at mid-radius is achieved by quenching in 38°C (100°F) oil. The effective case depth developed was 0.76 mm (0.003 inches), and the load weight was 68 kg (150 pounds).

FIGURE 20.8 | Load of die-cutting punches (courtesy of AmeriKen) [14]

Large marine transmission couplings (Fig. 20.9) of SAE 8620 weighing approximately 18 kg (39 pounds) each are oil quenched after vacuum carburizing to develop a surface hardness of 62-64 HRC and a core hardness greater than 25 HRC (typically 40-44 HRC). The effective case depth (measured at 50 HRC) is 1.27-1.78 mm (0.050-0.070 inches), and the load weight is approximately 385 kg (850 pounds).

FIGURE 20.9 | Marine transmission gears (courtesy of Midwest Thermal-Vac) [15]

Locking pliers (Fig. 20.10) of 12L14 material are oil quenched after a combination vacuum (copper) brazing followed by vacuum carbonitriding cycle. The load weight is approximately 300 kg (650 pounds), and the effective case depth is 0.254-0.381 mm (0.010-0.015 inches).

FIGURE 20.10 | Locking pliers (courtesy of C.I. Hayes) [14]

Drive flange gears (Fig. 20.11) of SAE 4620 weighing 1.8 kg (4 pounds) each are oil quenched after vacuum carburizing to achieve a surface hardness of 61-63 HRC and a core hardness greater than 28 HRC. The effective case depth (measured at 50 HRC) is 0.76-0.89 mm (0.030-0.035 inches), and the load weight is approximately 350 kg (780 pounds).

FIGURE 20.11 | Drive flange gears (courtesy of Midwest Thermal-Vac) [15]

Pinion gears (Fig. 20.12) of SAE 8822 weighing 0.75 kg (1.7 pounds) each were vacuum carburized then oil quenched to develop a surface hardness of 59-62 HRC and a core hardness greater than 35 HRC. The effective case depth (measured at 50 HRC) is 0.63-0.89 mm (0.025-0.035 inches), and the load weight is approximately 360 kg (800 pounds).

FIGURE 20.12 | Pinion gears (courtesy of Midwest Thermal-Vac) [15]

Future Needs

In-situ sensors need to be developed to monitor quench intensity, and devices and sensors need to be designed to measure the heat transfer inside a quench bath. These devices can then be used to monitor quench intensity in the quench tank in real time as a quality-control tool. This ensures that the quench conditions are known and corrective action or repairs can be implemented before more parts are quenched.

Summary

Oil quenching is a common practice in vacuum furnaces designed for this purpose. As with all quenching, the key is to understand and control the vital process variables. Proper selection of the type of oil and use of that oil under ideal conditions in a well-designed and well-maintained quench tank will assure consistent and repeatable results.

Oil quenching should be applied in those applications where its advantages outweigh its disadvantages. As with all technologies, it should be as completely understood as possible with respect to the performance requirements of the product so as to meet the application end-use.

REFERENCES

1. Herring, D.H., "A Review of Factors Affecting Distortion in Quenching," *Heat Treating Progress*, December 2002
2. Herring, D.H., M. Sugiyama and M. Uchigaito, "Vacuum Furnace Oil Quenching – Influence of Oil Surface Pressure on Steel Hardness and Distortion," *Industrial Heating*, June 1986

3. Herring, D.H., M. Sugiyama and M. Uchigaito, "Controlling Oil Surface Pressure in Vacuum Oil Quenching," *Heat Treating Magazine*, July 1987
4. Unpublished data, C.I. Hayes Inc.
5. "Cooling Rate Curves for Various Liquid Quenchants," *Advanced Materials & Processes*, February 1998
6. Totten, G.E. and G.M. Webster, "Quenchant Fundamentals: Condition Monitoring of Quench Oils," *Machinery Lubrication*, January/February 2003
7. Herring, D.H., Quenching Webinar, *Industrial Heating*, September 2010
8. Herring, D.H., "Oil Quenching Part One: How to Interpret Cooling Curves," *Industrial Heating*, August 2007
9. Liscic, Bozidar, Sasa Singer and Harmut Beitz, "Dependence of the Heat Transfer Coefficient at Quenching on Diameter of Cylindrical Workpieces," *The Journal of ASTM International*, STP 1532, contains IFHTSE Conference Proceedings, 2010
10. *ASM Handbook*, Volume 4: Heat Treating, ASM International, 1991
11. Totten, G.E. and K.S. Lally, "Proper Agitation Dictates Quench Success, Part 1," *Heat Treating Magazine*, 1992
12. Wachter, D.A., G.E. Totten and G.M. Webster, *Quenching Fundamentals: Quench Oil Bath Maintenance*
13. Hebauf, T. and A. Goldsteinas, "Experience in Low Pressure Carburizing," *Industrial Heating*, September 2003
14. Herring, D.H. and Steven D. Balme, "Oil Quenching Technologies for Gears," *Gear Solutions*, July 2007
15. Otto, Frederick J. and Daniel H. Herring, "Gear Heat Treatment, Part I & II," *Heat Treating Progress*, May and June 2002

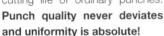

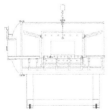

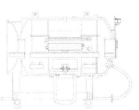

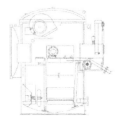

CHAPTER 21

BASKETS, FIXTURES AND GRIDS FOR VACUUM SERVICE

The selection of materials for use as grids, baskets and fixtures in vacuum service – whether they be high-temperature alloys, graphite, composite materials or combinations thereof – is an important consideration given the diverse range of processing applications, typically spanning from 65°C (150°F) to over 1650°C (3000°F).

More information can be obtained from a number of sources and standards from organizations such as SAE International (www.sae.org), ASTM International (www.astm.org), the Steel Founders' Society of America (www.sfsa.org) and the American Ceramics Society (www.ceramics.org) to name a few. There are other good sources of technical and research information for specific applications available from the many suppliers of these materials.

Design Considerations

Obtaining the most cost-effective design requires a thorough understanding of the service conditions under which the material will be exposed. Important considerations include:

- Normal operating (exposure) temperature as well as the maximum (and minimum) usage temperatures
- Metallurgical stability over the expected duty/thermal cycle (period, frequency and rate of heating/cooling)
- Thermal-expansion characteristics
- Fabrication (or casting) methods (with respect to development of thermal or chemical gradients in the material)

- Design with respect to applied load(s), repetitive force and transfer of the load to the load-bearing members
- Manner of loading, type of support and external constraints
- Required vs. desired life
- Environment(s) to which the alloy will be exposed
- Availability
- Cost vs. life

Designers are also concerned about some or all of the following factors:

1. Elevated-temperature properties
    - Hot strength and creep resistance
    - Hot ductility and thermal-shock resistance
    - Thermal-fatigue resistance and resistance to carbon pickup
    - Hot hardness

2. Room-temperature mechanical properties
    - Strength and ductility
    - Machinability
    - Weldability

3. Operating environment resistance
    - Reactive atmospheres – carburizing, nitriding, etc.
    - Oxidizing atmospheres
    - Sulfur- and chlorine-containing atmospheres
    - Quenchants – liquids, salts and gases

4. Cast properties
    - Fluidity
    - Freezing temperature

5. Physical properties
    - Coefficient of thermal expansion
    - Coefficient of thermal conductivity
    - Modulus of elasticity
    - Electrical resistivity

When selecting materials for vacuum applications, one must consider their interaction with the environment (Table 21.1) regardless of whether the attacking species is introduced as part of the process (e.g., carburizing or carbonitriding), a by-product of contamination (e.g., oils or contaminants from improperly cleaned parts) or other external factors (e.g., an air leak).

TABLE 21.1 | Environmentally induced influences on key mechanical properties of alloy systems [1]

Mechanical property affected (at use temperature)	Attacking species
High-temperature strength	Oxidation
Stress-rupture strength	Carburization
Creep strength	Nitriding
Fatigue strength	Sulfidation
Thermal stability	Halogenation
Thermal shock	Liquid-metal embrittlement
Toughness	Salt, molten and ash deposit

Alloy Selection Criteria

The subject of high-temperature alloys encompasses both cast and wrought products that are available from a number of qualified suppliers. Heat-resistant alloys are primarily selected for their ability to perform a function over a wide range of application temperatures. However, the measure of performance is different in each case and varies as the use factors change.

In many applications, high-temperature creep data (e.g., 1% creep in 10,000 hours) and/or stress-rupture data provided by manufacturers is a method for comparison of the differences between alloys used for heat-treating applications (Table 21.2a, 21.2b). Over time, however, creep may lead to excessive deformation and even failure (fracture) at stress levels much lower than those determined at room temperature or at elevated-temperature short-duration tensile conditions. When the degree (rate) of deformation is the limiting factor, the design stress at temperature should be chosen below that which produces that limiting rate (limiting creep stress) or below the stress that will produce that limiting degree of deformation in a given time.

TABLE 21.2a | Average rupture strength for selected wrought alloys (10,000-hour, psi) [2]

Alloy	Rupture strength, 10,000-hour, psi					
	Temperature, °C (°F)					
	538 (1000)	593 (1100)	649 (1200)	704 (1300)	760 (1400)	816 (1500)
446		3,500	2,700		1,100	
304L	25,000	15,600	9,700	6,000	3,700	2,300
304 304H	36,000	22,200	1,300	8,500	5,300	3,250
316L	39,000	23,500	14,200	8,500	5,100	3,050
321		23,500	12,900	7,200	4,000	2,280
347 347H	48,000	27,500	15,600	9,000	5,100	2,900
253MA		22,000	14,000	8,500	5,200	3,750
309			17,000	8,000	4,800	2,700
310			14,400	7,400	4,500	2,800
330	29,000	17,000	11,000	7,200	4,300	2,700
800AT			17,500	11,000	7,300	5,200
353MA		19,300	12,200	7,880	5,400	3,600
333		25,000	16,500	12,000	9,200	5,700
600		21,500	13,500	9,000	6,200	3,700
601	42,000	30,000	22,000	13,500	7,000	3,600
602CA			31,200		11,300	
625			42,500	22,500	12,000	
718	128,000	98,000	70,000			

TABLE 21.2b | Average rupture strength for selected wrought alloys (10,000-hour, psi) [2]

Alloy	Rupture strength, 10,000-hour, psi						
	Temperature, °C (°F) [a]						
	871 (1600)	927 (1700)	982 (1800)	1038 (1900)	1093 (2000)	1449 (2100)	1204 (2200)
446	450		230				
304L	1,400						
304 304H							
316L							
321							
347 347H							
253MA	2,500	1,650	1,150	860	680		
309	1,600	1,000	560				
310	1,500	940	660				
330	1,700	1,050	630	400	(280)		
800AT	3,500	1,900	1,200				
353MA	2,600	1,860	1,300	930	680	(450)	(320)
333	3,100	1,800	1,050	630	360		140
600	2,350	1,650	1,150				
601	1,850	1,200	820		(330)	(200)	
602CA	3,200	2,180	1,490	990	670	440	290
625							
718							

Notes:

[a] Values in parentheses are estimated.

Types of Carrier Grids

Carrier or base grids are available in a number of styles, including serpentine (Fig. 21.1), cast alloys (Fig. 21.2) and fabricated wrought alloy (Fig. 21.3) designs. Choice of design is a function of application and is primarily dependent on hot strength (i.e. the measure of strength at the operating temperature and the ability to resist deformation or attack of various gases present in the vacuum system in a given thermal environment) and creep (a measure of a combination of failures such as distortion, thermal fatigue, thermal shock, carburization and elemental attack on the metal matrix).

FIGURE 21.1 | Serpentine grid style

FIGURE 21.2 | Cast grids (courtesy of AFE North American Cronite)

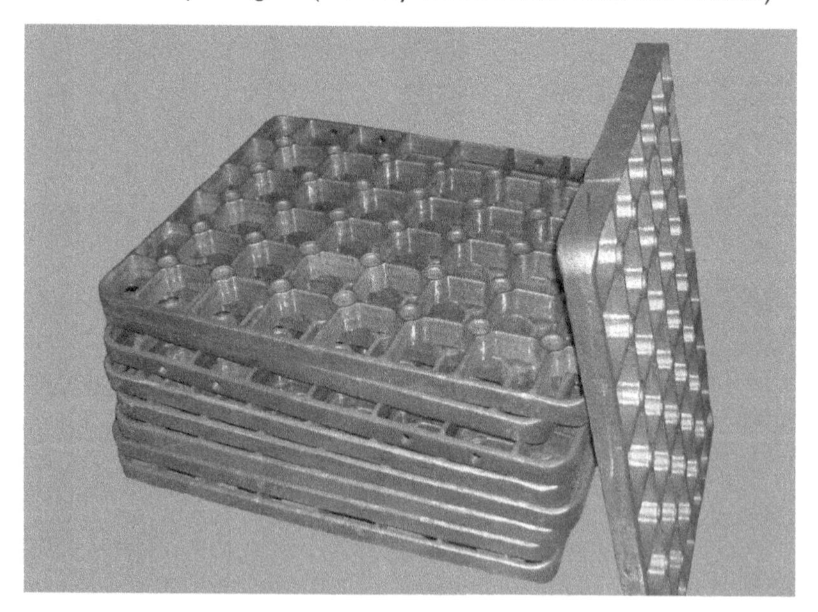

FIGURE 21.3 | Molybdenum-fabricated grid design (courtesy of PLANSEE USA)

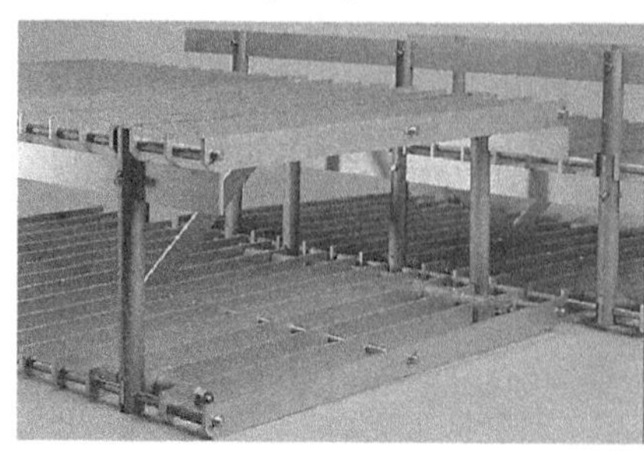

Types of Baskets

Not only must grid materials and designs be appropriate for the application, but selection, loading and maintenance of baskets and fixtures is equally important. Selection of a basket material is often (but not always) influenced first by cost, then service environment and finally compatibility with the workpiece and furnace hearth.

Common are rod framework baskets constructed from alloy bars typically 9.5-16 mm (0.375-0.625 inches) in diameter with woven wire-mesh liners (Fig. 21.4). For vacuum-service temperatures to 980°C (1800°F), austenitic stainless steels (e.g., 304, 309, 310 and 330) are commonly used. However, these alloys can become embrittled from long exposure to temperatures in the range of 595-815°C (1100-1500°F) due to carbide precipitation and sigma-phase formation. In this temperature range, more expensive alloys such as 300 or Inconel® 600 have been reported to be more stable and may justify the extra cost. For even higher temperatures, alloys like Haynes 230, MA 956 (a nickel-based oxide-dispersed alloy) or molybdenum alloys are reported to provide good life. Care must be taken in handling any baskets, particularly those made from molybdenum since they become brittle after initial exposure to high temperatures.

A fine grain size is preferred to resist thermal-fatigue damage (ASTM 5 or finer). Welded, fully penetrated joints are necessary, such as those produced by pressure welding (cross-wire resistance welding) the frame members.

In vacuum heat treating of tool and stainless steels, for example, temperatures above 1040°C (1900°F) are often required. Strength is important to minimize fixture weight. Alloys such as RA 353 MA, 602 CA and others are worthwhile considerations to retain adequate strength at such temperatures. Alloy selection for vacuum hardening and carburizing may include 330, Inconel® 600, Inconel® 601, Haynes 230 and Mancellium®, with other specialty alloys also used. Initial grain size is of somewhat less concern as temperatures are high enough that most alloys will grain-coarsen in service.

FIGURE 21.4 | Typical rod mesh baskets (courtesy of Solar Atmospheres Inc.)

Types of Fixtures

Various types of fixturing are also used to support parts being run in vacuum. This can consist of parts loaded directly on carrier grids (Figs. 21.5, 21.6), hung on rods (Fig. 21.7), saddle supports (Fig. 21.8), plates upon which parts are placed (Fig. 21.9) or other support designs (Figs. 21.10-21.12). The design, construction and material depend on the type and material of the part being supported. When designing fixtures that clamp or restrain the workpiece, careful consideration must be given to differences in thermal expansion between the workpiece and fixture. When possible, the coefficient of thermal expansion of the fixture material should closely match that of the workpiece.

Before being used in production, it is recommended that a new fixture be subjected to a vacuum conditioning cycle at a temperature at least 25°C (45°F) higher than its maximum operating temperature. This will degas the fixture to rid it of any contaminants formed in fabrication, such as welding oxides. Fixtures may also require periodic grit blasting if they become discolored during service. Regular inspections for straightness, cracking, distortion or other thermal-cycling damage should be performed. To avoid catastrophic failures, damaged fixtures should be repaired or replaced immediately.

FIGURE 21.5 | Parts placed directly onto carrier grid (courtesy of SECO/WARWICK Corp.)

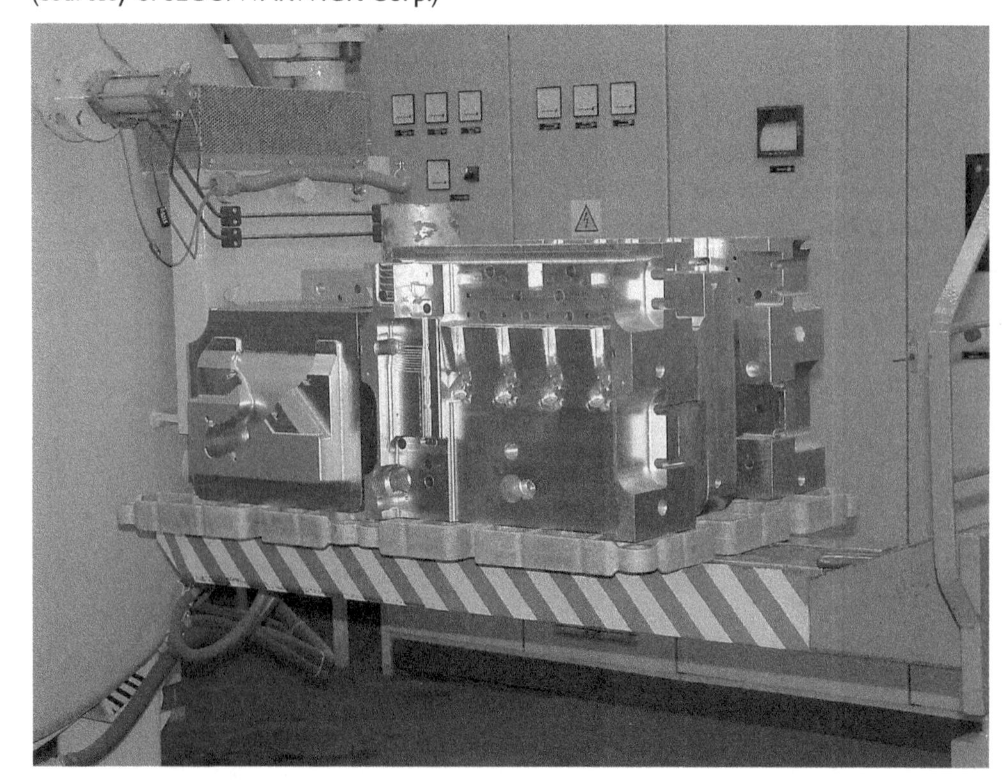

FIGURE 21.6 | Parts on multi-tiered carrier grids (courtesy of SECO/WARWICK Corp.)

FIGURE 21.7 | Parts supported on alloy rods (courtesy of Solar Atmospheres Inc.)

FIGURE 21.8 | Saddle supports for heavy ingots (courtesy of Solar Atmospheres Inc.)

FIGURE 21.9 | Parts supported by ceramic plates (courtesy of Solar Atmospheres Inc.)

FIGURE 21.10 | Tool-steel parts supported by mesh basket liner (courtesy of Solar Atmospheres Inc.)

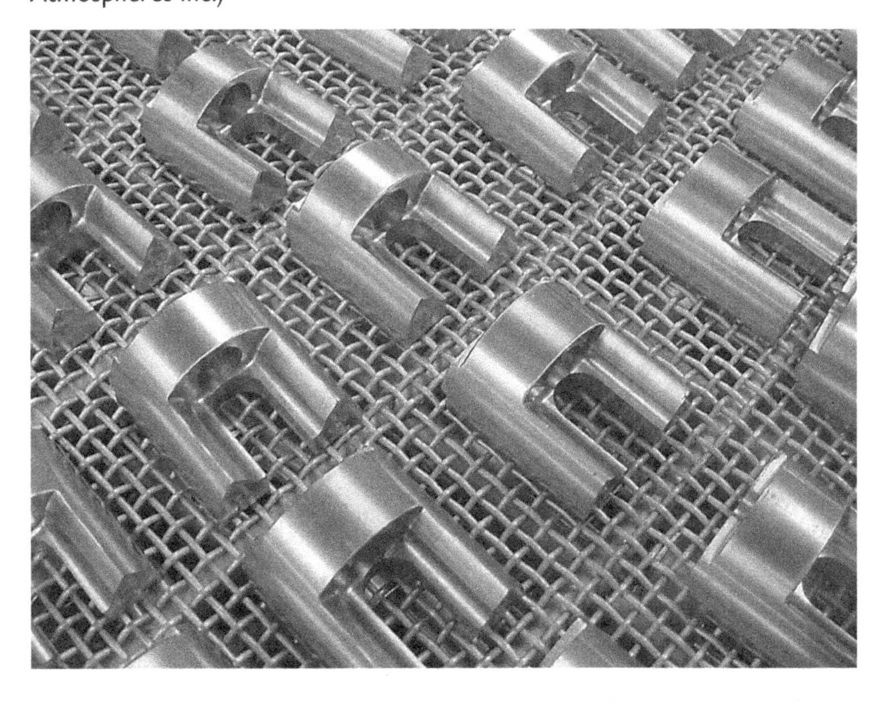

FIGURE 21.11 | Parts supported on expanded metal mesh (courtesy of Solar Atmospheres Inc.)

FIGURE 21.12 | Various types of supports for multiple-job load (courtesy of Solar Atmospheres Inc.)

The primary objectives in designing any alloy support system are to minimize thermal mass (weight) and cost and achieve a long service life. This always involves trade-offs. For example, certain materials (such as molybdenum) have great strength at temperature but are not weldable, which requires fabrication using mechanical fastening methods. This will add to fabrication and maintenance costs but may allow for replacement of individual component parts. In other applications, a low thermal mass rather than strength may be important for efficient heating and cooling (especially in gas-quench applications).

Effect of Individual Alloying Elements

The basic alloying elements and the primary reason for their selection as components of cast and wrought alloys can help the user understand how these alloys affect the variety of chemistries offered by manufacturers. Here is a summary of the general effect of the various alloying elements.

- **Nickel** is used in most grades of heat-resistant alloys and can range from a fraction of a percent to grades where the percentage exceeds 75%. Nickel serves as a strengthener and austenite stabilizer. It also improves resistance to oxidation, carburization, nitriding and thermal fatigue. Nickel increases ductility and also improves weldability. The use of high-nickel alloys should be avoided in the presence of sulfur and chlorine.

- **Chromium** provides the basic oxidation resistance for heat-resistant alloys. It is normally present in the range of 10-30%. Chromium improves high-temperature strength and carburization resistance and also promotes carbide precipitation. These carbides account for much of the strength of these alloys at high temperatures. Chromium also promotes the formation of ferrite and, in high percentages, sigma phase.
- **Carbon** accounts for the strength differences that exist between cast and wrought heat-resistant alloys. It is undoubtedly the most important of the minor alloying elements used in cast heat-resistant alloys. It can be present from 0.15-0.75%. As a generalization, carbon can be expected to increase the hot strength and creep resistance while reducing the cold ductility. Carbon is a potent austenite stabilizer.
- **Silicon** is normally present in heat-resistant alloys in the range of 0.50-2.50%. Silicon is a ferrite former, and it is primarily used for improvements to carburization and oxidation resistance. Silicon serves several purposes in the melting of heat-resistant alloys. Excessive levels of silicon reduce creep strength and shorten creep-rupture time. Silicon levels exceeding 2.5% also reduce weldability.
- **Nitrogen** is known to confer hot strength and promote formation of austenite. Nitrogen tends to prevent ferrite and sigma-phase formation. It is normally present in amounts less than 0.15% to prevent embrittlement effects.
- **Manganese** is normally added for steelmaking and casting purposes. It has negligible effect on physical properties at the normally used levels of less than 1.5%.
- **Molybdenum** improves high-temperature strength in alloys, and it is a ferrite stabilizer. Molybdenum will increase high-temperature creep-rupture strength and promote sigma formation unless counterbalanced by austenitizing elements such as nickel or cobalt.
- **Tungsten** improves high-temperature strength similar to molybdenum. However, tungsten does not have the adverse effects on oxidation resistance that molybdenum does. Tungsten is a carbide former and tends to retard the coalescence of strengthening carbides. Tungsten can be added at levels up to 14.0% or greater.
- **Niobium (Columbium)** is a strengthening and ferritizing element, a strong carbide former and promotes sigma-phase formation. Niobium carbides are stable at high temperature and offer improved thermal-

fatigue resistance. Weldability is also improved as niobium modifies the primary carbide geometry. Optimal levels are 0.5-1.0% and not to exceed 1.7% as niobium is subject to high oxidation rates. Elevated levels may reduce the oxidation resistance of an alloy.

- **Cobalt** is a very effective contributor to hot strength and resistance to creep, and it is an austenite stabilizer like nickel. Cobalt is an expensive addition and normally added in the 0.15-0.50% range.
- **Iron** is the economical base element in most heat-resistant alloy grades and may be present from 8-75%. It can act as a strengthening agent, but it is easily oxidized and carburized.
- **Titanium** is a strong carbide former and can be added to achieve improvements in strength or to promote precipitation in age-hardenable alloys.
- **Zirconium** is a strong carbide former and is added to certain high-temperature alloys in small amounts (typically less than 0.1%).
- **Boron** increases creep-rupture strength, and it is used at rather low concentrations (typically around 0.002%). Boron is an interstitial element and tends to concentrate at the grain boundaries. High boron levels may reduce the formation of continuous carbide network in austenitic alloys.
- **Aluminum** is typically added to improve oxidation resistance. It is a ferritizing element and promotes sigma-phase formation.
- **Cerium, lanthanum** and **yttrium** are added to improve oxidation resistance in high-temperature alloys.

Cast vs. Wrought

Both cast and wrought production methods have advantages and disadvantages (Table 21.3). In many designs, either cast or wrought alloys may be used, so both should be considered. Since similar compositions in cast or wrought form vary in physical properties and initial costs, their advantages and disadvantages for the intended application are important considerations.

TABLE 21.3 | Comparison of advantages and disadvantages of various heat-resistant alloys [2]

Condition	Advantages	Disadvantages
Cast alloys	Creep strength at elevated temperature; availability of shapes that are impossible or uneconomical to fabricate; custom chemistries not available in wrought alloys; initial cost (per pound)	Pattern cost; in-service issues such as porosity, shrink and surface integrity; cracking (embrittlement) over time; (in general) heavier and thicker sections than equivalent fabrications
Wrought alloys	Form (shape); availability; section size is practically unlimited; smooth surfaces; avoidance of casting defects; thermal-fatigue resistance (inherently fine grained microstructures)	Creep strength; compositional chemistry range; hot corrosion and carburization resistance limitations in certain applications

Words of Caution

Technical data describing the properties of heat-resistant alloys provides important guidelines to alloy selection, but predicting the behavior of an alloy over long-term exposure to the temperature environment is a complex process. This behavior is not always predictable, and looking at room-temperature properties in most cases has little relevance to the performance of the actual design under elevated-temperature operating conditions. Furthermore, the room-temperature properties may be subject to considerable change after even brief exposure to service temperatures and cyclic conditions. In addition, selection and use of acceptable but not optimized alloys may be a necessity, but the search for and trial of better materials is an important undertaking.

Care must also be taken to properly interpret creep and stress-rupture data because values are often extrapolated from shorter-time or lower-temperature tests, and this fact must be considered when comparing alloys. Also, emphasis is often placed on only one or two mechanical properties, while others (of equal or greater importance) may be overlooked. If creep and stress rupture were the only mechanical properties of importance, all heat-resistant alloys would be supplied in the solution-annealed condition. However, thermal fatigue and thermal conductivity are responsible for many failures in furnace applications. Failure to understand these factors in the design of alloy components may result in thermal strains far in excess of those experienced due to mechanical loading.

Room-temperature damage is another example. One of the inherent advantages of cast alloys is high-temperature strength, but strength alone is rarely the only factor to consider. Rough handling (such as pounding on or tossing/dropping cast grids or fixturing) may result in failure due to the brittle nature of the casting rather than from thermal fatigue brought about from stress rupture or creep. Wrought-alloy baskets are not necessarily the solution. These are often pounded back into shape after repeated cycling, which results, over time, in failure at weld joints that have become sensitized in service not due to strength or fabrication issues.

Graphite Selection Criteria

Graphite fixtures are used for certain applications since the strength of graphite increases with temperature. That, combined with good thermal-shock resistance and conductivity and the fact that it is relatively inexpensive, often makes it an attractive choice. Graphite fixtures, however, can be heavy.

Graphite fixtures are typically used by the heat-treating industry in the form of solid flat plates, plates with milled cavities (Fig. 21.13) or plates with holes for passage of cooling gases. Normally, the thickness for solid plates will range from 3-19 mm (0.125-0.75 inches), while milled plates can be up to 100 mm (4 inches) or thicker. Many graphite fixtures are designed to be stackable.

FIGURE 21.13 | Machined graphite for part placement (courtesy of Solar Atmospheres Inc.)

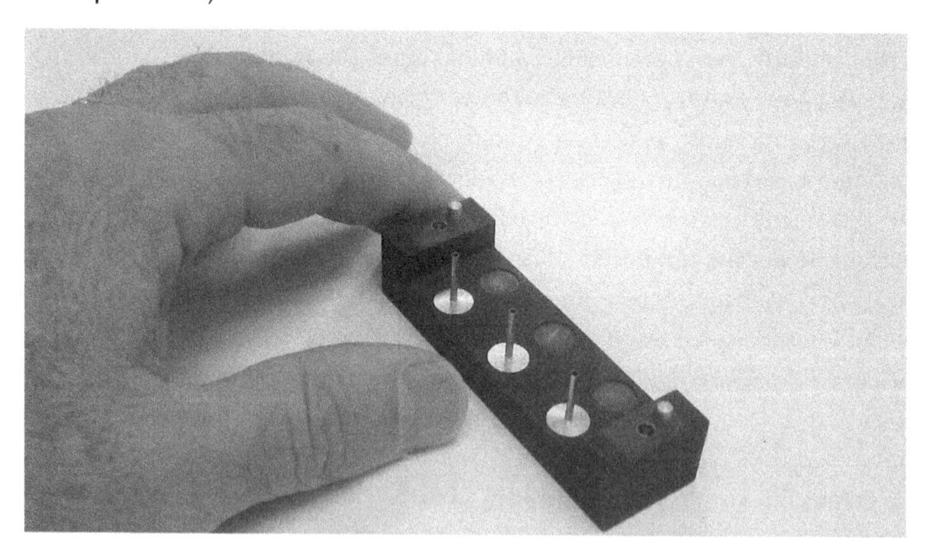

Graphite fixtures should be made of high-purity, high-density graphite. It is difficult to cite specific property levels for commercially available carbon and graphite materials since each manufacturer offers a variety of sizes, shapes and grades (Table 21.4). These different grades of graphite and their use depends on the application, thermal conditions, fixture design, loading, life expectancy and the amount of labor required for loading. When price is an issue and application does not require extended life, medium grades or isostatically pressed lower grades of graphite may be considered. Ultra-fine (ultrahigh purity) and superdense grades of graphite can significantly extend the life of a graphite fixture.

TABLE 21.4 | Properties of manufactured carbon and graphite

Property	Carbon		Graphite	
	Fine-grained	Coarse-grained	Fine-grained	Coarse-grained
Filler particle size, mm (inch)	0.025-0.76 (0.001-0.03)	1.27-6.35 (0.05-0.25)	0.025-0.76 (0.001-0.03)	1.27-6.35 (0.05-0.25)
Bulk density, g/cm³	1.60-1.70	1.40-1.60	1.70-1.90	1.50-1.70
Total porosity, %	20-25	30-35	10-20	20-30
Total ash, wt %			0.0001-0.6	0.0001-1.0
Tensile strength, kPa (ksi)				
parallel to grain			11,032-20,684 (1.6-3.0)	4,137-5,515 (0.6-0.8)
perpendicular to grain			8,963-16,547 (1.3-2.4)	2,757-4,137 (0.4-0.6)
Flexural strength, kPa (ksi)				
parallel to grain	13,789-41,369 (2.0-6.0)	4,826-13,789 (0.7-2.0)	20,684-48,263 (3.0-7.0)	6,895-11,032 (1.0-1.6)
perpendicular to grain	6,875-20,684 (1.0-3.0)	2,758-10,342 (0.4-1.5)	17,237-34,474 (2.5,5.0)	5,516-6,875 (0.8-1.0)
Electrical resistivity, ohm-cm				
parallel to grain	0.002	0.002	0.00025-0.00049	0.0003
perpendicular to grain	0.003	0.003	0.00035-0.0008	0.0005
Coefficient of thermal expansion, 25-600°C (in/in/°C)				
parallel to grain			2.5×10^{-6} to 5.0×10^{-6}	2.0×10^{-6} to 2.5×10^{-6}
perpendicular to grain			3.5×10^{-6} to 5.0×10^{-6}	2.5×10^{-6} to 3.5×10^{-6}
Thermal conductivity, W/m-°C (BTU-feet/hour-feet²)				
parallel to grain			161-184 (93-106)	
perpendicular to grain			105-107 (61-62)	

Graphite has a low coefficient of thermal expansion. For example, graphite will expand 0.13 mm/mm (0.005 in/in) at 1095°C (2000°F), so grooves in milled plates must be larger to account for a differential in expansion between the parts and the graphite.

Depending on the component part material, process temperature and vacuum level, some materials can come in direct contact with the graphite fixture (Fig. 21.14) while others should not. Alumina powder is often used to prevent sticking or surface reactions.

Remember that graphite fixtures are slow to heat and slow to cool, so this must be taken into account when designing a cycle using these materials.

FIGURE 21.14 | Tungsten carbide sintering of parts loaded on flat graphite trays

Carbon/Carbon Composite Selection Criteria

Carbon/carbon composite (C/C) fixtures and grids are constructed from two primary components: carbon fibers and a carbon matrix (or binder). Carbon fibers are extremely thin, typically 0.005-0.010 mm (0.0002-0.0004 inches) in diameter, and are strands of carbon atoms. They are interlaced in such a way as to provide mechanical strength, stiffness and thermal conductivity. The carbon matrix allows for uniform weight transfer and chemical resistance to attack. Properties can vary depending on whether they are measured parallel or perpendicular to the surface.

Now a word of caution. Eutectic melting is possible at temperatures exceeding 1050°C (1922°F), but this is highly dependent on the alloy being treated. For example, it has been reported that while melting of certain tool steels (e.g., M50) has occurred when placed in direct contact with C/C materials, other materials such as superalloy turbine blades processed at temperatures in the range of 1290°C (2350°F) show no reaction. Ceramic barrier layers or alloy-mesh screens can be used to avoid this problem. Care must also be taken when attempting to unload these materials into open air at temperatures above 350°C (662°F) because C/C readily oxidizes over time, thus destroying or severely degrading mechanical properties.

C/C material has low thermal mass (specific heat), high strength-to-weight ratio at temperature and negligible thermal deformation, which creates favorable net/tare weight ratios. This allows for rapid heating and cooling rates, heavy part loading and improvements in part distortion. C/C also has excellent fatigue resistance, minimizing issues with crack propagation. As with graphite materials, each manufacturer offers a variety of sizes, shapes and grades. Purity levels in the range of 300 ppm total impurities are acceptable for general-purpose heat treating, with specialty applications available with impurities as low as 10 ppm. A fully densified C/C material will be approximately 1.50 g/cm^3 or higher. Once again, it is important to view all properties together when evaluating the quality of a composite material.

Vacuum carburizing furnaces with high-pressure gas quenching commonly utilize C/C fixtures (Fig. 21.15) and/or combination designs (Fig. 21.16). The oxygen-free vacuum environment, coupled with inert quenching gases, avoids surface reactions with the fixtures. In this service environment, one can take full advantage of the material properties of carbon/carbon composites.

FIGURE 21.15 | Load of transmission gears on C/C fixtures (courtesy of ALD Thermal Treatment)

Construction techniques for C/C components include unitized construction (Fig. 21.17) and press-fit joints (Fig. 21.18).

FIGURE 21.16 | Combination C/C and alloy fixture design (courtesy of ALD Thermal Treatment)

FIGURE 21.17 | UniGrid® composite grid system (courtesy of Schunk Graphite Technology)

FIGURE 21.18 | Press-fit "egg crate" C/C grid construction [11]

Final Thoughts

More sharing of practical information about high-temperature alloy performance under the broad spectrum of heat-treating applications is needed throughout the industry. For their part, the heat treater must keep better records of the service history of his grids, baskets, fixtures and internal furnace components, including a history of duty cycles as a function of application, performance life and failure modes of the alloys. In addition, the alloy fabricators and casters must help interpret this field data, add their technical expertise on material design and help design more meaningful tests. As a whole, the industry must better educate itself to understand how to apply the alloys we have and work as partners in the development of new alloys to keep pace with the changing nature of the heat-treating industry.

REFERENCES

1. Agarwal, D.C., "Material Degradation Problems in High Temperature Environments (Alloys-Alloying Effects-Solutions)," *Industrial Heating*, October 1994
2. Kelly, James and Jason Wilson, *Alloy Selection Guide for Thermal Process Equipment*, Rolled Alloys
3. Kelly, James, *Heat Resistant Alloys*, Rolled Alloys
4. Bob Hill, Solar Atmospheres Inc, private correspondence
5. Jones, Roger A., "Assessing Work-Basket Alloys for Vacuum Furnaces," *Heat Treating*, 1992
6. Herring, D.H., "Heat & Corrosion Resistant Materials/Composites: A Survey of High Temperature Alloy Selection in Heat Treating," *Industrial Heating*, September 2006
7. *Encyclopedia of Industrial Chemical Analysis*, Vol. 14, John Wiley & Sons, 1971
8. Loeser, K., V. Heuer and D.R. Faron, "Distortion Control by Innovative Heat Treating Technologies in the Automotive Industry," *Gear Technology*, August 2008
9. Heuer, V. and K. Loeser, "Low Distortion Heat Treatment of Transmission Components," AGMA Technical Paper, AGMA, October 2010
10. Weiss, Roland, Chapter 4: "Carbon/Carbons and Their Industrial Applications," *Ceramic Matrix Composites Fiber Reinforced Ceramics and Their Application*, Walter Krenkel, editor, John Wiley & Sons, 2008
11. Bill Warwick and Thorsten Scheibel, Schunk Graphite Technology, private correspondence
12. *Vacuum Brazing & Heat Treating, Training Manual*, VAC AERO International Inc.

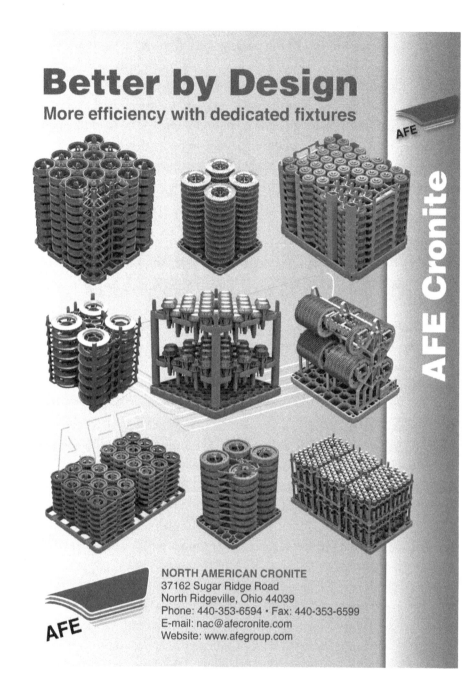

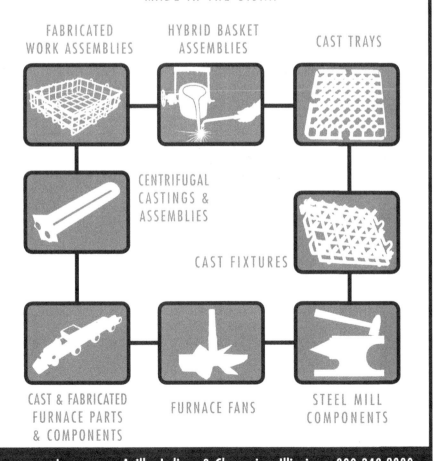

Straight as an Arrow.
Light as a Feather. Stronger than Steel.
schunk
888 560-7744
schunkgraphite.com
Carbon/Carbon Composites

CHAPTER 22

BACKFILL GASES, SURGE TANKS AND DISTRIBUTION PIPING FOR INSIDE/OUTSIDE GAS STORAGE SYSTEMS

Backfill Gases

The most common gases used in vacuum processing are nitrogen, argon, helium and hydrogen (Tables 22.1-22.4). Other common specialty gases include various hydrocarbons and ammonia (for vacuum carburizing/carbonitriding) and carbon dioxide. Each of these gases and how they are used will be discussed.

TABLE 22.1 | Relative gas supply cost [a]

Gas	Relative cost per 2.83 m^3 (100 cubic feet) [b]
Argon	6
Helium	21
Hydrogen [c]	4
Nitrogen	1

Notes:

[a] Based on a minimum usage of 2,830 cubic meters (100,000 cubic feet) per month.

[b] All gases compared to nitrogen, whose relative cost is unity.

[c] Based on liquid supply.

TABLE 22.2 | Liquid properties of common backfill gases [2]

Gas type	Liquid density, g/cm³ (lb/ft³)	Normal boiling point, °C (°F)
Argon	1.394 (87.020)	-185.87 (-302.56)
Nitrogen	0.807 (50.410)	-195.79 (-320.42)
Helium	0.125 (7.802)	-268.93 (-452.07)
Hydrogen	0.0708 (4.419)	-252.87 (-432.17)

Notes:

1. Appropriate health and safety measures should be observed at all times. Contact your gas supplier for more details regarding safe handling and usage of any gases present in the plant.
2. Refer to the HAZARDS IDENTIFICATION section of the Material Safety Data Sheet (MSDS) for the gas in question.
3. The inert gases, for example, are all asphyxiants. Hydrogen is flammable and is also an asphyxiant. Gas pressure is also a potential concern.

TABLE 22.3 | Physical properties of common backfill gases (@ 25°C, 1 bar) [2]

Gas type and property	Argon	Helium	Hydrogen	Nitrogen
Density @ 15°C, kg/m3	1.6687	0.167	0.0841	1.170
Density ratio to air	1.3797	0.138	0.0695	0.967
Molar mass, kg/kmole	39.948	4.0026	2.0158	28.0
Specific heat capacity (Cp), kJ/kg-K	0.5024	5.1931	14.3	1.041
Thermal conductivity (λ), W/m-K	0.0177	0.1500	0.1869	0.0259
Dynamic viscosity (h), N-s/m²	0.0000226	0.00001968	0.00000892	0.00001774

TABLE 22.4 | Conversion between common pressure and vacuum units

Atmospheres	bar	psia	mbar	torr	microns
250	253.3	3674.0			
200	202.6	2939.2			
100	101.3	1469.6			
50	50.7	734.8			
40	40.5	587.8			
20	20.3	293.9			
15	15.2	220.4			
12	12.2	176.3			
10	10.1	147.0			
6	6.1	88.2			
5	5.1	73.5			
2	2.02	29.4			
1	1.01	14.7	1013.2	760.0	760,000
0.5	0.51	7.35	506.6	380.0	456,000
0.0330	0.0334	0.48	33.4	25.08	25,080
0.0065	0.0066	0.096	6.56	4.94	4,940
0.0013	0.0013	0.191	1.32	0.99	988.0
0.00060	0.00061	0.009	0.608	0.456	456.0
0.00024	0.00024	0.0035	0.2432	0.182	182.4

NITROGEN (N_2)

Nitrogen makes up 78.03% of air (by volume), has a gaseous specific gravity of 0.967 and a boiling point of -195°C (-320.5°F) at atmospheric pressure. It is colorless, odorless and tasteless. Commercially, nitrogen is produced by a variety of air-separation processes, including cryogenic liquefaction and distillation, adsorption separation and membrane separation. Nitrogen used for vacuum applications is, in most cases, adequately supplied with industrial-grade gas. The typical impurity levels in the specification for industrial-grade nitrogen are 10 ppm oxygen (maximum) and a minimum dew point of -68°C (-90°F) or lower (3.4 ppm by volume). The actual levels are usually in the 2 ppm range for both oxygen and water vapor. It is often the case that there is more "pick up" of impurities in the piping to the equipment than in the supply product itself, primarily from leaks.

Nitrogen is the most common backfill gas (Fig. 22.1) used in vacuum heat treatment due in large part to its cost and the fact that it is generally considered to be inert because of its nonreactive nature with many materials. As a function of temperature, however, nitrogen is reactive to certain materials, especially alloys containing chromium and molybdenum. It is commonly used as a partial-pressure gas in the range of 0.67-6.67 mbar (500-5,000 microns) or in positive-pressure convection heating, where it has been found to be effective in reducing heat-up/cycle time. In most instances, it is not recycled.

FIGURE 22.1 | Cryogenic nitrogen storage systems (courtesy of ALD Thermal Treatment)

ARGON (Ar)

Argon is a monatomic, chemically inert gas comprising slightly less than 1% of the air (by volume). Its gaseous specific gravity is 1.38, and its boiling point is -185.9°C (-302.6°F). Argon is colorless, odorless, tasteless, noncorrosive, nonflammable and nontoxic. Commercial argon is the product of cryogenic air separation, where liquefaction and distillation processes are used to produce a low-purity crude argon product that is then purified to the commercial product.

Argon is used primarily for its properties as an inert gas, especially where it is important to prevent embrittlement in materials such as titanium, tantalum, niobium and zirconium. Argon is used as a partial-pressure gas in the range of 0.67-6.67 mbar (500-5,000 microns) during brazing to minimize volatilization. It is also used as a purge or sweep gas (due to the large size of the argon molecule) for double pump-down techniques used prior to the onset of heating.

HELIUM (He)

Helium, the second-lightest gas behind hydrogen, is a chemically inert gas and has a gaseous specific gravity of 0.138. It's a colorless, odorless, tasteless gas with a boiling point of -268.9°C (-452.1°F) at atmospheric pressure. Helium is present in air at a concentration of 0.0005% (by volume). The principal supply source is natural gas deposits, where the crude helium is extracted from the natural gas stream and then purified. Helium can be stored and shipped either as a compressed gas or cryogenic liquid.

Helium is used for its properties as an inert gas. Due to cost, helium is only used as a partial-pressure gas in specialized applications such as those in the electronics industry. Argon, which is also totally inert, is often substituted. Helium is used in high-pressure gas quenching for its heat-transfer properties in the range of 2-20 bar to increase the cooling rate of a load during quenching. This allows proper transformation to take place in some steels. In others, such as the case of very large, thick parts or very dense loads, it allows for shorter overall cycle times.

Helium is expensive and in limited supply. As such, recycling technology is cost effective (Fig. 22.2). The components of a helium gas-recovery system include:

- Low-pressure blower
- Low-pressure storage
- Recompression
- High-pressure storage
- Filtration
- Purification
- Gas analysis
- Interface with vacuum-furnace gas supply

FIGURE 22.2 | Helium gas-recovery system [1]

SEE FULL COLOR IMAGE ON PAGE C

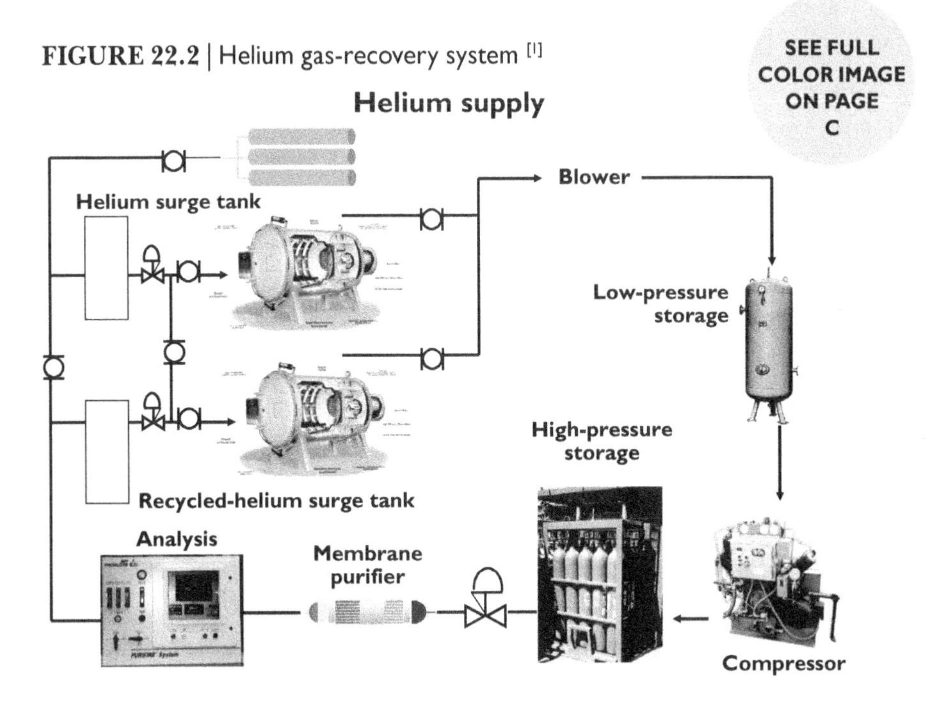

HYDROGEN (H)

Hydrogen is the lightest gas, having a gaseous specific gravity of 0.0695 and a boiling point of -252.8°C (-423°F) at atmospheric pressure. It's a colorless, odorless, tasteless, flammable gas found at concentrations of about 0.0001% in air (by volume). Hydrogen is produced by several methods, including steam/methane reforming, dissociation of ammonia, electrolysis and recovery from by-product streams during chemical manufacturing and petroleum reforming. Hydrogen is stored and transported as either a gas or a cryogenic liquid.

Hydrogen is used in heat treating primarily for its properties as a reducing gas. It is commonly used as a partial-pressure gas in the range of 0.67-6.67 mbar (500-5,000 microns) in many applications, including brazing (e.g., oxide reduction), hydriding (e.g., titanium, tantalum), sintering and even during the heat-up portion of the cycle in low-pressure vacuum carburizing, where it acts as a scrubbing (reducing) agent and shortens heat-up times. Another common use of hydrogen is for cleanup cycles involving either the furnace or the baskets/grids/fixtures. Several companies report that they are using hydrogen in over-pressure (+0.5 psig) conditions, not for heating but for sintering of powder-metal green compacts to aid in lubricant removal.

Hydrogen can also be used in high-pressure gas-quenching systems due to its high heat-transfer properties. There are safety concerns, however, and special precautions must be observed. (Note: The use of hydrogen in any type of furnace can be extremely dangerous.) The original equipment manufacturer (OEM) should be consulted, and the end user must comply with all appropriate federal, state, local and company codes, rules and regulations prior to attempting to do so (e.g., NFPA 86, latest edition). Also, some metals (e.g., titanium, tantalum) have adverse embrittlement reactions in the presence of hydrogen.

The following is advice gathered from a number of industry experts on what you should watch out for and what precautions are necessary when using a hydrogen atmosphere. *These are only suggestions and do not necessarily constitute NFPA compliance.*

1. The vacuum-furnace leak rate must be 10 microns per hour or less prior to any processing with hydrogen.
2. The pressure-relief and/or chamber-release valve must be vented outdoors via hard pipe on an upward angle to ensure no trapped hydrogen gas. A rain cap is required. Use of an inert gas added to this venting line for purging will help ensure that there is no trapped hydrogen gas.
3. If the amount of gas venting from the furnace is minimal, it is possible to burn the gas at the exit. Be sure to use a check valve.
4. Ensure there are no leaks in the hydrogen supply lines. A vacuum should be pulled on the hydrogen supply lines to check for leaks. Bubble checking all fittings while under pressure should be part of routine maintenance.
5. Before opening the furnace door, pump down to less than 100 microns and backfill with inert gas. It would be a good idea to perform this

operation two times to dilute any hydrogen gas trapped in the pumping exhaust system.

6. Institute a good preventive maintenance program/schedule that includes inspecting furnace integrity. Items should include such things as O-rings (not only if they are dirty but also if they are flat), disassembly of solenoid valves to clean their surfaces and components, and oil changes on pumps as well as the condition of the belts so as to avoid a pump failure.
7. Intrinsically safe and redundant control systems to ensure operator mistakes – such as venting, door opening, air releasing, etc. – are minimized.
8. Compliance with NFPA 86 (latest edition) is mandatory.
9. Provide an inert gas flow, typically 2.83-4.24 m^3/h (100-150 cfh), to the mechanical-pump plenum at all times when hydrogen is flowing. Install an inert gas line to the pumping exhaust line to purge hydrogen gas out at any time.
10. Have a manual safety shutoff valve added to the hydrogen supply line that is easily accessible and prominently labeled.
11. All hydrogen supply lines should be metal, not flexible plastic. No cast iron components are allowed.
12. The use of an interlock is mandatory before the hydrogen circuit is enabled. The interlock will not allow the addition of hydrogen until the furnace has been evacuated to below 0.13 mbar (100 microns).
13. Hydrogen supply lines shall be equipped with high- and low-pressure switches to prevent operation above and below specified pressure ranges.
14. Components that use hydrogen shall be designed and approved for hydrogen use.
15. Consider the use of a flow-limiting valve on the hydrogen supply line to limit the maximum allowable flow.
16. Mechanical roughing pumps on all furnaces should have "run monitors" installed and interlocked to ensure that the pump itself (not just the pump motor) is operating.
17. Ballast valves on all mechanical roughing pumps should be plugged to prevent the accidental introduction of air into the pump while hydrogen is being used.
18. Change all vacuum-line air-release valves to normally open style. Pipe those valves to an argon supply equipped with a low-pressure regulator (with pressure gauge) set no higher than 6,900 Pa (1 psig).
19. Remove site glasses from furnace.
20. Always have an operator nearby when operating with a hydrogen atmosphere.

21. Have five times the volume of inert gas standing by at all times ready to purge the furnace of hydrogen in an emergency or a power-failure situation.

A full tertiary diagram (Fig. 22.3) for hydrogen, oxygen and nitrogen includes the flammability envelope for ambient conditions. Mixtures inside the envelope are flammable. Temperature corrections (Fig. 22.4) have been determined assuming oxygen is the contaminant.

FIGURE 22.3 | Tertiary diagram for hydrogen, oxygen and nitrogen (with flammability envelope for ambient conditions) [4]

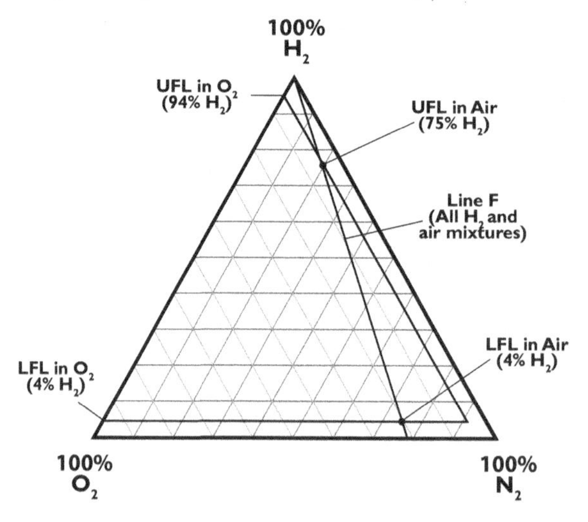

FIGURE 22.4 | Portion of right side of tertiary diagram for hydrogen, oxygen and nitrogen at elevated temperatures [4]

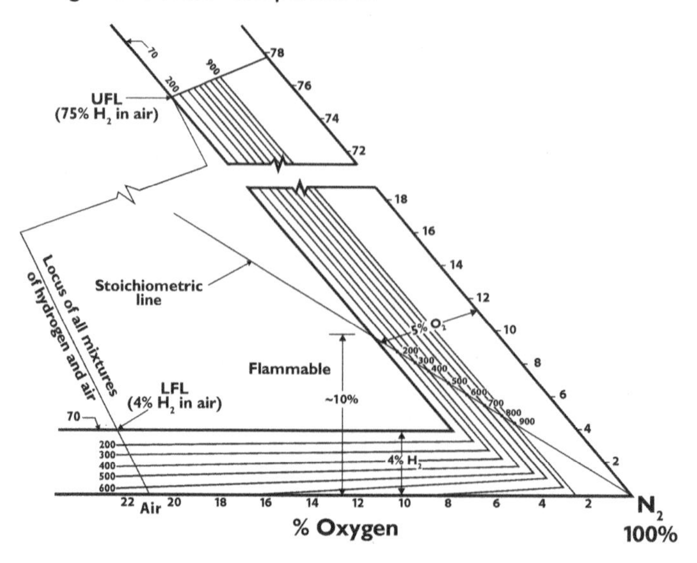

HYDROCARBONS

Various hydrocarbon gases such as acetylene (C_2H_2), cyclohexane (C_6H_{12}), ethylene (C_2H_4), methane (CH_4), propane (C_3H_8) and propylene (C_3H_6) are used as a carbon source in processes such as low-pressure vacuum carburizing and carbonitriding (in combination with ammonia) or plasma carburizing/nitriding. All are flammable and most have a detectable odor. These gases will be covered in more detail in Chapter 26.

OTHER GASES

Gas mixtures can be used for high-pressure gas quenching. These include:

1. Nitrogen/hydrogen

One of the methods used by the heat-treating industry to increase the cooling rate (i.e. heat-transfer coefficient) of nitrogen is to add a small percentage of hydrogen into the gas. Typically, 3% hydrogen is added to ensure that the mixture is noncombustible, but up to 12% hydrogen has been reportedly used (with appropriate safety precautions) to enhance the results with marginal materials. For example, vacuum carburized SAE 8620 can be gas quenched at lower pressures (depending on part section thickness) in nitrogen/hydrogen mixtures to achieve results equivalent to those produced by oil quenching or higher nitrogen gas pressures (e.g., case hardness, surface and core hardness, microstructure).

2. Helium/carbon dioxide*

Having a specific gravity of 1.53 (air = 1.0), carbon dioxide is a nonflammable, colorless and odorless gas found in air at concentrations of about 0.03% (by volume). Although not an inert gas, carbon dioxide is nonreactive with many materials and can be combined with helium for use in high-pressure gas quenching (Figs. 22.5, 22.6).

**Caution: Use of helium/carbon dioxide above 980°C (1800°F) may result in partial breakdown of the carbon dioxide into its individual constituents.*

FIGURE 22.5 | Reported improvement in heat-transfer coefficient using helium/carbon dioxide mixtures [5]

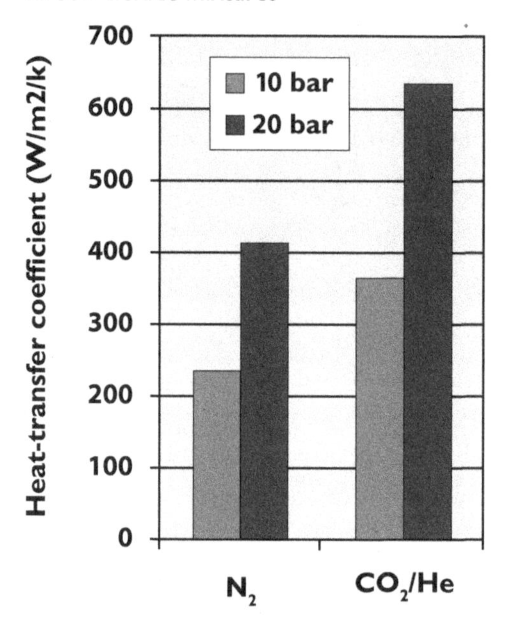

FIGURE 22.6 | Hardness response as a function of helium/carbon dioxide mixture concentration (20-bar pressure quench) [5]

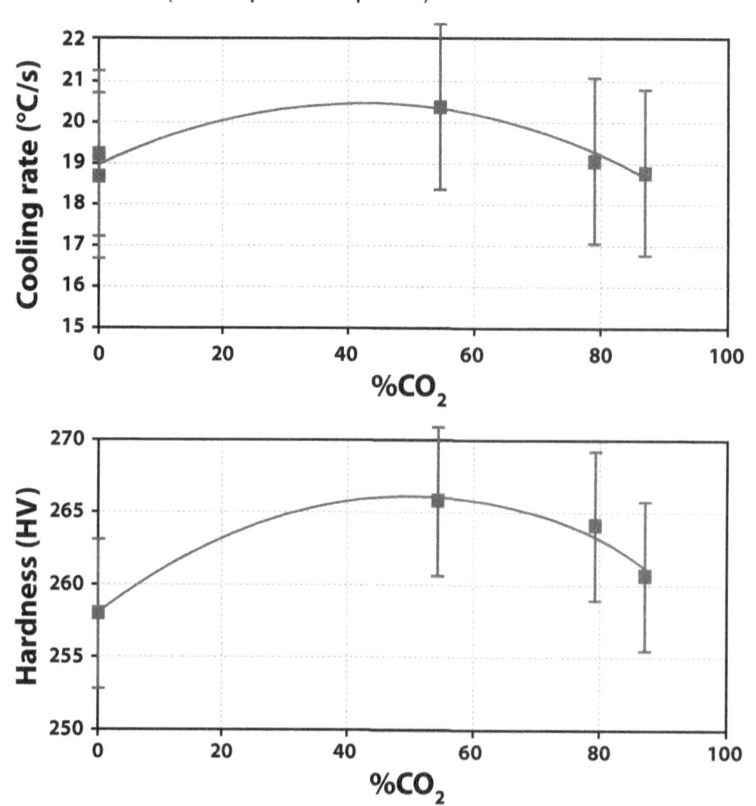

3. Argon/helium or argon/hydrogen

The cooling properties of argon can be enhanced by the addition of either helium or hydrogen. Mixtures of argon and helium in the range from 50% helium/50% argon to 95% helium/5% argon have been used to enhance cooling rates as a function of pressure and material. Mixtures in the range of 10% hydrogen/90% argon to 40% hydrogen/60% argon have been reported to improve hardness of marginal materials. Again, all appropriate safety precautions must be observed. An added benefit to using either helium or hydrogen gas mixtures is that less powerful quench fans are required (due to decreased gas density) and less power is consumed (Table 22.5).

TABLE 22.5 | Electrical costs using different backfill gases [6]

Gas type	kW requirement (300-HP motor)	Relative cost per hour [a]
Nitrogen/Argon	300	3
Helium	100	1

Notes:

[a] Based on a utility cost of $0.10 per kWH.

Surge (Accumulator) Tanks

The following are considerations in selecting a vacuum-furnace surge tank.

Many gas-supply systems for a vacuum furnace simply consist of a surge (accumulator or gas receiver) tank (Figs. 22.7, 22.8) sized to contain sufficient quench gas to backfill the furnace quench system to its maximum operating pressure within 5-30 seconds. In applications where rapid gas quenching is required (particularly for effective pressure quenching), it is important that the furnace chamber be backfilled quickly. Otherwise, the cooling that occurs during backfill may sufficiently slow the overall cooling rate so that optimum properties will not be achieved in the material being processed.

These tanks are typically designed to be connected to a cryogenic bulk gas supply. Purification may be required if contaminated quench gas is suspected. Contamination may result from leaks in the supply system or when maintenance requires that the supply system be opened to the ambient environment. Purification is accomplished by first venting any residual gas in the surge tank and quench system. Backfill valves are then opened, and the furnace pumping system is used to evacuate the entire quench system to 0.13 mbar (100 microns). New, uncontaminated gas can then be pumped into the system.

FIGURE 22.7 | Typical surge (accumulator) tank installed on a high-pressure-quench vacuum furnace (courtesy of SECO/WARWICK Corp.)

FIGURE 22.8 | Surge-tank schematic (courtesy of VAC AERO International)

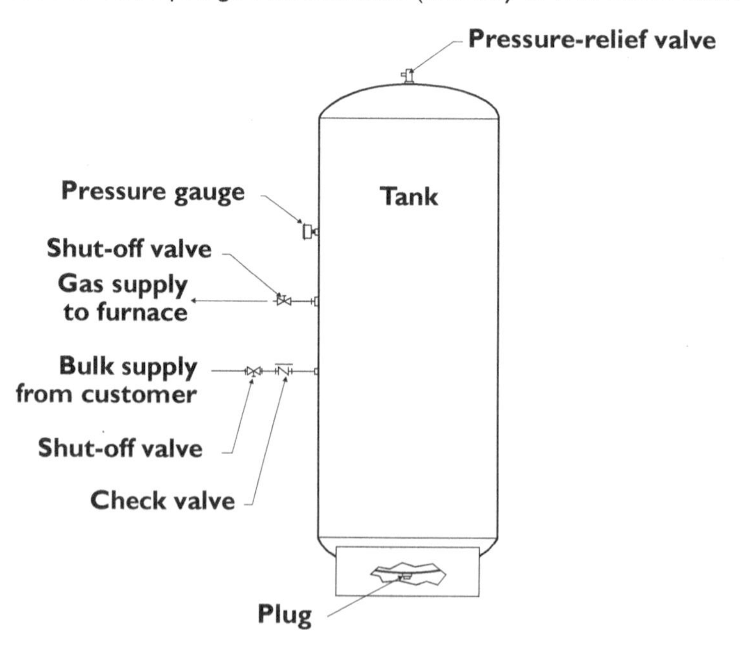

The shift toward faster quenching through higher-pressure backfills has made surge-tank selection (size and pressure rating) more critical. To size a surge tank, you first need to determine the required tank operating pressure that will provide the necessary furnace backfill pressure and time to backfill. There are trade-offs between tank size, its pressure rating, the resulting stored volume of gas and the cost of the tank. The gas-supply system must also be able to provide adequate pressure to refill the tank. There are natural pressure-level breakpoints for standard cryogenic-based supply systems like, for example, 200 psig for a standard 250-psig-rated liquid cryogenic tank.

Be sure that the ASME-approved surge tank is rated for the pressure that you are using and that it is adequately protected from over-pressurization. If you're using a cryogenic supply system, make sure it has adequate vaporization and a low-temperature shut-off valve to prevent embrittlement of the carbon-steel surge tank.

The volume of a surge tank is usually referred to in terms of its gallons of water displacement. Since there are 0.0038 m^3 (0.134 $feet^3$) per 3.78 liters (1 gallon), a 3,785-liter (1,000-gallon) surge tank has a volume of 3.79 m^3 (134 $feet^3$). Therefore, for each atmosphere of pressure, there are 3.79 m^3 (134 $feet^3$) of gaseous volume available for the backfill. For example, 3.79 m^3 (134 $feet^3$) of gaseous volume is available at 1 bar (14.7 psig), 7.59 m^3 (268 scf) is available at 2 bar (29.4 psig) and so on.

A surge tank needs to be able to store the proper volume of gas at an adequate pressure level above the backfill pressure of the furnace. Using simple ideal gas laws, if 2.83 m^3 (100 $feet^3$) is required for a 5-bar quench pressure (approximately 72 psig), it would require 17 m^3 (600 scf) of gas for a backfill from full vacuum. That's assuming a minimum pressure of 6 bar is required to provide an adequate flow rate to backfill within the desired time. The resulting surge tank would need to be about 2,840 liters (750 gallons) with a minimum operating pressure level of approximately 12 bar (175 psig). A tank with a 13.8-bar (200-psig) maximum allowable working pressure (MAWP) rating would be recommended, and the actual size would be based on how much overdesign might be desired. A smaller tank with a much higher operating pressure could be used.

Sizing of Surge Tanks for Vacuum Furnaces [7]

Note: The information that follows is intended as a general guideline only. A registered professional engineer should be consulted to verify all formulas and calculations prior to their use.

To size a surge (accumulator) tank for a vacuum furnace, solve Equation 22.1 for V_t.

22.1) $$\frac{V_t \times P_t}{T_t} = \frac{V_f \times P_f}{T_f} = \frac{V_t \times P_f}{T_t}$$

where:

V_t is the volume in cubic feet.

P_t is the initial backfill gas pressure in the surge tank in psia (where psia = psig + 14.7).

T_t is the temperature of backfill gas in the surge tank (ambient temperature) in °R (where °R = °F + 460).

V_f is the volume of furnace vessel in cubic feet.

P_f is the final backfill gas pressure in the furnace (and surge tank) in psia (where psia = psig + 14.7).

T_f is the temperature of the furnace at the start of the gas backfill in °R (where °R = °F + 460).

FIGURE 22.9 | Surge-tank sizing

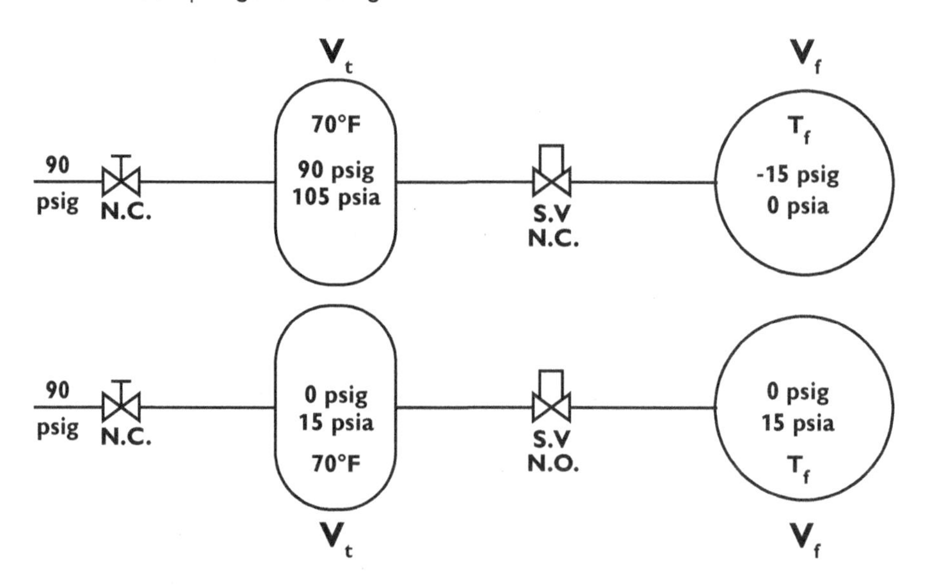

Conclusion

The choice of backfill gas and the size of surge tank selected for a particular vacuum-furnace system remain important considerations from a metallurgical as well as fiscal standpoint. Properly sized, installed and leak-tested backfill-gas piping is mandatory for proper system performance.

REFERENCES

1. Don Bowe, Air Products and Chemicals Inc., technical review and private correspondence
2. The Linde Group
3. Trevor Jones, Solar Atmospheres Inc., private correspondence
4. Dwyer Jr., John, James G. Hansel and Tom Philips, *Temperature Influence on the Flammability Limits of Heat Treating Atmospheres,* Air Products and Chemicals Inc., 2003

5. "Performance of Gas Quenching," ECM-USA, white paper
6. Robert Hill, Solar Atmospheres Inc., private correspondence
7. Alex Jarema, Lindberg, private correspondence
8. ALD Vacuum Technologies GmbH
9. John Anderson, Airgas, private correspondence

PRAXAIR
When you need reliability and experience, choose Praxair.

PRAXAIR
Making our planet more productive

CHAPTER 23

GETTER MATERIALS

Experience tells us that sensitive materials in the presence of minute quantities of unwanted gaseous contaminates can destroy the integrity and shorten the life expectancy of parts made from these materials. It is natural to ask ourselves what can be done to further protect the work in a vacuum environment after the pumps have done their job in reducing the chamber pressure as low as is economically feasible in a production environment. This task falls to getter materials.

What is a getter?

A getter is simply a reactive material that is deliberately placed inside a vacuum system or furnace hot zone for the purpose of improving the efficiency of that vacuum by scavenging unwanted contaminates. Essentially, when gas molecules strike the getter material, they combine with it chemically or by adsorption so as to be removed from the environment. In other words, a getter eliminates even minute amounts of unwanted gases from the evacuated space.

Getter materials fall into three broad categories: bulk getters, coating getters and flash getters. Examples vary from simple foils, wraps and stamped forms to machined turnings placed in and around the work. Getters can also be applied to the surface in the form of coatings, or they can be placed in the gas stream in the form of pellets. The hot zone of most vacuum furnaces, whether graphite or metallic (e.g., molybdenum and/or stainless steel shields), acts as a getter (metallic to a lesser extent).

Getters are especially important when processing reactive metals such as titanium or tantalum and in sealed systems such as vacuum tubes, cathode ray tubes and the like during the initial use period.[5] The sophistication of the getter is in direct relation to the task at hand.

FIGURE 23.1 | Titanium discs used as a getter material in brazing of oxidation-sensitive components (courtesy of California Brazing)

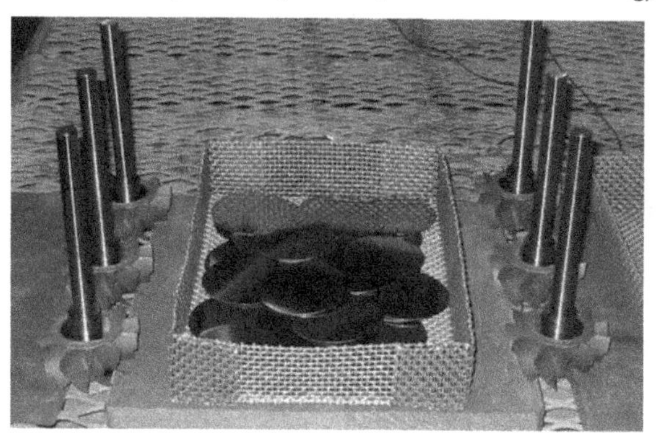

Getter Properties

The action of a getter material depends on:

- Adsorption (i.e. accumulation of gas molecules at the surface)
- Absorption (i.e. diffusion of gas molecules in the solid)
- Chemical binding (i.e. reaction with the surface atoms)

A getter material is designed to react with the gas species present, which creates a chemical reaction. Typical gases present in vacuum systems are carbon monoxide (CO), carbon dioxide (CO_2), nitrogen (N_2), hydrogen (H_2) and water vapor (H_2O). Most metal surfaces have a protective oxide on them, which must dissolve and diffuse into the getter material under vacuum and at high temperature before the process can be effective. Therefore, a getter material must also have this diffusion ability for contaminating gases to be absorbed.

Getters bind gases on their surfaces. Thus, the greater the surface area, or the more porous the material, the better it will perform. In some cases, getters can be reused provided a bake-out process is employed (typically performed in vacuum) so as to unbind the absorbed gas species. Certain types of surfaces (e.g., nitrides) require diffusion rather than extraction of the gas to return the surface to a useful condition.

Getter Materials

For most heat-treating applications, stainless steels and metals in Group IV of the periodic table (titanium, zirconium, hafnium) are particularly well suited as getter materials. For more sophisticated applications, tantalum, niobium, thorium and many other materials have been successfully used.

The amount of gas that a getter can absorb is often referred to as its "getter capacity" (Table 23.1). In more scientific terms, it is the number of atoms or molecules of the contaminating species bound up inside the material.

TABLE 23.1 | Getter capacity of common materials [1]

Getter material	Gas species	Getter capacity, Pa-l/mg
Aluminum	Oxygen (O_2)	1.00
Barium	Carbon dioxide (CO_2)	0.69
	Hydrogen (H_2)	11.50
	Nitrogen (N_2)	1.26
	Oxygen (O_2)	2.00
Magnesium	Oxygen (O_2)	2.70
Rare-earth elements (cerium, lanthanum)	Carbon dioxide (CO_2)	0.29
	Hydrogen (H_2)	6.13
	Nitrogen (N_2)	0.43
	Oxygen (O_2)	2.80
Titanium	Hydrogen	27.00
	Nitrogen	0.85
	Oxygen	4.40

By way of example, the getter capacity of titanium for hydrogen from Table 23.1 is 27 Pa-l/mg. Since the molar mass of titanium is 48 g/mol, we find that 1 mg of titanium contains 1.25 x 10^{19} titanium atoms. The Ideal Gas Law ($PV = nRT$) tells us the state of any amount of gas is determined by its pressure, volume and temperature. As a result, 27 Pa-l contains 6.7 x 10^{18} particles (i.e. 6.7 x 10^{18} hydrogen molecules or 1.34 x 10^{19} hydrogen atoms). These can be distributed over the 1.25 x 10^{19} titanium atoms so that each titanium atom in the getter corresponds to (approximately) one hydrogen atom.

The chemical reactions (in simplified form) for getter materials (GM) are as follows:

23.1) **GM** + $O_2 \rightarrow$ 2**GMO**
23.2) **GM** + $N_2 \rightarrow$ 2**GMN**
23.3) **GM** + CO $\rightarrow$ **GMC** + **GMO**
23.4) **GM** + $CO_2 \rightarrow$ CO + **GMO** $\rightarrow$ **GMC** + **GMO**
23.5) **GM** + $H_2O \rightarrow$ H + **GMO** $\rightarrow$ **GMO** + H (bulk)
23.6) **GM** + $H_2 \rightarrow$ **GMH**+ H (bulk)
23.7) **GM** + $C_xH_y \rightarrow$ **GMC** + H (bulk)
23.8) **GM** + inert gas (He, Ne, Ar, Kr, Xe)$\rightarrow$ No reaction

Getter capacity is affected by temperature because diffusion rates of the surface-bound gas atoms in the bulk of the getter material increase with temperature. This keeps the getter surface continuously active, and getter capacity rises for gases that bind only to the surface due to chemical reactions. Adsorption also continues for longer periods of time (until saturation).

Titanium can be used as an effective getter material when running titanium parts in vacuum (Fig. 23.2). For example, to keep part surfaces clean (i.e. to avoid oxidation and discoloration) during annealing at a temperature in the 650-760°C (1200-1400°F) range, titanium scrap (often in the form of clean, dry machine turnings) is commonly included with the load.

FIGURE 23.2 | Titanium sponge used as a getter material in sintering metal injection molded (MIM) tensile bars (courtesy of Elnik Systems)

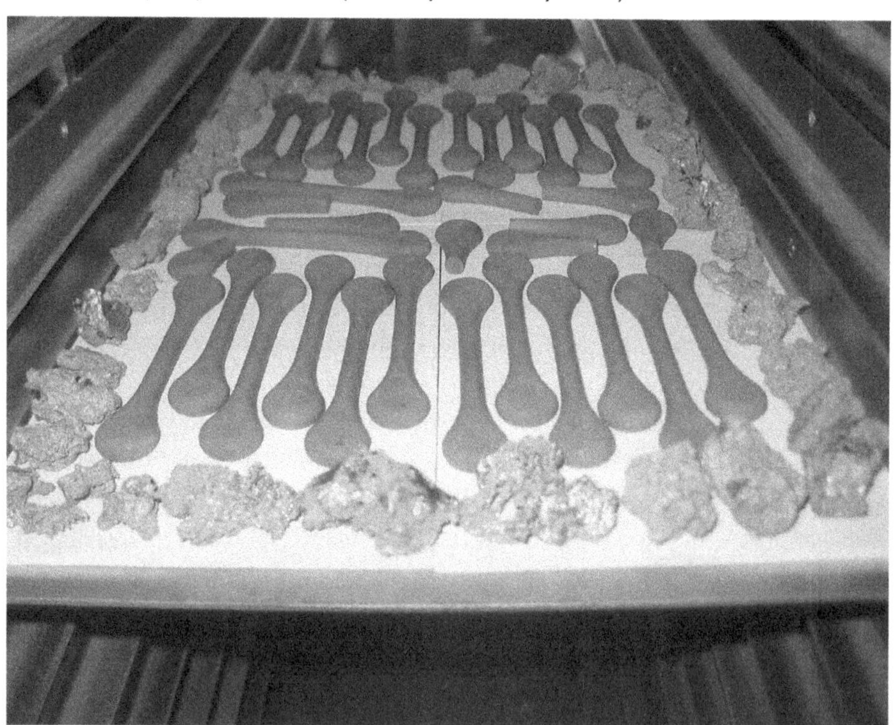

Non-evaporative Getters

Non-evaporable getters have become an integral part of many ultrahigh-vacuum environments due mainly to their unique surface properties, which are conducive to achieving extremely high vacuum conditions, in the order of 10^{-10} mbar (10^{-9} torr) or lower. Binary, ternary and multi-component alloys from Group IV and Group V (vanadium, niobium, tantalum) elements are most often used. These generally consist of a film of a special alloy, often zirconium-based. The requirement is that the alloy materials must form a passivation layer at room temperature that disappears when heated. Many of these alloys have names of the form St (Stabil®) followed by a number:

- St 707 – a 70% zirconium, 24.6% vanadium balance iron alloy
- St 787 – an 80.8% zirconium, 14.2% cobalt balance mischmetal alloy
- St 101 – an 84% zirconium and 16% aluminum alloy

For vacuum tubes used in electronics, the getter material coats plates within the tube, which are heated in normal operation. When getters are used within more general vacuum systems, such as in semiconductor manufacturing, they are introduced as separate pieces of equipment in the vacuum chamber and turned on when needed.

Final Thoughts

For heat treaters, getters are often considered a last resort to help keep parts bright and clean. In fact, they play an important role in the successful vacuum processing of many highly sophisticated products and materials. As a result, we need to do a better job of understanding their role and how and where they can help.

REFERENCES

1. Jousten, Karl, *Handbook of Vacuum Technology*, Wiley-VCH, 2008
2. Kohl, Walter Heinrich, *Handbook of Materials and Techniques for Vacuum Devices*, 1967
3. Danielson, Phil, *A Journal of Practical and Useful Vacuum Technology*, The Vacuum Lab, 2001
4. Wikipedia
5. Dushman, Saul, *Vacuum Technique*, John Wiley and Sons Inc., Third Printing, May 1955, Copyright 1949, General Electric Co.

CHAPTER 24

VACUUM APPLICATIONS: HARDENING

In general, applications involving vacuum heat treating can be broken down into four main categories:

- Processes that can be done in no other way than in vacuum
- Processes that can be done better in vacuum from a metallurgical standpoint
- Processes that can be done better in vacuum from an economic viewpoint
- Processes that can be done better in vacuum from a surface-finish perspective

Hardening is one of the most common practice performed on steel components throughout the heat-treating industry. Hardening is a process that produces an increase in the strength of a material by a suitable thermal treatment usually involving heating followed by cooling at a rapid rate. Other common process names and related subjects are through-hardening, neutral hardening, quench hardening, direct hardening and surface hardening. For the purposes of this discussion, hardening will not include case-hardening techniques (c.f. Chapter 26).

The Hardening Process

The sequence of operation used for a typical vacuum-hardening operation can be divided into two types: those used in single-chamber vacuum furnaces and those used in multi-chamber vacuum furnaces. What follows is a generic sequence of operations that is intended as an example. An exact sequence of operation should be created by the user, checked for accuracy and approved by the original equipment manufacturer.

SINGLE-CHAMBER VACUUM-FURNACE OPERATION

The steps outlined below are typically involved for vacuum hardening a workload in a single-chamber (horizontal or vertical) gas-quench vacuum furnace.

1. Backfill the vacuum furnace to atmospheric pressure with nitrogen or another inert gas. (Note: This assumes that the vacuum vessel is empty and has been pumped down at the end of the last cycle and left under vacuum.)
2. Inspect the furnace interior, especially the heating chamber, for signs of damage and do repairs as necessary.
3. (Optional) Attach workload thermocouples either onto parts or in slugs of predetermined size representative of the parts being processed.
4. Load the parts into the hot zone of the furnace. (Note: Use caution to prevent damage to the furnace interior due to load configuration or placement.)
5. Inspect (for damage), wipe down and (if necessary) re-grease the door O-ring seal.
6. Close the load door and ensure that it is properly locked in position. *(CAUTION: Never operate a vacuum furnace if the outer door is not properly secured.)*
7. Select the proper furnace recipe and initiate the automatic cycle sequence. The process typically involves the following:
   - ❖ Ramp to setpoint(s) as dictated by the recipe selected. Ramp rates exceeding approximately 17°C/minute (30°F/minute) should be used only if the consequences of faster heating have been taken into consideration by the recipe designer.
   - ❖ Soak at setpoint temperature.
   - ❖ Backfill to a preset pressure level.
   - ❖ Fan cool-down to unload temperature.
   - ❖ Hold for an extended period of time prior to unloading to ensure that the furnace chamber and workload are less than 65°C (150°F). This will prevent damage to the furnace interior unless it is specifically designed with materials that can withstand exposure to higher temperatures.
8. Silence the end-of-cycle alarm.
9. Backfill the vacuum furnace to atmospheric pressure. Unload.
10. Reload or close the outer door and pump the unit down to an intermediate vacuum level until another load is ready for processing.

MULTI-CHAMBER VACUUM-FURNACE (INCLUDING OIL-QUENCH DESIGNS) OPERATION

The steps outlined below are typically involved for vacuum hardening a workload in a multi-chamber furnace (including oil-quench vacuum furnaces).

1. Ensure that the load/unload chamber of the furnace is empty so that the loading sequence will not initiate a transfer into a chamber that already has a load in it (Note: For in-out oil-quench designs, make sure that the previous workload has been removed from the quench tank prior to backfilling and attempt to place the next load in the chamber).
2. Backfill the load chamber of the vacuum furnace to atmospheric pressure with nitrogen or another inert gas. This assumes that the chamber is empty, isolated from the next chamber, has been pumped down and is currently under vacuum. In the case of oil-quench vacuum furnaces, the quench tank often doubles as the load/unload chamber.
3. Inspect the chamber for signs of damage and do not proceed with the load sequence if damage is observed until repairs have been made. (Note: This may require interrupting the automatic sequencing of the furnace.)
4. Place the parts into the load chamber. Be sure that the load is positioned in the chamber so as to allow the internal load-transfer mechanism to properly engage it.
5. Inspect for damage, wipe down and, if necessary, re-grease the door O-ring seal.
6. Close the chamber door and ensure that it is properly locked in position. *(CAUTION: Never operate a vacuum furnace if the outer door is not properly secured.)*
7. If required, select the proper furnace recipe and initiate the automatic cycle sequence. In the case of multi-chamber designs with workloads in process in other chambers, placing a load in a load chamber often initiates pump-down of the chamber automatically in anticipation of automatic sequencing.
8. Actions from this point on depend on the function and configuration of the load chamber. For an oil-quench vacuum furnace this may involve the following automatically triggered activities.
   - ❖ When the chamber reaches a preset vacuum level (equivalent in pressure to the chamber to which it is attached), an inner vacuum (or thermal) door opens and the workload is transferred from one chamber to the other.
   - ❖ The thermal-processing cycle begins in this next chamber (e.g., ramp to setpoint and soak).

❖ Backfill to a preset pressure level so as to allow equalization in pressure with the next chamber in the sequence of operation.
❖ Opening of a vacuum door and initiation of load transfer.

9. Silence the end-of-cycle alarm.
10. Backfill the vacuum furnace to atmospheric pressure. Unload.
11. Reload and/or close the outer door and pump the unit down to an intermediate vacuum level until another load is ready for processing.

Types of Vacuum Hardening

Vacuum-hardening methods are best discussed by focusing on each of the different types of materials being processed in vacuum furnaces.

PLAIN-CARBON AND ALLOY-STEEL HARDENING BY OIL QUENCHING

Oil quenching of plain-carbon and alloy steels often takes place in horizontal vacuum furnaces equipped with integral-quench tanks (Figs. 24.1, 24.2). The design of the quench tank is similar to its atmosphere counterpart – fixed or variable-speed oil circulation via agitators or pumps located on one or both sides of the tank and internal baffles to guide the respective oil flow around and through the load. Cold or preheated oil in the 50-65°C (120-150°F) range are the most common. Special (hot) oils run at 135-175°C (275-350°F) have also been used with success. Heaters control the oil temperature, and the oil is cooled via double-wall construction and/or external heat exchangers usually employing air for safety reasons.

FIGURE 24.1 | Two-chamber oil-quench vacuum furnace (courtesy of Specialty Heat Treating Inc.)

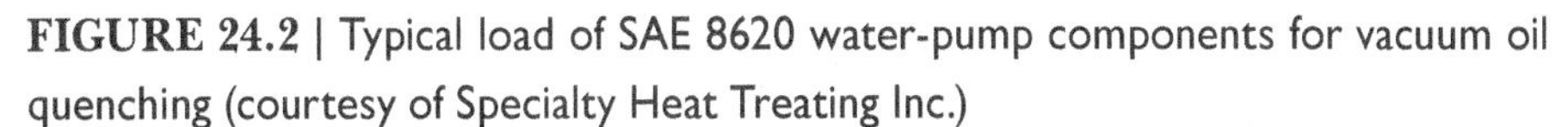

FIGURE 24.2 | Typical load of SAE 8620 water-pump components for vacuum oil quenching (courtesy of Specialty Heat Treating Inc.)

As covered in Chapter 20, a peculiarity of quenching in vacuum furnaces is that the low pressure above the oil causes standard quench oils to degas violently. The duration of this degassing process depends on the amount of air or nitrogen absorbed by the oil during the loading and unloading of the furnace. Vacuum oils have been created to minimize these problems. Oils that are not degassed properly have a worse quenching severity and produce discolored components.

Vacuum quench oils are distilled and fractionated to a higher purity than normal oils, which is important in producing the better surface appearance of quenched parts. In practice, quenching in vacuum furnaces is frequently done with a partial pressure of nitrogen above the oil between 540 mbar (400 torr) and 675 mbar (500 torr).

It is well known, however, that a pressure increase just before initiating the quench also changes the oil-cooling characteristics. The pressure increase shortens the vapor-blanket phase, thus increasing the quench severity at high temperatures (in the pearlite-ferrite transformation range). On the other hand, it lowers the quench rate in the convective cooling phase (in the bainite or martensite transformation range). Thus, high partial pressures above the oil can be advantageous in producing full hardness on unalloyed or low-alloy materials, whereas low pressures above the oil produce higher hardness and lower distortions on components made of medium- or high-alloy steels.

Very low pressures under 50 mbar (37.5 torr) above the oil and very high quenching temperatures in the area of 1095°C (2000°F) can lead to carbon deposition and/or pickup on the surface of parts. This is due to the thermal decomposition of the quench oil, which has been experienced in hardening certain tool steels. The carbon originates from the fractionation of the oil vapor in contact with the hot surface of the load in the initial phase of the quench process. High nitrogen pressures over the oil greater than 200 mbar (150 torr) tend to reduce or eliminate this effect.

Medium-alloy steels (Table 24.1) and most case-hardening steels (Table 24.2) are hardened either by oil quenching or high-pressure gas quenching (typically up to 20 bar).

TABLE 24.1 | Quench selection guide for plain-carbon and alloy steels [2]

Material grade (AISI/SAE)	Hardening temperature, °F (°C) [a]	Quench medium [b]	Typical surface hardness, HRC — Diameter values, mm (inches) [c]			
			12 (0.5)	25 (1)	50 (2)	100 (4)
1340	830 (1525)	OQ, PQ	58	57	39	32
3140	830 (1525)	OQ, PQ	57	55	46	34
4140	855 (1570)	OQ, PQ	57	55	49	36
4150	830 (1525)	OQ, PQ	64	62	58	47
4340	800 (1475)	OQ, PQ	58	57	56	53
4640	800 (1475)	OQ, PQ	57	54	39	30
6152	845 (1550)	OQ, PQ	61	60	54	42
8740	830 (1525)	OQ, PQ	57	56	52	42
9440	830 (1525)	OQ, PQ	56	51	41	28

Notes:

[a] Austenitizing temperature in vacuum is often 15-30°C (25-50°F) higher.

[b] OQ = oil quench; PQ = gas pressure quench

[c] As-quenched (oil) data shown.

ALLOY-STEEL HARDENING BY GAS QUENCHING

Hardening by inert-gas pressure quenching (2-20 bar) is the most popular method of quenching in vacuum furnaces (Figs. 24.3-24.5). In recent years, some systems have been designed for the use of hydrogen for cooling in the 25- to 40-bar range due to its extremely high heat-transfer rates. Appropriate cautionary steps are MANDATORY. Reference NFPA 86 when considering the use of hydrogen or hydrogen-containing mixtures (c.f. Chapter 22).

In gas quenching, it has been found that some part dimensional changes, although repeatable, are different than when quenching in oil. A current trend is to "dial in" the quench pressure (i.e. use only the highest pressure required to properly transform the material). Recent changes in material chemistry and pressure-quench design (e.g., alternating gas flows, directionally adjustable blades, variable-speed drives) have made this possible, and gas quenching is now used to produce full hardness in many traditional oil-hardening steels.

FIGURE 24.3 | Single-chamber gas pressure-quench vacuum furnaces (courtesy of Nevada Heat Treating)

FIGURE 24.4 | Core hardening of 4140 automotive transmission parts prior to ferritic nitrocarburizing (courtesy of Specialty Heat Treating Inc.)

FIGURE 24.5 | Hardening M4 roll-threading dies with a 6-bar nitrogen quench (courtesy of Solar Atmospheres Inc.)

Flow rate and density of the cooling gas blown onto the surface of the load are important factors for achieving high heat-transfer (i.e. high cooling) rates. In addition to high gas velocities, high gas pressures are needed to through-harden a wide variety of steel parts with appreciable dimensions (c.f. Chapter 19). Calculations of the heat-transfer coefficient alpha (α) show that it is proportional to the product of gas velocity and gas pressure. The heat transfer is an exponential function so that the heat transfer increases considerably with the first few bars of pressure but decreases with increasing pressure. Poor equipment design can contribute to non-uniformity of cooling or the use of higher pressures or velocities that can contribute to increased distortion.

The critical transformation range for most steel is between 800-500°C (1475-930°F). Time-temperature-transformation (TTT) curves or lambda (λ) values are available from steel suppliers for many material grades as a relative measure of the required cooling rate. Lambda (λ) values are numbers that represent the time required to pass through the stated temperature range divided by 100 seconds.

Cooling in argon produces the slowest heat-transfer rates, followed by nitrogen, then helium and finally hydrogen. All of these gases are popular (c.f. Chapter 22), but nitrogen is the most attractive purely from a cost standpoint. Theoretically, there is no limit to the improvement in cooling rate that can be obtained by increasing gas velocity and pressure. Practically, however, high-pressure and high-velocity systems are complex and costly to construct. In particular, the power required for gas recirculation increases faster than benefits may accrue.

There are pressure/gas combinations that achieve heat-transfer coefficients within the range of those produced by still and mildly agitated oil. Gas quenching offers potential advantages over liquid (e.g., oil or salt) quenching in that altering gas type, velocity or pressure can change the cooling rate. However, the effect of load weight on the resultant cooling speed during gas quenching is more pronounced than, for example, during liquid quenching. Maximum section size (ruling section) is an important consideration as well.

TABLE 24.2 | Gas-quench pressure for select carburizing steel grades [4]

Material grade (AISI/SAE)	Quench medium [b]	Gas pressure, bar [a]	Typical core hardness, HRC	Process
1018	Oil		20-24	Carburizing
1030	Oil		20-24	Carburizing
1117	Oil		22-26	Carbonitriding
12L14	Oil		22-26	Carbonitriding
3310	Gas	10-15	32-36	Carburizing
4027	Gas	15-20	33-35	Carburizing
4118	Gas	15-20	33-35	Carburizing
4120	Gas	15-20	33-35	Carburizing
4142M	Oil		50	Carburizing
4320	Gas	12-20	33-35	Carburizing
4615	Gas	12-20	28-34	Carburizing
4620	Gas	10-20	30-35	Carburizing
4820	Gas	12-20	32-38	Carburizing
5115	Oil/Gas	12-20	32-36	Carburizing
5120	Gas	15-20	28-32	Carburizing
8620	Gas	20	32-40	Carburizing
8822	Gas	20	38-42	Carburizing
9310	Gas	5-10	38-44	Carburizing
C-61	Gas	2-3	50-54	Carburizing
C-61S	Gas	2-3	50-52	Carburizing
C-62	Gas	2-3	50-52	Carburizing
C-69	Gas	2-3	50-52	Carburizing
Pyrowear 53	Gas	5	36-40	Carburizing
Pyrowear 675	Gas	20	40-44	Carburizing

Notes:

[a] Data for section thickness up to 25 mm (1 inch).

[b] Quench medium is nitrogen.

TOOL STEELS

For most tool steels (Fig. 24.6), equivalent end-product performance, surface hardness and mechanical and microstructural properties (e.g., carbide size and distribution) can be achieved by vacuum hardening. For example, the hardening of hot-work dies with high toughness (e.g., H11 and H13) in vacuum furnaces with high gas-quench capability and cooling rates of 20-40°C (50-100°F) per minute, thus avoiding issues with carbide precipitation, decarburization and unwanted carbon pickup from the furnace atmosphere. Controlling the heating and the quenching phases produces a martensitic microstructure with proeutectic carbides and reduced residual stress and distortion.

In general, an air-hardening tool-steel part such as A2 (Fig. 24.7) is hardened in much the same way as in atmosphere furnaces. It is preheated, heated to a high austenitizing temperature and cooled at a moderate rate. The medium-alloy air-hardening steels in the A-series and the high-carbon, high-chromium steels in the D-series are regularly hardened in gas-quench furnaces using nitrogen up to 6 bar.

A rough vacuum in the range of 1.3 to 1.3 x 10^{-1} mbar (1 to 1 x 10^{-1} torr) is used in the heat treatment of tool steels. This level of vacuum is required mainly because of the relatively high vapor pressures of chromium, manganese and other easily vaporized elements.

FIGURE 24.6 | Typical tool and die parts for vacuum hardening (courtesy of Nevada Heat Treating)

FIGURE 24.7 | Vacuum-hardened A2 tool-steel custom-shape dies (courtesy of AmeriKen)

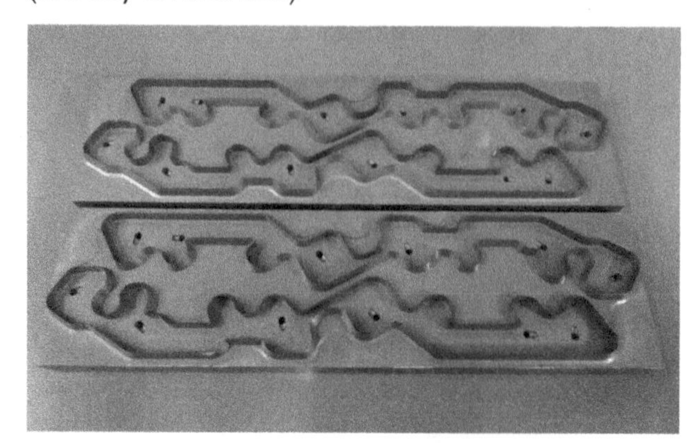

Many grades of hot-work tool steels such as H11 (Fig. 24.8a) and H13 are high-pressure gas quenched in nitrogen up to 10-bar pressure as opposed to oil quenched. In many cases, an isothermal hold (Fig. 24.8b) is introduced to minimize distortion. Gas quenching has been reported to achieve surface cooling rates in the 55°C/minute (100°F/minute) or higher range, which is adequate for most service applications. By contrast, oil quenching can typically achieve rates of up to about 165°C/minute (300°F/minute).

FIGURE 24.8a | H-11 tool-steel die for hardening by high-pressure gas quenching (courtesy of SECO/WARWICK Corp.)

FIGURE 24.8b | Process cycle with isothermal hold during quenching (courtesy of SECO/WARWICK Corp.)

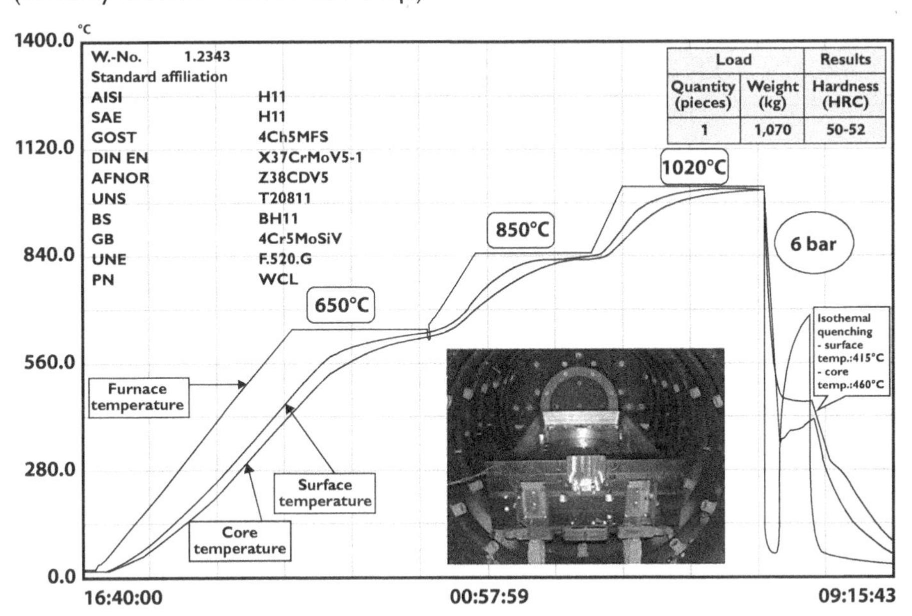

MARTENSITIC STAINLESS STEELS

All grades of martensitic stainless steels can be processed in vacuum furnaces using similar austenitizing temperatures and general considerations as those processed in atmosphere furnaces (Table 24.3). Since the austenitizing temperatures are usually below 1100°C (2000°F), vacuum levels in the range of 10^{-3} mbar (10^{-3} torr) are very often used, which result in clean and bright part surfaces. To avoid evaporation of certain alloying elements, processing is also done at vacuum levels ranging from 6.67 x 10^{-1} to 1.3 mbar (5 x 10^{-1} to 1 torr), with some sacrifice to brightness.

Due to the differences in the hardenability of the various martensitic stainless alloys, there is a limitation on the section sizes that can be fully hardened by recirculated nitrogen gas quenching. Other types of cooling gas (e.g., helium) can be used, but the economic benefits must be carefully considered. The actual values of section size limits depend on the type of cooling system and the capability of the specific furnace employed.

TABLE 24.3 | Typical hardening cycles for select martensitic stainless steels [5]

Material grade (AISI/SAE)	Preheat 1, °C (°F) [a]	Preheat 2, °C (°F)	Hardening temperature, °C (°F)	Quench medium [b]	Typical surface hardness, HRC [c]
410	540 (1000)	760-790 (1400-1450)	925-1010 (1695-1850)	OQ, PQ	40-44.5
420	540 (1000)	760-790 (1400-1450)	980-1065 (1795-1950)	OQ, PQ	47.5-55.5
440A	540 (1000)	760-790 (1400-1450)	1010-1065 (1850-1950)	OQ, PQ	52-57
440B	540 (1000)	760-790 (1400-1450)	1010-1065 (1850-1950)	OQ, PQ	56-59
440C	540 (1000)	760-790 (1400-1450)	1010-1065 (1850-1950)	OQ, PQ	60-62

Notes:

[a] Rapid heating rates can cause distortion and/or cracking. In vacuum, heating rates of 8-14°C/minute (15-25°F/minute) are recommended for small parts or intricate shapes.

[b] Certain parts will benefit from an initial preheat at the temperatures shown.

[c] OQ = oil quench; PQ = gas pressure quench

[d] As-quenched (oil) data shown.

PRECIPITATION HARDENING OF STAINLESS STEELS

The heat-treatment temperature for precipitation-hardened stainless steels (Fig.

24.9) depends on the particular alloy grade, the type of parts being treated and the required properties (Table 24.4). In several instances, multiple heat treatments are specified. In other cases, material is purchased in the so-called Condition "A," requiring only an aging operation to be performed (this is typically not done in vacuum). For optimum creep and creep-rupture properties, the high side of the solution-annealing temperature range is typically used. For optimum strength during relatively short-term service at high temperatures, a low-end annealing temperature is used. A final aging heat treatment produces a finely dispersed precipitate throughout the microstructure, significantly increasing the room-temperature yield strength.

A typical example is A-286, which is also identified as K-660, 286 and Type 660. This is an austenitic iron-nickel-chromium steel with good strength up to 705°C (1300°F) in the aged condition. It is used for gas-turbine wheels, blades and jet-engine components. Unlike other stainless steels, A-286 has very high titanium content, about 2%, along with about 15% Cr, 25% Ni, 1.3% Mo, 0.3% V, 0.2% Al, 0.05% C and 0.004% B. The usual solution-annealing temperature range is 900-1010°C (1650-1850°F). It is processed in high vacuum for one hour per inch of cross section, followed by rapid cooling.

TABLE 24.4 | Typical solution-treating and aging cycles for select precipitation-hardening stainless steels [5]

Material grade (AISI/SAE)	Solution heat-treat temperature, °C (°F)	Method of cooling [a]	Aging temperature, °C (°F)	Aging time, hours [b]	Method of cooling [a]
A-286	980 (1800)	OQ, PQ	720 (1325)	16	AQ
13-8 Mo	925 (1695)	PQ	510-620 (950-1150)	4	AQ
15-5	1040 (1900)	OQ, PQ	480-620 (900-1150)	4	AQ
17-4 PH	1040 (1900)	OQ, PQ	480-620 (900-1150)	4	AQ
17-7 PH	1050 (1925)	PQ	510-595 (950-1100)	4	AQ
Custom 455	830 (1525)	WQ, PQ	480-565 (900-1050)	4	AQ
Rene 41	1080 (1975)	WQ, PQ			
Udimet 700	1175 (2150)	PQ	845 (1550)	24	AQ
Waspaloy	1080 (1975)	PQ	845 (1550)	24	AQ

Notes:

[a] WQ = water quench; OQ = oil quench; PQ = gas pressure quench; AQ = air cool

[b] Under certain conditions, H900 aging operations only require a one-hour soak.

FIGURE 24.9 | 17-4 PH stainless steel die for age hardening after solution heat treating (courtesy of Solar Atmospheres Inc.)

SUPERALLOYS

Superalloys cover a wide range of materials, typically nickel, cobalt or iron-based, and are generally intended for high-temperature applications. Most of these alloys are hardened using a solution-treating and aging process (Tables 24.5, 24.6). Solution treating involves heating the alloy to a temperature in the range of 982°C (1800°F) or higher, followed by gas quenching. In most cases, a very fast quench speed is not required, and gas quenching with nitrogen at pressures of 2 bar or less is often sufficient. This is followed by aging at an intermediate temperature for extended periods of time. Normally, the complete solution-treating and aging cycles can be programmed into the furnace so that unloading is not required between cycles. Certain superalloys, however, require other special treatments to develop required properties.

TABLE 24.5 | Typical solution-treating and aging cycles for select wrought superalloys [6]

Alloy	Solution heat-treat temperature, °C (°F)	Solution heat-treat time, hours	Method of cooling [a, b]	Aging temperature, °C (°F)	Aging time, hours	Method of cooling [a, b]
A-286	980 (1800)	1	OQ, PQ	720 (1325)	16	AC
Incoloy 925	1010 (1850)	1	AC	730 (1350)	8	FC
Inconel 625	1150 (2100)	2	[c]			
Inconel 718	980 (1800)	1	AC	720 (1325)	8	FC
Inconel X-750	1150 (2100)	2	AC	845 (1550)	24	AC
Rene 41	1065 (1950)	0.50	AC	760 (1400)	16	AC
Udimet 700	1175 (2150)	4	AC	845 (1550)	24	AC
Waspaloy	1080 (1975)	4	AC	845 (1550)	24	AC

Notes:

[a] FC = furnace cooling; AC = air cooling; RAC = rapid air cool; OQ = oil quench; PQ = gas pressure quench

[b] Air cooling equivalent is defined as cooling at a rate not less than 40°F per minute to 1100°F and not less than 15°F per minute from 1100-1000°F. Any rate may be used below 1000°F.

[c] To provide adequate quenching after solution heat treatment, cool below 540°C (1000°F) rapidly enough to prevent carbide precipitation. Oil or water quenching may be required on thick sections.

TABLE 24.6 | Typical solution-treating and aging cycles for select cast superalloys [6]

Alloy	Solution heat-treat temperature, °C (°F)	Solution heat-treat time, hours	Method of cooling [a, b]	Aging temperature, °C (°F)	Aging time, hours	Method of cooling [a, b]
Hastalloy C	1220 (2225)	0.5	AC			
Inconel 100	1080 (1975)	4.0	AC	870 (1600)	12	AC
Inconel 625	1190 (2175)	1	AC			
Inconel 718	1095 (2000)	1	AC	[c]	[c]	[c]
Rene 41	1065 (1950)	3	AC	[d]	[d]	[d]
Rene 80	1220 (2225)	2	PQ	[e]	[e]	[e]
Udimet 700	1175 (2150)	4	AC	[f]	[f]	[f]
Waspaloy	1080 (1975)	4	AC	[g]	[g]	[g]

Notes:

[a] FC = furnace cooling; AC = air cooling; RAC = rapid air cool; PQ = gas pressure quench

[b] Air cooling equivalent is defined as cooling at a rate not less than 40°F per minute to 1100°F and not less than 15°F per minute from 1100-1000°F. Any rate may be used below 1000°F.

[c] Balance of heat treatment for Inconel 718: 1120°C (1750°F) for 1 hour, followed by air cooling, then 720°C (1325°F) for 8 hours, followed by furnace cool, then 620°C (1150°F) for 8 hours, followed by air cool.

[d] Balance of heat treatment for Rene 41: 955°C (2050°F) for 1.5 hours, followed by pressure quench, then 900°C (1650°F) for 4 hours, followed by air cool.

[e] Balance of heat treatment for Rene 80: 1095°C (2000°F) for 4 hours, followed by air cooling, then 1050°C (1925°F) for 4 hours, followed by air cool, then 845°C (1550°F) for 16 hours, followed by air cool.

[f] Balance of heat treatment for Udimet 700: 1175°C (2150°F) for 4 hours, followed by air cooling, then 1080°C (1975°F) for 4 hours, followed by air cool, then 845°C (1550°F) for 24 hours, followed by air cool, then 760°C (1400°F) for 16 hours, followed by air cool.

[g] Balance of heat treatment for Waspaloy: 845°C (1550°F) for 4 hours, followed by air cool, then 760°C (1400°F) for 16 hours, followed by air cool.

Final Thoughts

When considering vacuum hardening it is necessary to take a number of factors into consideration, not the least of which is adherence to print requirements and compliance to all appropriate specifications (e.g., AMS or other aerospace standards). Done properly, vacuum hardening can be expected to meet or exceed the quality being produced by other hardening methods.

REFERENCES

1. Herring, D.H., "Technology Trends in Vacuum Heat Treating, Part Two: Processes and Applications," *Industrial Heating*, November 2008
2. *Modern Steels and Their Properties,* Handbook 268, Bethlehem Steel, 1949
3. *Vacuum Brazing & Heat Treating Training Manual,* VAC AERO International Inc.
4. Beauchesne, Dennis and Aymeric Goldsteinas, "A Users Guide to Gas and Oil Quenching After Vacuum Carburizing," *Heat Treating Progress*, September/October 2004
5. *Heat Treaters Guide: Practices and Procedures for Ferrous Alloys,* Candler, Harry, editor, ASM International, 1995
6. *Heat Treaters Guide: Practices and Procedures for Nonferrous Alloys,* Candler, Harry, editor, ASM International, 1996
7. Patrick McKenna, Nevada Heat Treating, editorial review and private correspondence

World Class Heat Processing Equipment
innovation • quality • performance
SECO/WARWICK
www.secowarwick.com

VACUUM HEAT TREATMENT

CHAPTER 25

VACUUM APPLICATIONS: BRAZING

Vacuum brazing represents one of the larger application uses for vacuum furnaces. The transportation industry (e.g., automotive and aerospace) in particular has provided the impetus for increasing demand for vacuum-furnace brazing. In addition, the gain in popularity of lightweight, high-strength materials have also contributed to the popularity of vacuum brazing.

What is brazing?

Brazing is a joining process and not a heat-treatment process per se, although it requires the controlled application of thermal (heat) energy in order to be successful. Brazing takes place between materials of the same or dissimilar composition. It is achieved by heating to a temperature where a filler metal with a liquidus temperature above 450°C (840°F) and below the melting point of the base metal is distributed between closely fitted surfaces by capillary action.

Brazing has four distinct characteristics.

1. The joining, or coalescence, of an assembly of two or more pieces into a single structure is achieved by heating the assembly or the area to be joined to a temperature above 450°C (840°F).
2. Assembled parts and brazing filler metal (BFM) are heated to a temperature high enough to melt the filler metal but not the base metal.
3. The filler metal must melt and flow into the closely fitted surfaces or joint, and it must wet or spread out onto the base-metal surfaces.

4. The parts are cooled to solidify, or freeze, the filler metal, which is held in the joint by capillary attraction and anchors the part together.

There are many different types of brazing, including flame (torch) brazing, furnace brazing, induction brazing, dip brazing, resistance brazing and infrared brazing to name a few. These are typically distinguished by the method of heating. A successful braze joint results in a metallurgical bond that is generally as strong or stronger than the base metal(s) being joined. Brazing differs from welding, in which the joint is formed through melting of both the base and filler metals. It is similar to soldering but, by definition, is performed at higher temperatures.

In brazing, the filler metal can be placed within the joint as a foil or placed over the joint in the form of paste, preform or wire. Joint clearances must be very carefully controlled and usually do not exceed 0.12 mm (0.005 inches). Since capillary action draws the molten filler metal into the joint and holds it there, the base-metal components must be designed to enhance this action.

Brazing is a process well suited for vacuum furnaces given their method of heating and the prevention of unwanted oxidation. To achieve a good joint, the base-metal components must be clean and protected from excessive oxidation during heating so that the molten filler metal will flow unrestricted throughout the braze-joint area.

As with all brazing, variables that need to be controlled to produce mechanically sound braze joints include: base-metal selection and characteristics; filler-metal selection and characteristics; component design; joint design and clearance; surface preparation; filler-metal flow characteristics; temperature and time; source; and rate of heating so that the molten filler metal will flow unrestricted throughout the braze-joint area.

Some factors affect the ability to produce a metallurgically sound braze joint by influencing the behavior of the brazed joint. Others affect the base-metal properties, while still others influence the interactions between the base metal and filler metal. Effects on the base metal include carbide precipitation, hydrogen embrittlement, nature of the heat-affected zone, oxide stability and sulfur embrittlement. Filler-metal effects include vapor pressure, alloying, phosphorous embrittlement and stress cracking. Interaction effects include post-brazing thermal treatments, corrosion resistance and dissimilar metal combinations.

Vacuum furnaces can be either horizontal or vertical in design and have technical advantages, including:

- ❖ The process permits brazing of complex, dense assemblies with blind passages that would be almost impossible to braze and adequately clean using atmospheric flux-brazing techniques.

- Vacuum furnaces operating at 1.3 x 10^{-4} to 1.3 x 10^{-5} mbar (1 x 10^{-4} to 1 x 10^{-5} torr) remove essentially all gases that could inhibit the flow of brazing alloy, prevent the development of tenacious oxide films, and promote the wetting and flow of the braze alloy over the vacuum-conditioned surfaces.
- Properly processed parts are unloaded in a clean and bright condition, often avoiding additional processing.
- A wide variety of materials – aluminum, cast irons, stainless steel, steels, titanium alloys, nickel alloys and cobalt-based superalloys – are brazed successfully in vacuum furnaces without the use of any flux.

Key Considerations

To achieve a successful braze joint, it is essential to understand the factors that influence the outcome as well as how they interact with one another. These include:

- Wetting and flow
- Base-metal properties
- Filler-metal characteristics
- Part surface cleanliness
  - Part surfaces protected from chemical interaction prior to brazing
- Joint design and clearance
- Brazing parameters: time and temperature
  - Heating rate and source to proper brazing temperature
  - Proper heat distribution at brazing temperature
  - Cooling or solidification rate

We will investigate each of these in some detail.

WETTING AND FLOW

Wetting and flow of the base and filler metal are essential elements to successful brazing.

Wetting is the ability of the liquid filler metal to spread out and adhere to the solid base metal and, when cooled below the "solidus" temperature, to make a strong bond with that metal. Wetting is a function not only of the BFM but also the nature of the metal or metals to be joined. There is considerable evidence that in order to wet well, a molten metal must be capable of dissolving or alloying with some of the metal on which it flows.

The wetting characteristics of a BFM have varying effects on the reach or the rate of capillary action. For example, copper melts and flows easily, whereas aluminum filler metals aggressively alloy with their aluminum base metals and can act sluggishly. Indeed, the wetting characteristics determine whether the capillary action will be generated at all. In effect, the wetting characteristics are a go/no-go situation. Wetting, however, is only one of the important facets of the brazing process. If the molten BFM does not flow into the joint, the effectiveness of the filler metal is greatly reduced.

Flow is the property of a BFM that determines the distance it will travel away from its original position because of the action of capillary forces. Capillary action is an essential element in producing a good braze. To achieve a properly brazed joint, it is therefore necessary that the parts be designed so that when properly aligned, they afford a capillary for the filler metal. Capillary attraction is the result of greater adhesive attraction of closely spaced (parallel) solid surfaces for a liquid than the cohesive force of the liquid. In other words, the force that wants to hold the molecules of liquid together is overcome by surface-tension forces that want the liquid to adhere to the walls of the small space, or capillary, through which the liquid is flowing.

To flow well, a filler metal must not suffer an appreciable increase in its liquidus temperature, even though its composition is altered by the addition of the metal it has dissolved. This is important because brazing operations are carried out at temperatures above the liquidus.

In the real world, a variety of interactions alter wetting and flow characteristics. The rate at which these interactions take place depends on temperature, time at temperature and the materials involved. In practice, these interactions are minimized by:

- Selection of the proper BFM
- Keeping the brazing temperature as low as possible (but high enough to produce flow)
- Keeping the time at temperature short
- Cooling the braze joints as quickly as possible without causing cracking or distortion (in many cases this involves an initial slow cool to "set the braze," followed by rapid cooling)

BASE-METAL PROPERTIES

A wide range of metallic and non-metallic materials can be brazed. The strength of the base metal has a profound effect on the strength of the brazed joint. Therefore, this property must be clearly kept in mind when designing the joint for specific properties. The ease of brazing is partially determined by the choice of base metal and by the specific brazing process selected.

Choices for base metals are not limited to the common metals and their alloys. They include more exotic metals such as titanium and titanium alloys, cobalt and cobalt alloys, tantalum, and precious metals such as gold, silver and platinum. We are not limited to metals. Ceramics, cermets, glass and carbides can all be brazed to themselves or to various metals provided that proper filler metals and other factors are taken into consideration.

The oxides of the less reactive metals like iron, nickel and cobalt tend to dissociate (break down) under high vacuum and temperature. Therefore, alloys such as the 300- and 400-series stainless steels, carbon steels and many tool steels can be successfully brazed in vacuum at relatively high pressures, first by pumping the chamber down to the 1.3 x 10^{-4} mbar (1 x 10^{-4} torr) range (if diffusion pumping capability is available) or to 1.33 mbar (1 torr) followed by processing in partial pressure in the range of 0.001-0.067 mbar (1-50 microns). Conversely, alloys containing appreciable amounts of reactive elements (e.g., aluminum or titanium at levels of 0.4% or greater) singly or in combination with one another tend to form oxides at high temperatures, which impede the flow of the BFM. Many of the superalloys fall into this category, and the severity of the problem varies depending on alloy composition. These materials should be brazed at high vacuum levels of at least 2.67 x 10^{-4} mbar (2 x 10^{-4} torr).

There are several reliable techniques for improving the brazeability of difficult-to-braze materials. These include nickel plating of the joint surfaces, chemical etching techniques (to remove aluminum and titanium from a shallow layer at the joint surface) and special aggressive BFMs with self-fluxing characteristics.

Titanium- and zirconium-based alloys can be vacuum brazed using specially formulated BFMs. Because of their tendency to become contaminated by even small amounts of oxygen or moisture, these alloys must be brazed at high vacuum levels in a clean furnace. Aluminum and many of its alloys can also be brazed in vacuum. Aluminum alloys, however, must always be brazed in dedicated furnaces designed for extremely close temperature uniformity at temperatures up to approximately 650°C (1200°F). Outgassing of magnesium and aluminum during brazing can quickly contaminate furnace equipment and loads subsequently processed at higher temperatures. For these reasons, it is mandatory to remember that vacuum brazing of aluminum must always be performed in dedicated furnaces.

Also of critical importance is the coefficient of thermal expansion, especially of dissimilar materials that are used to make up the assembly. The gaps may open or close on heating, making it easier or more difficult to braze them.

FIGURE 25.1 | Thermal-expansion curves for common materials [2]

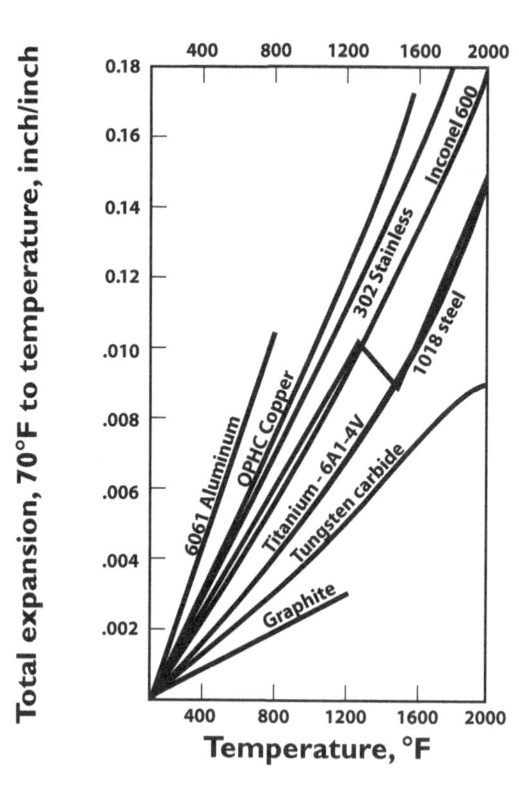

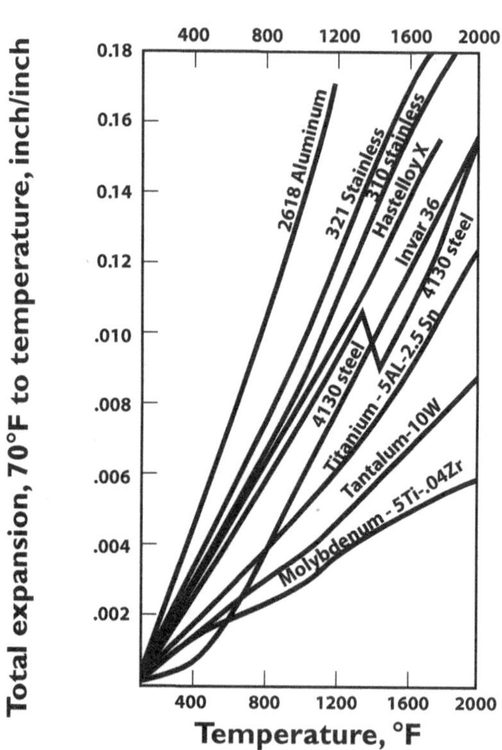

FIGURE 25.2 | Thermal expansion vs. temperature for selected materials [2]

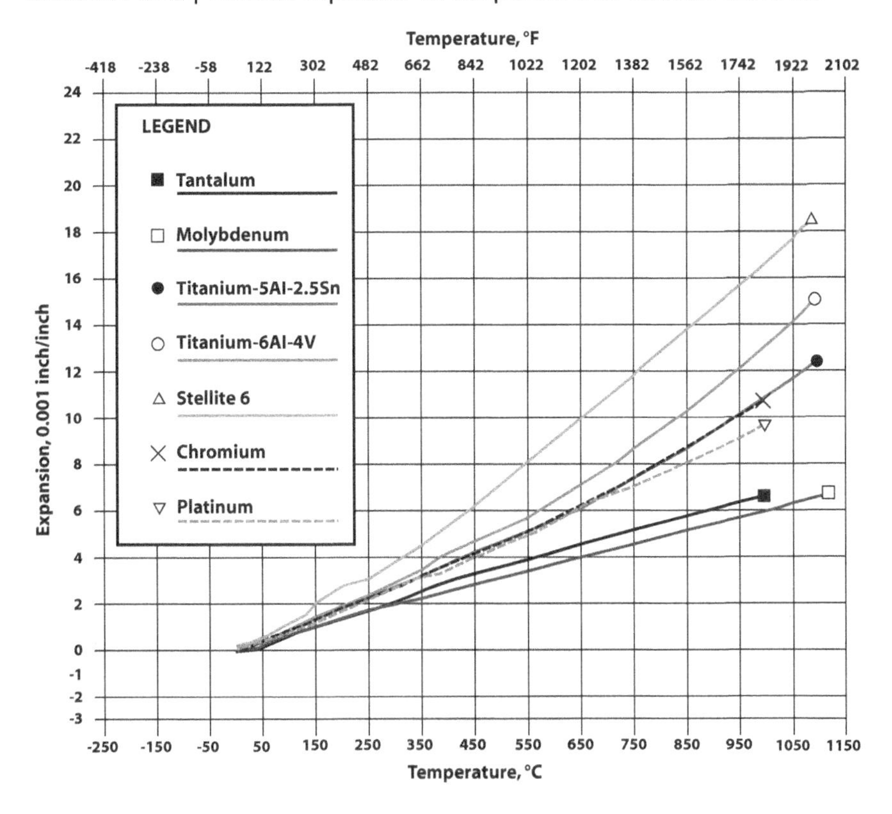

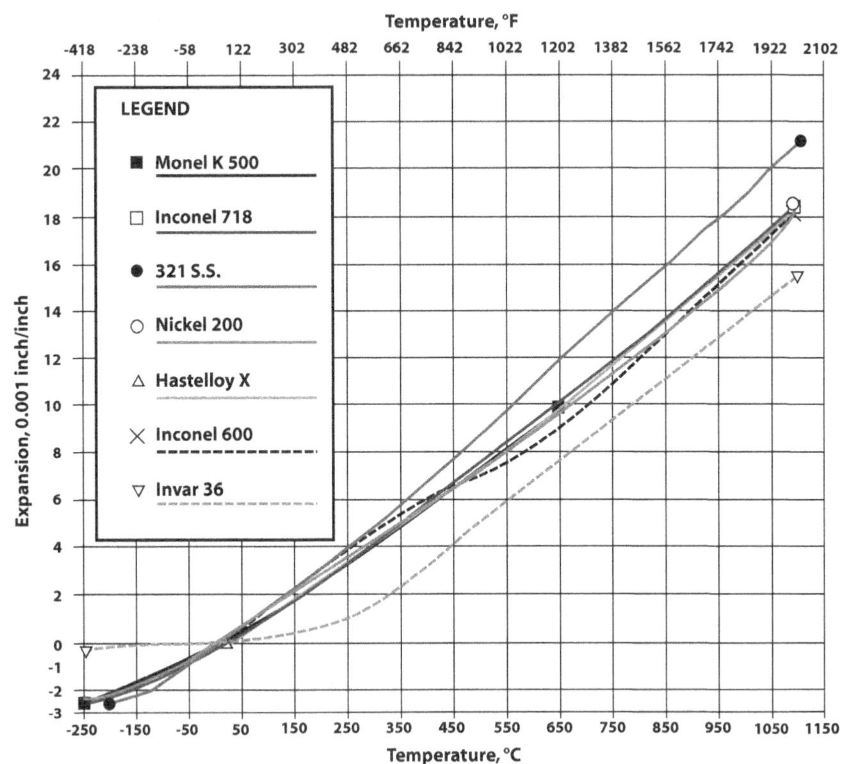

TABLE 25.1 | Select data on thermal expansion of materials [2]

Metal	Condition	Coefficient of thermal expansion x 10^6 at 21-100°C (70-212°F)	
		mm/mm-°C	in/in-°F
Aluminum, 1100 Series	Annealed (O)	7.3	13.1
Aluminum, 6061	Heat treated (T4)	7.3	13.1
Aluminum bronze (9.5%)	Extruded	5.0	9.0
Copper, oxygen free	Annealed		9.4
Ductile iron (+Mg)	Cast	4.2	7.5
Gold (pure)	Cold rolled	4.3	7.8
Gray cast iron	Cast	3.3	6.0
Hastaloy X	Cast, solution treated	4.3	7.7
K-Monel	Hot rolled	4.3	7.8
Kovar	Annealed	1.3	2.3
Magnesium	Cold worked	8.1	14.5
Molybdenum	Stress relieved, 870°C (1600°F)	1.9	3.4
Nickel (pure)	Annealed	8.1	14.5
Phosphor bronze (8%)	½ Hard	5.6	10.1
Rene 41		4.2	7.5
Stainless steel, 304/304L	Annealed	4.4	8.0
Stainless steel, 310	Cold rolled (50%)	4.4	8.0
Stainless steel, 316	Cold rolled (49%)	4.9	8.8
Stainless steel, 330	Cold drawn		3.5
Stainless steel, cast HT alloy (35% nickel-15% chromium)	Cast	5.4	9.7
Stainless steel, 410	Heat treated	2.8	5.1
Stainless steel, 420	Heat treated	3.1	5.5
Stainless steel, 440A	Hardened, tempered	3.1	5.6
Steel, 1020	Annealed	3.6	6.5
Steel, 4340	Annealed	3.5	6.3
Steel, A286	Solution treated	5.2	9.4
Titanium	Commercial purity	2.6	4.7
Zirconium	Extruded	1.6	2.9

Finally, the effect of the brazing cycle on the base metal and final joint strength must be taken into consideration. By its very nature, the brazing process will usually anneal the base metal unless the brazing temperature is very low and the time at heat is very short. When it is essential to design a brazement having strength above the annealed strength of the base metal, it will be necessary to specify a heat-treatable base metal capable of being either air or oil quenched that can be brazed and hardened in a combined operation. Selecting a precipitation-hardening grade (in which the brazing cycle and the solution-treatment cycle may be combined) is another possibility.

FILLER-METAL CHARACTERISTICS

Many types of BFMs (or braze alloys as they are commonly referred to) are in use today, each chosen for a particular characteristic that makes it ideally suited for joining a certain combination of base metals. These are available from a broad range of manufacturers. All of them however, must have these common characteristics:

- Ability to wet the base-metal joint surfaces and form a strong, sound bond with them
- Have suitable melting and flow characteristics at brazing temperature to ensure flow by capillary action and provide full alloy distribution
- Stability to avoid premature release of low melting-point elements from the filler metal
- Alloying elements of the BFM must have low volatility and/or avoid environmentally objectionable constituents (without special precautions being taken)
- Have the ability to produce a joint with needed service capabilities such as strength, corrosion resistance and electrical conductivity
- Depending on the requirements, have the ability to produce or prevent base-metal/braze filler-metal interactions

In addition, it is desirable that the filler-metal alloy diffuses into (i.e. combines with) the parent metal to form an alloy with a higher melting temperature than the original BFM. This is particularly noteworthy when attempting to re-braze components, since this operation must often be attempted at a higher temperature than the original brazing operation.

The degree to which the BFM penetrates the alloys with the base metal during brazing is referred to as diffusion. In applications requiring strong joints for high-temperature, high-stress service conditions (common for most aerospace components), it is generally good practice to specify a BFM that has high diffusion and solution properties with the base metal. Diffusion is a normal part of the metallurgical process that can contribute to good brazed joints in these demanding applications.

The compatibility of the braze alloy with the base metal must also be considered. There is almost always some interaction between the braze alloy and base

TABLE 25.2 | Base-metal/filler-metal combinations [2]

	Al and Al alloys	Mg and Mg alloys	Cu and Cu alloys	Carbon and low-alloy steels	Cast iron
Al and Al alloys	BAl-Si				
Mg and Mg alloys	[a]	BMg			
Cu and Cu alloys	[a]	[a]	BAg, BAu, BCuP, BNi, RBCuZn		
Carbon and low-alloy steels	[a]	[a]	BAg, BAu, RBCuZn BNi	BAg, BAu, BCu, RBCuZn, BNi	
Cast iron	[a]	[a]	BAg, BAu, RBCuZn BNi	BAg, BNi, RBCuZn	BAg, BNi, RBCuZn
Stainless steel	BAl-Si	[a]	BAg, BNi, BAu	BAg, BAu, BCu, BNi RBCuZn	BAg, BAu, BCu, BNi RBCuZn
Ni and Ni alloys	BAl-Si	[a]	BAg, BAu, RBCuZn	BAg, BAu, BCu, BNi RBCuZn	BAg, BCu, BNi RBCuZn
Ti and Ti Alloys	BAl-Si	[a]	BAg [c]	BAg [c]	BAg [c]
Be, Zr, V and alloys (reactive metals)	[b]	[a]	BAg	BAg	BAg
W, Mo, Ta, Cb and alloys (refractory metals)	[a]	[a]	BAg	BAg, BCu, BNi	BAg, BCu, BNi
Tool steels	[a]	[a]	BAg, BAu, BNi, RBCuZn	BAg, BAu, BAu, BCu, RBCuZn	BAg, BAu, RBCuZn,BNi

Notes:

[a] Not recommended. Special techniques may be practicable for certain dissimilar metal combinations.

[b] Generalizations on these combinations cannot be made.

[c] Special BFMs are available and are used successfully for specific metal combinations.

Stainless steel	Ni and Ni alloys	Ti and Ti alloys	Be, Zr, V and reactive alloys	W, Mo, Ta, Cb and refractory alloys	Tool steels
BAg, BAu, BCu, BNi					
BAg, BAu, BCu, BNi	BAg, BAu, BCu, BNi				
BAg [c]	BAg [c]	BAg, BAl-Si [c]			
BAg [c]	BAg [c]	[b]	[b]		
BAg, BCu, BNi, BAu	BAg, BCu, BNi, BAu	[b]	[b]	[b]	
BCu, BNi, BAg, BAu	BAg, BAu, BCu, BNi, RBCuZn	[a]	[a]	[a]	BAg, BAu, BCu, BNi, RBCuZn

metal, the extent of which varies depending on compositions and the thermal cycle. Under certain conditions, molten braze alloy can dissolve base metals, resulting in erosion or embrittlement. This can be devastating to the component, especially if the base metal is thin.

In choosing BFM, the first criterion is the melting points of the base metal and braze alloy. Pure metals and eutectic alloys transform from the solid state to the liquid state at one temperature when heated. However, other alloys melt over a temperature range. As the alloy melts, there is a range where both the solid and liquid phases coexist. The highest temperature at which a metal or alloy is completely solid is called the solidus; the lowest temperature at which a metal or alloy is completely liquid is called the liquidus. In brazing, therefore, the solidus is considered the melting point of the filler metal and the liquidus is considered its flow point.

BFMs in which the solidus and liquidus are close together flow readily and should be used for designs with small joint clearances. As the solidus and liquidus band widens, the mixture of solid and liquid metal can aid in gap filling. As a general rule, the solidus of the base metal should be at least 55°C (133°F) higher than the liquidus of the braze alloy. Table 25.2 lists typical base-metal/filler-metal combinations for a number of materials.

Factors other than the need to melt below the solidus of the base metal are important as well. It may have to melt below the temperature at which the assembly loses strength or above the temperature at which oxides are reduced. Oxidation resistance, corrosion resistance and electrochemical potential with other parts of the assembly may also be important considerations. Color match to the base metal, electrical and thermal conductivity, joint filling capability, hardness, machineability and the ability to provide visual fillets (especially important in automotive applications) are other considerations.

BFMs are available in a variety of shapes and sizes, including rod, ribbon, powder, paste, wire, sheet, foil and preforms (stamped shapes, washers, rings, shaped wires). For any given part, single or multiple choices, either alone or in combination, may prove to be the best solution. The choice is dependent on the joint design, heating method and level of automation desired. The major types of filler metals include:

- ❖ Copper and copper alloys
  - Pure copper is used extensively for brazing ferrous metals. Molten copper is free-flowing and is often used in furnace brazing with a reducing atmosphere. Because of its flow characteristics, copper penetrates joints with small clearances or with interference fits. Brazing temperatures typically exceed 1100°C (2000°F).
  - The most common form of brazing is in an atmosphere or vacuum furnace.

- Certain copper alloys, such as phosphor bronze, are often flame hardened rather than vacuum brazed.

❖ Aluminum and aluminum alloys
- Certain aluminum alloys can be brazed readily with aluminum-alloy BFMs. Alloy 3003 is common.
- The most common heating methods are flame, dip and furnace brazing, including vacuum.

❖ Nickel and nickel-based alloys
- Base alloys most commonly brazed with these alloys are 300- and 400-series stainless steels, and nickel- or cobalt-based alloys. Nickel-based filler metals retain their heat resistance up to 980°C (1800°F) and higher, depending on the specific alloy.
- The most common form of brazing is in a vacuum furnace.

❖ Silver alloys and other precious metals
- Silver alloys are probably among the most versatile filler metals and may be used to join ferrous and nonferrous metals, except aluminum and magnesium. Their melting range is between 600-870°C (1100-1600°F).
- The most common form of brazing is in an atmosphere or by induction.
- All heating methods can be used with these alloys.

❖ Gold and palladium filler metals are used for brazing iron-, nickel- and cobalt-based metals when resistance to oxidation or corrosion is required and high operating temperatures are anticipated.

TABLE 25.3 | Brazing filler metals for refractory metals [4]

Brazing filler metal	Liquidus temperature	
	°C	°F
Ag	960	1760
Cu	1052	1980
Ni	1454	2650
Pd-Mo	1571	2860
Pt-Mo	1774	3225
Ag-Cu-Mo	779	1435
Ni-Cu	1349	2460
Mo-Ru	1899	3450
Pd-Cu	1204	2200
Au-Cu	885	1625
Au-Ni	949	1740

When selecting a braze alloy, it is also important to consider the manner in which the alloy will be introduced into the joint and the form in which it is commercially available. Ductile metals such as copper-, silver- and gold-based braze alloys are available in the form of wire, shim, sheet paste and powder. Shim and sheet can be pre-placed directly in the joint during assembly of the components to be brazed.

The alloying ingredients in nickel-based BFMs, particularly those added for temperature depression, are such that they tend to make these materials hard and non-ductile. Therefore, nickel-based filler metals cannot be formed by rolling, drawing or similar manufacturing operations. They can be mixed with binders to form a paste that is easily applied over the joint. Joint design has some influence over which form of braze alloy is preferred. For thick joints, preplacement of the braze alloy may be necessary to ensure the joint is completely filled. Also, different braze alloys often require different joint clearances for effective capillary flow.

Braze alloys containing appreciable amounts of volatile elements may need to be brazed under a partial pressure of a gas such as hydrogen, nitrogen or argon. Copper-based braze alloys and those containing zinc and/or cadmium as melting-point suppressants are typical examples. Some silver-based and even nickel-based braze alloys also require the use of partial pressures to prevent vaporization of key alloying elements.

Finally, the placement of the BFM is an important consideration. Several very general rules apply.

- Wherever possible, place the filler metal on the most slowly heated part of the assembly (in order to ensure complete melting of the filler metal).
- Use gravity to assist filler-metal flow (particularly for those filler metals having wide ranges between their solidus and liquidus temperatures).
- BFMs may be chosen to fill wide gaps or gaps of variable size (such as around a corner).
- The filler-metal placement is often preference-dependent outside the joint (unless movement between components is unimportant or can be corrected) or inside the joint (if possible) for solid performs.
- The BFM should be placed on heavier sections (which heat up more slowly) if erosion of thin members is a concern.
- The type and amount of braze alloy (preform or paste) needs to be carefully controlled for economy, reproducibility and to regulate and maintain joint properties and configuration.

PART SURFACE CLEANLINESS

A clean, oxide-free surface is imperative to ensure uniform quality and sound brazed joints. Uniform capillary action will occur only when all grease, oil, dirt and oxides have been removed from both the braze alloy and base metal prior to brazing. The length of time that cleaning remains effective depends on the material involved, atmospheric conditions, storage techniques and the amount of handling that may be involved. It is recommended that brazing be performed as soon as possible after the material has been cleaned.

Cleaning is commonly divided into two major categories: chemical and mechanical. Chemical cleaning is a function of chemistry, temperature, time and energy. The selection of the chemical cleaning agent depends on the nature of the contaminant, base metal, surface condition and joint design. Regardless of the cleaning agent or method used, it is important that all residue or surface film be removed from the cleaned parts by adequate rinsing to prevent the formation of an equally undesirable film. A common mistake is to not pay attention to the degree of contamination being transferred to the cleaning agent and then attempting to clean parts with dirty chemistry. Mechanical cleaning techniques include blasting, grinding, filing, wire brushing and machining.

Apart from cleanliness and freedom from oxides, surface roughness is important in determining the ease and evenness of the braze-alloy flow. Generally, a liquid that wets a smooth surface will wet a rough surface even more. A rough surface will modify the filler-metal flow from laminar to turbulent, prolonging flow time and increasing the possibility of alloying and other interactions.

Mechanical cleaning methods such as grinding, wire brushing, machining and blasting are often used to remove objectionable surface conditions. They can also be used to roughen joint surfaces, which may promote braze-alloy flow. Care must be taken to ensure that no residues from the cleaning media are left on joint surfaces. In particular, do not use non-metallic media (such as aluminum oxide, glass beads or sand), wheels coated with these materials or boron nitride as blast media. They do contaminate. After the materials to be brazed are thoroughly cleaned, the braze alloy is applied. Proper application of braze alloy is best learned through practice, but a few general rules apply.

1. Excessive application of braze alloy should be avoided, particularly when brazing thin sections with aggressive fillers. The volume of braze alloy applied should be carefully considered, especially when using braze alloys in the form of paste. Pastes may contain more than 50% binder, so the size of the bead applied compared to the actual amount of braze alloy delivered can be deceiving.

2. Although braze alloy will flow uphill due to capillary action, it should be positioned over the joint to take advantage of gravitational forces whenever possible. During application of pastes, joints should not be completely sealed. Each joint must be allowed to vent during pump-down of the vacuum furnace. Stop-off paints can be used to limit the flow of braze alloy into unwanted areas, but care must be taken not to overuse these materials.
3. Components to be joined by brazing must be assembled in a fixed position relative to each other, and this position must be maintained throughout the brazing cycle. During assembly, care should be taken to ensure that proper joint clearances are also maintained. Whenever possible, parts should be designed so they are self-fixturing. If this is impractical, tack welding is the next best alternative. However, some assemblies may require auxiliary fixturing. Again, fixturing materials with coefficients of thermal expansion similar to the base metal should be used. Auxiliary fixtures should be low in mass and simple in design. If the use of screws or bolts is considered, careful consideration should be given to their effects on relative exposure and contact with the part before considering their use.

JOINT DESIGN AND CLEARANCE

The design engineer must determine how and where the components in an assembly are to be joined. Since brazing relies on capillary action, the design of the joint must provide an unobstructed and unbroken capillary path to allow the BFM to get into the joint or to allow air or flux, if used, to escape from the area. Small clearances and correspondingly thin filler-metal films make sound joints, and sound joints are strong joints. Some other important considerations include:

- Base-metal properties (composition, ductility, strength)
- Filler-metal properties (viscosity, surface tension, gravity)
- Tendency of the filler metal and base metal to alloy with one another during capillary flow of BFM
- Type of stress – shear (desirable) or tensile (undesirable)

The actual joint design should be based on a consideration of factors such as:

- Strength
- Corrosion resistance
- Electrical conductivity
- Thermal conductivity
- Materials to be joined

- Mode of application of the BFM
- Post-joining inspection requirements

The two most important design factors that have a strong influence on brazing results are the type of joint (Fig. 25.3) and the amount of clearance, or gap, between the members of the joint. Although there are many kinds of brazed joints, they are all variations of two basic types: butt joints and lap joints (Fig. 25.4).

FIGURE 25.3 | Various types of joint designs [3]

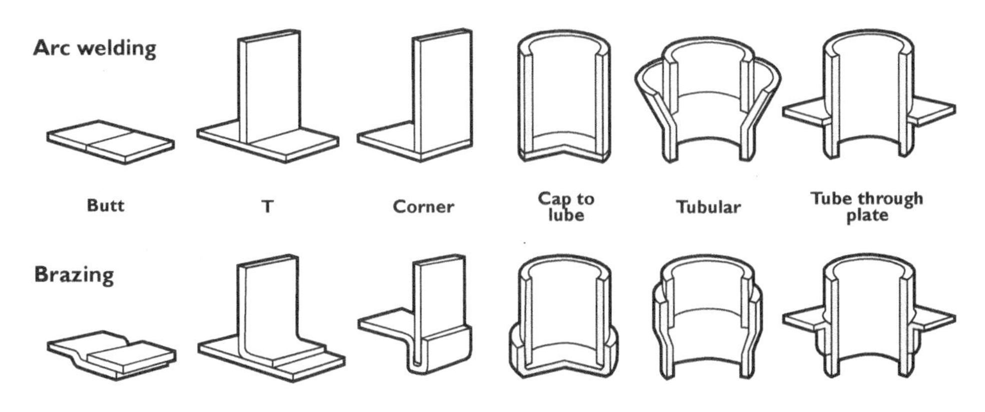

FIGURE 25.4 | Butt- and lap-type joints [3]

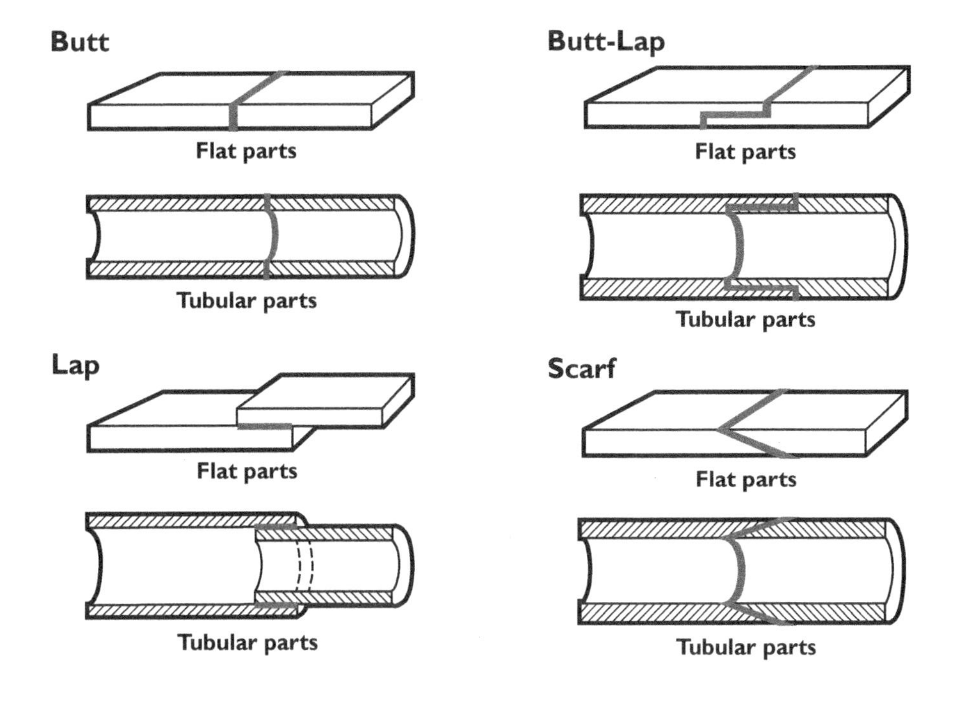

The butt joint has the advantage of a single thickness at the joint. The braze area is determined by the thinnest member of the joint, however, which dictates the maximum strength of the joint. Thus, butt joints should be chosen only when the thickness of the joint is a critical consideration. When properly made, butt joints should be stronger than the base metal being joined. By contrast, the lap joint has a much larger braze area and, consequently, is preferred for brazing operations. For maximum strength, lap-joint length should equal three to four times the thickness of the thinner member.

Selection of joint type is influenced by:

- Configuration of the parts
- Stress considerations
- Fabrication techniques
- Production quantities
- Service requirements (such as electrical conductivity, pressure tightness and appearance)
- Method of feeding BFM

Joint clearance is probably one of the most significant factors in brazing. As previously stated, the gap or distance between the surfaces to be joined may widen or close during heating, which causes problems when the BFM begins to melt and move. It is always desirable to design the joint so that the solidifying filler metal is exposed to compressive rather than tensile stresses.

Table 25.4 provides information on recommended joint clearances based on joints having members of similar metal and equal mass. When dissimilar metals and/or widely differing masses are joined for brazing, other problems may arise that necessitate more specialized selection among the various BFMs, with joint clearances empirically determined for the job at hand.

The temperature of the BFM has an important effect on the wetting action. In general, wetting and alloying improve as the temperature increases. Also, avoiding metal evaporation is critical to braze quality (Fig. 25.5).

Lower brazing temperatures may be beneficial for:

- Minimizing heat effect on the base metal (annealing, grain growth, warpage)
- Minimizing base-metal/filler-metal interactions
- Avoiding stress cracking
- Increasing the life of fixtures, jigs and other tooling
- Energy conservation

Higher brazing temperatures may be beneficial for:

- Use of a more economical BFM
- Combined processing steps (heat treatment and brazing)
  - Hardening
  - Case hardening (e.g., carbonitriding)
- Promoting base-metal/filler-metal interactions in order to modify BFM (increasing remelt temperature of the joint)
- Removal of surface contaminants (oxides)

The time at brazing temperature also affects the wetting action, particularly with respect to the distance the BFM may creep. If the filler metal has this tendency, the distance generally increases with time. Similarly, the vacuum level is important to prevent vaporization and change the ultimate brazing temperature.

TABLE 25.4 | Recommended gap clearances for different brazing filler metals [2]

Filler-metal group (AWS Classification)	Joint clearance		Remarks
	mm	inches	
BAlSi	0.000-0.050	0.000-0.002	Recommended for vacuum brazing
	0.050-0.203	0.002-0.008	Lap length <6.3 mm (<0.25 inches)
	0.203-0.254	0.008-0.10	Lap length >6.3 mm (<0.25 inches)
BCuP	0.025-0.127	0.001-0.005	Joint length <25.4 mm (1 inch)
	0.178-0.381	0.007-0.015	Joint length >25.4 mm (1 inch)
BAg	0.000-0.050	0.000-0.002	For maximum strength, a press fit of 0.025 mm/mm (0.001 in/in) should be used.
BAu	0.000-0.050	0.000-0.002	For maximum strength, a press fit of 0.025 mm/mm (0.001 in/in) should be used.
BCu	0.000-0.050	0.000-0.002	For maximum strength, a press fit of 0.025 mm/mm (0.001 in/in) should be used.
BCuZn	0.050-0.127	0.002-0.005	
BMg	0.050-0.127	0.002-0.005	
BNi	0.050-0.127	0.002-0.005	
	0.000-0.050	0.000-0.002	Free-flowing type

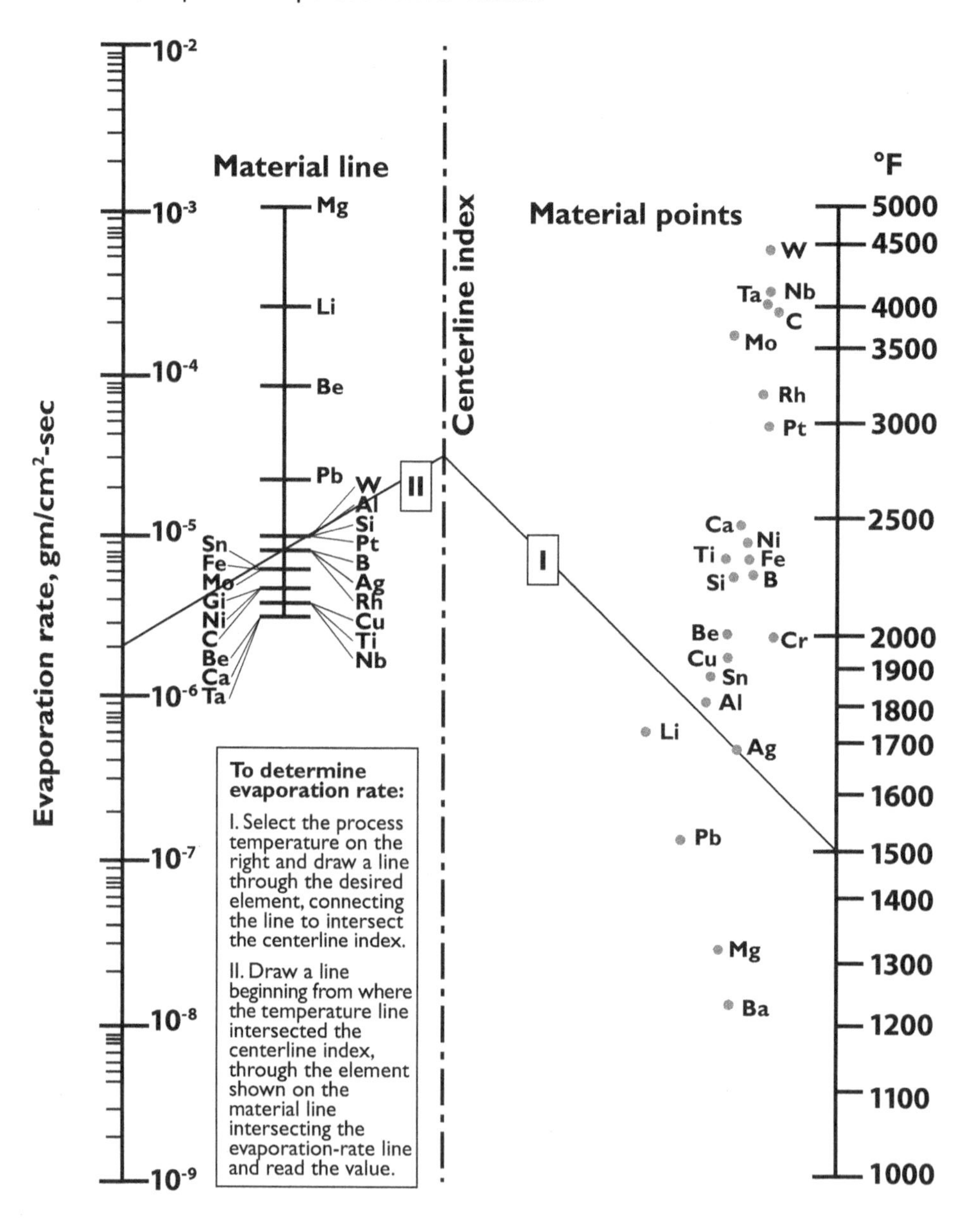

FIGURE 25.5 | Metal evaporation under vacuum [6]

In general, for production work, both temperature and time are kept to a minimum once a good quality part has been produced.

Solidification of the brazed assembly is accomplished by first slow cooling to set or solidify the braze followed by cooling rapidly from brazing temperature. The rate of cooling can create movement of the braze joint prior to solidification and induce stress. In all applications, the brazed parts must be cooled slowly to set the braze prior to rapid cooling.

Brazing Cycles

A properly designed time-temperature vacuum cycle is a critical step in producing a successfully brazed component. A typical brazing cycle consists of an initial pump-down (or a pump-down/backfill to near atmospheric pressure/repump-down), initial heating ramp, preheating/stabilization soak, heating ramp to brazing temperature, soak, initial cooling for braze solidification and final cool-down.

There are a number of factors that influence the development of a brazing cycle. These include such things as base-metal and braze-alloy composition, mass of the assembly and joint design. However, each cycle is comprised of a number of common steps (Fig. 25.6). For relatively easy-to-braze materials, evacuating to a level of 10.7 x 10^{-4} mbar (8 x 10^{-4} torr) is usually sufficient, while the initial pump-down should be below 1.33 x 10^{-4} mbar (1 x 10^{-4} torr) for difficult-to-braze materials (e.g., nickel-based superalloy with appreciable amounts of aluminum and titanium).

A vacuum safety interlock should be programmed into the cycle to ensure these levels are reached. The initial heating ramp should not exceed 10-15°C/min (20-30°F/min). Faster rates are not recommended due to possible part distortion, spalling of the braze paste or slurry, and the likely occurrence of excessive outgassing with subsequent pumping issues. During the initial pump-down, water vapor absorbed by the parts and furnace is driven off.

For difficult-to-braze materials, particularly for those with heavily applied braze alloys, a furnace bake-out cycle following the brazing run is highly recommended as part of a good maintenance program to avoid excessive pressure rise on heating. For most brazing applications, a pump-down before heating to a vacuum level of 1.33 x 10^{-4} mbar (1 x 10^{-4} torr) or better is often recommended.

Heating continues to a temperature at about 25°C (45°F) below the solidus temperature of the braze alloy. If needed, the load is then soaked at this temperature to achieve temperature uniformity and to allow vacuum levels to recover. A soak time of 15-30 minutes is usually sufficient, though the incorporation of a second vacuum safety interlock in the braze-cycle program may be desirable.

The final heating rate to brazing temperature is the most critical. A fast ramp rate in the order of 15-25°C/min (27-45°F/min) is recommended to heat the load to the braze temperature. The rate must be fast enough to prevent liquation of the braze alloy, where lower melting-point constituents begin to melt separately. A faster heating rate also reduces the risk of erosion of the base metal. For thin metals less than 0.25 mm (0.010 inches), heating rates of 28-40°C/min (50-75°F/minute) are essential.

FIGURE 25.6 | Typical vacuum brazing cycles

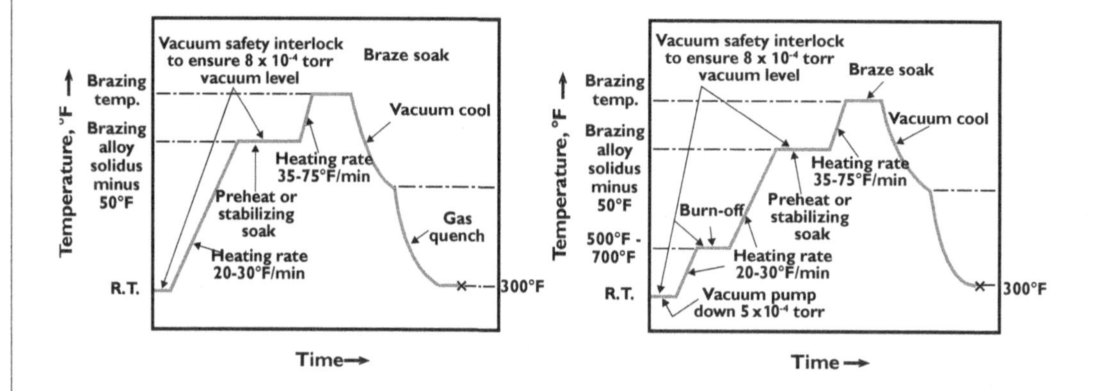

The brazing temperature selected should be the lowest possible within the recommended range. For many braze alloys, the minimum brazing temperature will be at least 25°C (45°F) above the liquidus temperature. Minimum brazing temperatures are essential when using free-flowing braze alloys, when trying to braze large gaps and when brazing thin materials. At these lower temperatures, molten braze alloy will be more sluggish and less reactive with the base metal. The time at brazing temperature should be just long enough to ensure that all sections of a part and all parts within the load reach the desired temperature. This time normally ranges between 5 and 10 minutes, but it may be longer for heavy loads. When the braze soak is complete, the cooling cycle can begin. It is strongly recommended that the load be cooled to a temperature at least 25°C (50°F) below the solidus temperature of the braze alloy before gas quenching is initiated. This will ensure that the molten braze alloy has re-solidified and will not blow away from the joint during the quench.

Despite taking all the necessary precautions, joint imperfections will occasionally occur. Fortunately, re-brazing can often repair defects. Because of diffusion and mixing of constituents between braze alloy and base metal, most braze alloys tend to develop a higher remelt temperature after the initial braze. Rather than attempting to repair the joint defect by remelting the existing joint, it is better to apply a small amount of additional alloy in the defective area. To prevent remelting of the existing joint, a re-braze temperature lower than that used in the first braze is preferable, particularly if wide gap joints are involved. The defective area should be re-inspected for cleanliness before additional braze alloy is applied. The brazing cycle can then be repeated as before with modifications to the brazing temperature.

Brazing Examples

Many different types of nickel, nickel-base, copper, copper-base, gold-base, palladium-base, aluminum-base and silver-base brazing alloys are used for the filler metal. Generally, alloys that contain easily vaporized elements for lowering the melting points are avoided for vacuum brazing. With respect to the heat treatment of steel, the copper and the nickel-base brazing alloys are the most widely used filler metals.

ALUMINUM AND ALUMINUM ALLOYS

In the brazing of aluminum components, it is important that vacuum levels be maintained in the 10^{-5} mbar (10^{-5} torr) range or better. Parts are heated to 575-590°C (1070-1100°F) depending on the alloy. Temperature uniformity is critical, typically ±5°C (±9°F) or better, and multiple-zone temperature-controlled furnaces are always required. Cycle times are dependent on furnace type, part configuration and part fixturing. Longer cycles are required for large parts and very dense loads.

COPPER AND COPPER ALLOYS

Copper-filler metal applied to the base metal either as paste, foil, clad or solid copper can be vacuum brazed, recognizing that the high vapor pressure of copper at its melting point causes some evaporation and undesirable contamination of the furnace internals. To prevent this action, the furnace is first evacuated to a low pressure 10^{-2} to 10^{-4} mbar (10^{-2} to 10^{-4} torr) to remove residual air. The temperature is then raised to approximately 955°C (1750°F) to allow outgassing and to remove any surface contamination. Finally, the furnace is heated to a brazing temperature normally between 1100-1120°C (2000-2050°F) under a partial pressure of inert gas up to 1 mbar (7.5×10^{-1} torr) to inhibit evaporation of the copper.

When brazing is completed, usually within minutes after the brazing temperature has been reached, the work is allowed to slow cool to approximately 980°C (1800°F) so that the filler metal will solidify. Parts can then be rapidly cooled by gas quenching, typically in the range of 2 bar, even when vacuum furnaces capable of higher quenching pressures are being used.

NICKEL-BASED ALLOYS

Brazing with nickel-based alloys is usually done without any partial pressure at the vacuum levels in the range of 10^{-3} to 10^{-5} mbar (10^{-3} to 10^{-5} torr). Normally, a preheat soak at 920-980°C (1700-1800°F) is used to ensure that large workloads are uniformly heated. However, filler metals such as BNi2 start to melt at 970°C (1780°F). After brazing, the furnace temperature can be lowered for additional solution or hardening heat treatments before gas cooling and unloading.

FIGURE 25.7a | Turbine blades brazed in a horizontal vacuum furnace (courtesy of Solar Atmospheres Inc.)

FIGURE 25.7b | Various aerospace components brazed in a bottom-loading vacuum furnace (courtesy of VAC AERO International)

Aerospace Brazing Process Examples

SILVER BRAZING

In development of the scram jet engine, NASA conducted various tests (Fig. 25.8) requiring a silver-brazed (BAg 8 wire) test box brazed in a partial pressure of argon at 800°C (1475°F). Total braze time was 15 hours.

FIGURE 25.8 | Brazed box hoisted into position for testing (courtesy of Solar Atmospheres Inc.)

SUPERALLOY BRAZING

A honeycomb seal (Fig. 25.9) is a jet-engine component designed to increase engine efficiency by surrounding the airfoil or turbine blade and preventing airflow around the blade tips. Honeycomb seals are made from a variety of nickel and cobalt superalloys (Table 25.5) designed to withstand the harsh service environments of jet-engine applications.

FIGURE 25.9 | Typical honeycomb seals

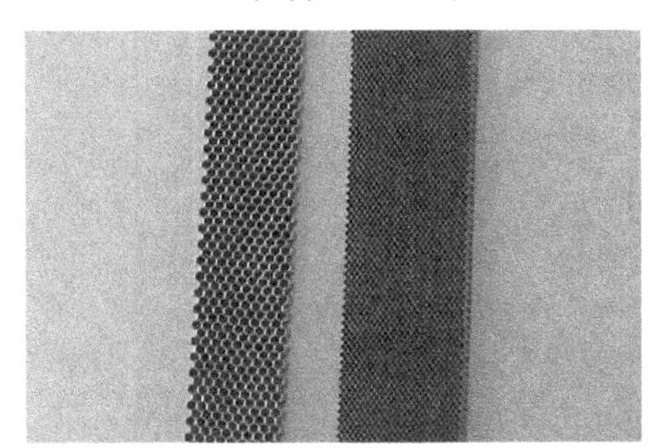

TABLE 25.5 | Honeycomb materials

Component	AMS Specification			Major constituents
Honeycomb	5536	5878		Ni, Cr, Co, Mo
Rings and details	5662	5596	5706	Ni, Cr, Mo, Cb, Ti
Filler metal	4777	4779	4782	Ni, Si, Cr, B

Rolling and tacking are two of the most important steps in the pre-braze assembly process. Firm tack is necessary to ensure intimate contact and achieve a sound joint during brazing. All dimensions are set during the tacking process. Braze tolerances of 0.25-0.50 mm (0.010-0.020 inches) are typical. Proper cleaning is another critical pre-braze step. Every effort should be made to ensure the part is clean and free of all oxides, contaminants and oils prior to braze preparation.

The furnace cycle is just as important as part preparation to the success of the brazing operation. Parts that ramp too fast are at risk for distortion and uneven temperature throughout the assembly. Parts that are not stabilized will not see proper braze flow. If the assembly quenches too rapidly, there is risk for distortion, quench cracking of the braze joint and splatter.

Brazing of these high-temperature nickel alloys is typically performed at 1040-1200°C (1900-2200°F) in a vacuum level of 10^{-4} to 10^{-5} mbar (10^{-4} to 10^{-5} torr). Brazing is performed 40-65°C (72-117°F) above the braze-alloy melting point.

Common problems include splatter of the braze alloy, quench cracking and distortion. All of these problems can be prevented by controlling cleanliness of the part, using proper setup technique, designing a proper brazing recipe and operating the furnace properly. Multiple re-brazes can be performed using shorter brazing cycles at slightly higher temperatures. Stop-off paints such as aluminum oxide (preferred) can be applied to reduce the risk of unwanted braze flow.

SPECIAL ALLOY BRAZING

Reusable rocket engines are a practical necessity facing NASA and other space agencies. Joint technology programs have incorporated Russian technology, most notably rotational compression brazing in combination with channel-wall construction, to enhance nozzle reliability via machine-controlled processes. Special gold-palladium (Au-Pd) alloys (Fig. 25.10) are used as filler metals when joining engine components such as the liner, jacket and stiffeners together.

During the brazing process, a vacuum of 10^{-5} mbar (10^{-5} torr) is pulled on a sealed inner assembly, while a positive pressure of 5-6 bar is applied to the outer surfaces. This pressure differential, in combination with rotating the part during heating and brazing to counteract the forces of gravity on the braze joint, produces a superior part.

FIGURE 25.10 | Section of channel-wall braze joints (Au-Pd alloy) [19]

NICKEL BRAZING

An example of a nickel-brazed assembly is a three-piece spool assembly of Inconel 718 material (Fig 25.11). The diameter is 1,400 mm (55 inches), the overall height is 1,525 mm (60 inches) and the total weight is 7,700 kg (17,000 pounds). The braze was intended to join the top and bottom spools onto the center hub. Filler metal was placed into pre-machined grooves, and the end spools were heated to several hundred degrees prior to assembly. The braze gap was 0.076 mm (0.003 inches) on the diameter with shims used to center the spools onto the hub. Total braze time was 30 hours.

FIGURE 25.11 | Nickel-brazed Inconel assembly (courtesy of Solar Atmospheres Inc.)

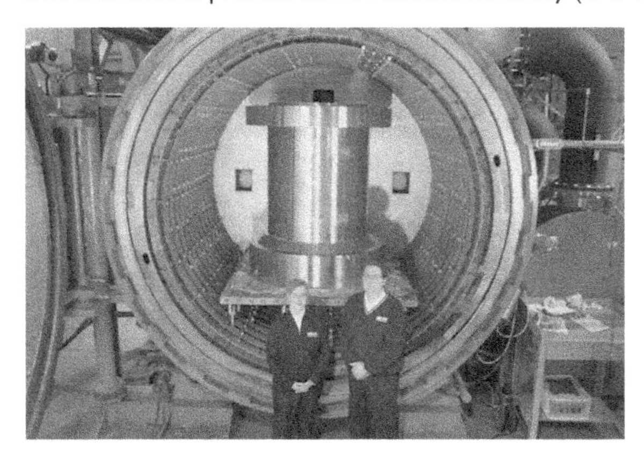

Common Brazing Problems and Solutions

Common brazing imperfections include:

- Voids and porosity
  - Lack of fill can be the result of improper cleaning, extensive clearances, insufficient filler metal, entrapped gases, insufficient temperature, leaks or oxidizing atmospheres, stop-off misplacement, movement of the mating parts and improper fixturing.
  - These imperfections reduce the strength of the joint by reducing the load-carrying area. They may also be a path for leakage.
- Flux entrapment
  - In any brazing operation, there is a danger of entrapped flux preventing flow of the filler metal into a particular area.
  - This problem can create a variety of issues, including corrosion of components with external flux on outside surfaces, and may severely reduce service life.
- Non-continuous fillets
  - This type of imperfection is evidenced by a large void(s) in the fillet.
  - These imperfections may be a path for leakage and create other problems.
- Base-metal erosion
  - This phenomenon is caused by alloying of the filler metal with the base metal during brazing. This results in melting of some of the base-metal constituents, which causes undercuts or the disappearance of mating surfaces.
  - These imperfections compromise joint integrity by reducing the cross-sectional area of the base metal.
- Unsatisfactory surface appearance
  - Excessive flow of the filler metal, roughness, excessive quantity of filler metal or unmelted filler metal (i.e. skulls) are common reasons for poor surface appearance.
  - In addition to aesthetic considerations, these imperfections may act as stress concentrators or corrosion sites.
- Cracks
  - These may reduce strength and service-life expectancy. They may act as stress raisers, causing premature fatigue failures as well as lowering the mechanical strength of the braze.
  - Cracks in braze joints should be considered defects, requiring repairs or in severe cases, scrapping of the component part.

Some metals and alloys also exhibit metallurgical phenomena that influence the performance of the brazed joint and may necessitate special procedures. These phenomena can be classified as:

- Base-metal effects
  - Carbide precipitation
  - Oxide stability
  - Hydrogen embrittlement
  - Heat-affected zone
  - Sulfur embrittlement
- Filler-metal effects
  - Vapor pressure
- Base-metal/filler-metal interactions
  - Alloying
  - Phosphorous embrittlement
  - Stress cracking

We will briefly consider some of these problems.

CARBIDE PRECIPITATION

This is a concern, primarily when brazing stainless steel components. If stainless steels are heated to temperatures between 425-815°C (800-1500°F), the carbon in the base metal combines with chromium to form chromium carbides, usually at the grain boundaries. This chromium depletion affects the corrosion resistance of the stainless steel. In addition, if the cooling rate is too slow from brazing temperature, carbide precipitation will occur.

A short brazing cycle will keep the chromium carbide precipitation to a negligible level in most stainless steels. When this is not possible, one of the special grades of stainless steel may have to be used or a special heat-treating cycle to redissolve the carbides may have to be considered. If (due to the mass of the part) rapid cooling is not possible, one of the stabilized stainless steel compositions or an extra-low carbon grade should be used.

HYDROGEN EMBRITTLEMENT

Hydrogen, if used as a partial-pressure or backfill gas, can be a source of trouble. Hydrogen can diffuse rapidly into many metals, and the rate of diffusion increases with temperature. There are three main types of hydrogen embrittlement: water-vapor formation, void entrapment and hydrides. We will review each of these phenomena.

If oxygen is present in the base metals being joined, the hydrogen may combine with it to produce water vapor. This will not diffuse out of the metal, and the results are stresses that can quite literally tear or create cracks, typically at the grain boundaries. This form of embrittlement is common in some copper and copper-based alloys, silver and palladium. The use of oxygen-free copper, or deoxidized copper, is one solution to this problem. If electrolytic tough-pitch copper must be brazed, the atmosphere should be free of hydrogen.

Another form of hydrogen embrittlement occurs in steel and certain ferrous alloys when the diffusing hydrogen accumulates in small voids, such as areas around non-metallic inclusions and at the grain boundaries. Water vapor is not formed, but the increased concentration of molecular hydrogen can lower the ductility, especially in high-stress applications. The solution is to perform a hydrogen bake-out procedure by heating to 95-205°C (200-400°F) and soaking for an extended period (24 hours at temperature is recommended) until the ductility is regained.

The third type of embrittlement is due to the formation of metal hydrides. These reduce notch toughness and affect the strain rate of the material. Titanium, zirconium, niobium, tantalum and their alloys are subject to this type of phenomena.

Avoidance of hydrogen in the vacuum furnace will prevent the forms of embrittlement due to hydrogen.

HEAT-AFFECTED ZONE

The heating of base metals may cause changes to their properties. The extent of these changes will vary with the process used. In vacuum-furnace brazing, it is important to remember that the entire assembly will be affected by the heating rate.

This requires the designer to pay close attention to how the base-metal properties were obtained and what is required after brazing. If mechanical properties were obtained by cold working during brazing, for example, they may undergo softening and an increase in grain size. If the mechanical properties were obtained by heat treatment, these may be altered by brazing. Materials in the annealed condition will generally experience no appreciable change due to brazing.

OXIDES

When metals are heated and oxygen is present either from a furnace leak or from being entrained in the partial-pressure or backfill gas, surfaces may form oxides. Oxidized surfaces are usually difficult to wet, and they act as barriers to BFM flow. Once they are formed, oxides are difficult to remove. In the case of stainless steels, a transparent oxide film exists on the part at room temperature. In certain applications, it is necessary to run individual components separately through a highly reducing atmosphere before brazing to minimize this oxide layer. This is followed almost immediately by assembly and brazing. Oxide formation can be prevented by brazing in vacuum or in a low dew-point atmosphere.

SULFUR EMBRITTLEMENT

If heated in the presence of sulfur or sulfur compounds, nickel and certain alloys containing appreciable amounts of nickel either form a low-melting eutectic or precipitate nickel sulfides at the grain boundaries, which significantly weakens the material or causes cracks to occur, particularly if the material is stressed.

It is imperative to clean all sulfur-containing compounds from the surface of nickel-bearing parts prior to brazing.

VAPOR PRESSURE

At room temperature, the vapor pressure of most materials is so small it can be considered non-existent. As the temperature increases and/or the pressure is reduced (such as in the case of vacuum brazing), however, a material with a relatively high vapor pressure will boil or volatilize gases at normal brazing temperatures. This includes constituents in the BFM.

In the case of vacuum brazing, special vacuum-grade filler metals are commercially available and designed to minimize these de-alloying effects.

An example of a material that causes problems in brazing is zinc and zinc alloys. Even at atmospheric pressure, "dezincification" (i.e. the evolution of zinc vapors when parts are heated to brazing temperatures) occurs. Cadmium and cadmium alloys have similar characteristics. Therefore, these materials are unsuitable for vacuum brazing.

BASE-METAL/FILLER-METAL INTERACTIONS, ALLOYING

Some detrimental effects may occur due to the melting of the BFM and the interaction with the base metal. These include:

- Formation of non-ductile intermetallic compounds that may lower joint strength
- Diffusion of the filler metal into the base metal, producing color changes
- Creation of new alloys with higher melting points that cause the flow of the original filler metal to cease

These effects can be minimized by proper selection and consideration of both the base-metal and the filler-metal alloy. Whenever excessive base-metal dissolution and diffusion are likely to occur, brazing should be done in the shortest time and at the lowest possible temperature. Sufficient BFM should be present to fill the joint completely, but excessive filler metal is both undesirable and uneconomical.

PHOSPHOROUS EMBRITTLEMENT

Phosphorous combines with many metals to form hard, non-ductile compounds known as phosphides. In general, copper-phosphorous filler metals are not used with iron- and nickel-based alloys.

STRESS CRACKING

In a highly stressed condition, many high-strength materials (e.g., stainless steels and nickel alloys) may display a tendency to crack during brazing. Materials with high annealing temperatures – particularly those that are age hardened – are susceptible to these phenomena. Stresses also occur during cooling from brazing temperature in the form of cracks in the braze joint area. These stresses will weaken this area or cause parts to leak or fail in service.

When stress is encountered, the usual remedy is to remove the source of stress. This can be done by:

- Using annealed rather than hardened material
- Annealing cold-worked parts prior to brazing
- Remove the source of externally applied stress
  - Improper fit
  - Unsupported weight
  - Poorly designed jigs
- Redesigning parts or revising joint design
- Heating and cooling at a lower rate
- Selecting the most compatible BFM

In Conclusion

Vacuum brazing is a complex process capable of producing high-quality, repeatable results. The problems associated with vacuum brazing are, in general, well understood by the industry. They can be overcome by the design engineer working closely with the brazing operation to understand the particular process and equipment variability that exists. Material selection and application are critical to success.

REFERENCES

1. Dan Kay, Kay & Associates, technical and editorial review
2. *Brazing Handbook, 4th Edition*, American Welding Society, 1991 and 2007
3. *Fundamentals of Brazing*, American Machinist, April 1981

4. *Brazing Manual,* American Welding Society, 1976
5. "Designing with Refractory Metals," white paper, Rembar
6. Kowalewski, Janusz and Janusz Szczurek, "Issues in Vacuum Brazing," *Heat Treating Progress,* May/June 2006
7. Herring, D.H., "Technology Trends in Vacuum Heat Treating, Part Two: Processes and Applications," *Industrial Heating,* November 2008
8. *Vacuum Brazing & Heat Treating Training Manual,* VAC AERO International
9. Herring, D.H., "Fundamentals of Brazing," white paper
10. *A Complete Guide to Successful Silver Brazing,* Wolerine Industries
11. *Brazing,* Air Products & Chemicals Co.
12. Carey, Howard B., *Modern Welding Technology,* Prentice-Hall
13. Kay, Dan, "Ten Reasons to Select Brazing," *AWS Welding Journal,* September 2002, p. 33-35
14. *Metals Handbook, 9th Edition,* Volume 6, "Welding, Brazing, and Soldering," ASM International, 1983
15. *The Brazing Book,* Handy & Harmon
16. Kay, Dan, "Liquation of Brazing Filler Metals, " *Industrial Heating,* November 2009
17. Schwartz, Mel, *Brazing,* ASM International, 1987
18. Peaslee, Robert L., *Brazing Footprints: Case Studies in High-Temperature Brazing,* Wall Colmonoy, 2003
19. Herring, D.H., "Understanding the Differences Between Soldering, Brazing and Welding," *Heat Treating Progress,* June/July 2003
20. Mark Bulaw, Advanced Thermal Processing, editorial review

CHAPTER 26

VACUUM APPLICATIONS: LPC AND OTHER CASE-HARDENING METHODS

Carburizing of a component surface is a function of the rate of carbon absorption at the surface and the diffusion of carbon away from the surface and into the material. While carburization can be performed at a fixed or variable carbon potential, the latter technique is most commonly associated with low-pressure (vacuum) carburizing. Once a high concentration of carbon has developed on the surface (during what is commonly called the boost stage), the process normally introduces a diffuse stage whereby the surface carbon concentration is reduced by diffusion into the interior. The result is a reduction of the carbon concentration at the surface while the depth of carbon absorption increases.

In the carburization process, residual compressive stresses result from the delayed transformation and volume expansion of the carbon-enriched surface. This induces a highly desirable stress state throughout the case-hardened layer.

Why vacuum carburizing?

Companies are interested in using vacuum carburizing due in part to considerations involving:

- Engineering designs focused on enhancing performance while reducing the total package size, weight and unit cost
- Metallurgical factors to enhance property development (mechanical, physical, metallurgical) from existing or new materials

- Engineering solutions requiring extending service-duty and/or enhanced product performance

Historical Perspective

The history of vacuum carburizing is a fascinating one. The process was invented in late 1968 and subsequently patented (U.S. Patent No. 3,796,615, U.S. Patent RE 29,881) by Herbert W. Westeren, director of research and development for C.I. Hayes, Inc. of Cranston, R.I. The process was commercialized in early 1969 (Fig. 26.1). Full acceptance of the process, however, involved three decades of work and contributions from all over the world, including the discovery and patenting of acetylene technology in the former Soviet Union (USSR Patent No. 668978) by V.S. Krylov, V.A. Yumatov and V.V. Kurbatov in 1977. It culminated with the application and patenting (U.S. Patent No. 5,702,540) of acetylene-based carburizing by K. Kubota of Japan's JH Corporation (formerly Japan Hayes Corporation). Since that time, a significant number of individuals and companies have made patentable inventions that have helped advance the technology.[1]

Developmental efforts involved areas such as:

- Improvements in the design and construction of vacuum furnaces
- Use of low-pressure carburizing methods
- Process optimization, particularly the selection of hydrocarbon gas
- Optimized gas-injection methods and flow/pressure controls
- Creation of empirical databases and design of process simulators
- Development of high-pressure gas-quenching technology
- Availability of low-cost carburizing alloys specifically designed to take advantage of vacuum carburizing and gas quenching (including high-temperature capability)

Modern-day carburizing installations (Fig. 26.2) have taken advantage of these developmental efforts.

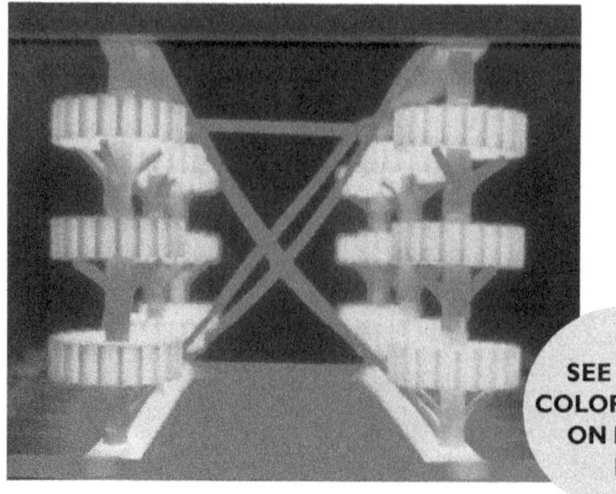

FIGURE 26.1 | First commercial heat-treat load, February 1969: gears carburized at 930°C (1700°F), 13 mbar (10 torr), methane (CH_4) (courtesy of C.I. Hayes)

SEE FULL COLOR IMAGE ON PAGE D

FIGURE 26.2 | Present-day commercial heat-treat load, February 2012: gears carburized at 930°C (1700°F), 14.5 mbar (11 torr), acetylene (C_2H_2) (courtesy of ALD Thermal Treatment)

LOW-PRESSURE (VACUUM) CARBURIZING METHODS

It is generally agreed that low-pressure carburizing (LPC) occurs at pressures of 27 mbar (20 torr) or less, typically at temperatures between 830-980°C (1525-1800°F) for carburizing and 800-900°C (1475-1650°F) for carbonitriding. In the past several years, higher carburizing temperatures, up to 1200°C (2200°F) in several instances, have been used for certain advanced materials.

Vacuum carburizing is a method of so-called pure carburizing and pure diffusion in that carbon is allowed to penetrate into the surface being processed without interference from external influences such as gas chemistry interactions or surface contaminants.

Vacuum carburizing is a modified form of the gas carburizing process, in which the carburizing is done at pressures far below atmospheric pressure, which is approximately 1013 mbar (760 torr). The typical pressure range for LPC is 1.3-26.7 mbar (1-20 torr).

This method of carburization takes advantage of a clean steel surface, and the vacuum environment allows faster transfer of carbon to the steel surface (i.e. higher carbon flux values) since atmosphere carburizing interactions such as those found in the water gas reaction do not take place. In addition, no intergranular oxidation can occur if the correct hydrocarbon gas is used.

The carbon produced by the breakdown of the hydrocarbon gas introduced into the chamber is free to penetrate into the surface of the steel, while the hydrogen and residual hydrocarbon by-products are removed from the system by the vacuum pump(s).

The hydrocarbons currently being used in vacuum carburizing (Table 26.1) are: acetylene (C_2H_4), propane (C_3H_8), methane (CH_4), ethylene (C_2H_4) and cyclohexane (C_6H_{12}), a liquid either solely or in combination and with or without dilution by hydrogen (H_2) or nitrogen (N_2).

TABLE 26.1 | Hydrocarbon combinations for low-pressure (vacuum) carburizing

Primary gas species	Mixture types
Acetylene	100% [a]
	Acetylene + Nitrogen [b]
	Acetylene + Hydrogen [c]
	Acetylene + Ethylene + Hydrogen [d]
Cyclohexane	100%
	Cyclohexane + Acetylene
Methane	100% [e]
	Methane + Propane [f]
Propane	100%
	Propane + Hydrogen or Propane + Nitrogen
	Propane + Methane [f]

Notes:

[a] DMF acetylene (without acetone) preferred (Fig. 26.3), though not mandatory

[b] Typical dilutions up to 50%

[c] Typical dilution 7:1 (U.S. Patent 7,514,035, Solar Atmospheres Inc.)

[d] Typically, ratios of acetylene to ethylene to hydrogen are 3:2:1 or 2:2:1 (U.S. Patent 7,550,049, SECO/WARWICK Corporation)

[e] Temperatures above 955°C (1750°F) recommended unless plasma assisted

[f] Typical dilution: 40/60 to 60/40 (methane/propane)

FIGURE 26.3 | Typical trailer setup (DMF acetylene)

In vacuum carburizing, the breakdown of hydrocarbon gases involves non-equilibrium reactions. This means that the surface of the steel is very rapidly raised to the saturation level of carbon in austenite. Repeated boost and diffuse steps (Fig. 26.4) achieve the desired carbon profile and case depth.

FIGURE 26.4 | Schematic representation of time/temperature and boost/diffuse process for LPC and gas pressure quenching

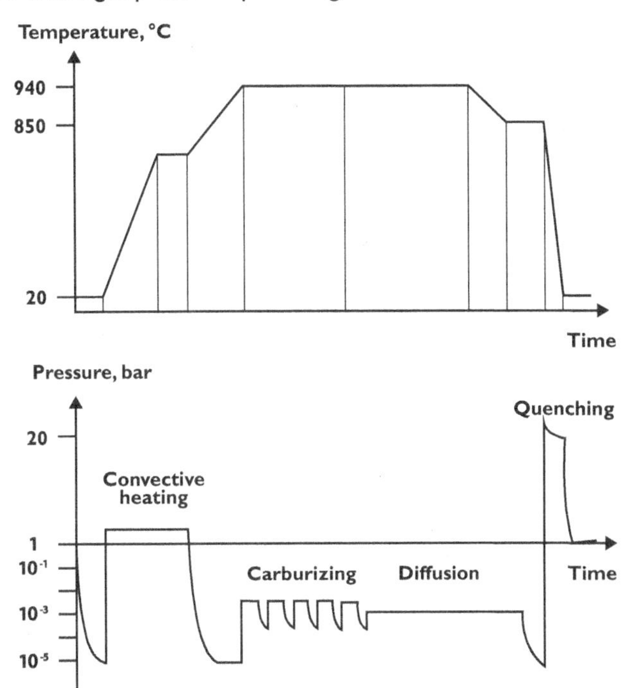

Depending on the type of hydrocarbon gas used, carbon is delivered to the surface via pyrolysis reactions (Equations 26.1-26.3). The various hydrocarbons are either catalytically or thermally decomposable into their constituent gases.

26.1) $C_2H_2 \rightarrow 2C + H_2$

26.2a) $C_3H_8 \rightarrow C + 2CH4$

26.2b) $C_3H_8 \rightarrow C_2H_4 + CH_4 \rightarrow C + 2CH_4$

26.2c) $C_3H_8 \rightarrow C_2H_2 + H_2 + CH_4 \rightarrow C + 2CH_4$

26.3) $C_2H_4 \rightarrow C + CH_4$

Process control is achieved through the use of simulation programs applied to the respective kinetic and diffusion models to determine the boost and diffuse times for a given case depth. Carbon transfer rates have been established as a function of temperature, gas type, gas pressure and flow rate. Material chemistry and surface area must be taken into consideration in these programs as well as initial and final surface carbon levels. Prediction of case depth and hardness profiles is the most obvious output of these programs. Research continues into prediction of microstructural results such as carbide size and distribution and retained-austenite levels.

CONTROL OF CASE DEPTH [2]

LPC is a recipe-controlled process. By contrast, atmospheric gas carburizing is controlled via carbon potential. In vacuum carburizing, process-related parameters such as temperature, carburizing-gas flow, time and pressure are adjusted and controlled to achieve a certain case profile in the parts.

One method of recipe development involves solving the following three equations:

26.4) $D = k\sqrt{t}$

26.5) $c = r \cdot t$

26.6) $t = c + d$

where:

D = effective case depth (50 HRC equivalent)
k = carburizing constant (Table 26.2)
t = total time, in hours
c = carburizing time, in hours
r = boost/diffuse ratio (Table 26.2)
d = diffusion time, in hours

TABLE 26.2 | Carburizing parameters

Carburizing constant			Boost/diffuse ratio
Temperature,°C (°F)	k value		r value
	D (mm)	D (inches)	
840 (1550)	0.25	0.010	0.75
870 (1600)	0.33	0.013	0.65
900 (1650)	0.41	0.016	0.55
930 (1700)	0.51	0.020	0.50
950 (1750)	0.64	0.025	0.45
980 (1800)	0.76	0.030	0.40
1010 (1850)	0.89	0.035	0.35
1040 (1900)	1.02	0.040	0.30

While this method is still in use, recipe development by means of a simulation program dominates the industry. The result is a recipe comprising a sequence of carburizing and diffusion pulses, in which carbon profile as a function of depth can be predicted. In addition, work is under way to use the software to predict such parameters as hardness values and residual-stress levels versus depth below the surface.

Simulation programs are available from any number of suppliers of vacuum carburizing equipment and are designed to create recipes and test scenarios for process development. These programs are based on a mathematical description of the carbon dissociation and adsorption of the carbon at the surface of the parts and equations, which describe the diffusion of the carbon into the material. While the carbon transport to the surface in LPC differs significantly from that in atmospheric gas carburizing, the same diffusion laws apply for the carbon transport within the material.

Input parameters of the software are material, carburizing temperature, targeted carburizing depth, targeted surface carbon content, surface carbon-content limit and load surface area. The simulation program can be used in two modes. The first mode allows one to enter a desired surface carbon content and case depth. Then the program calculates the required recipe consisting of different carburizing and diffusion pulses (Fig. 26.5). Furthermore, the program shows diagrams of carbon flux versus time and carbon profile versus time and distance from surface.

FIGURE 26.5 | Screen shot – LPC simulation program (courtesy of ALD Vacuum Technologies GmbH)

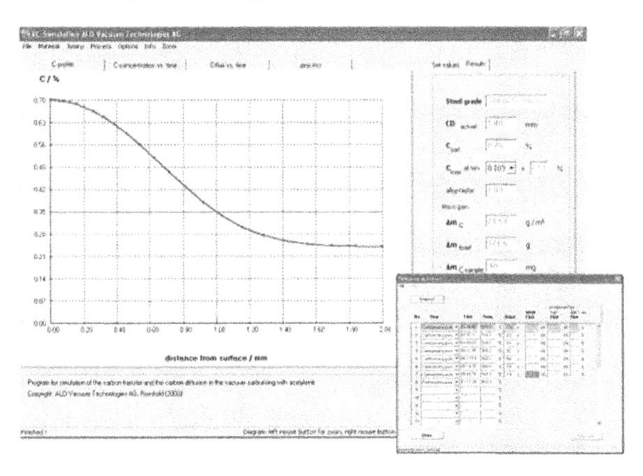

The second mode allows one to enter a certain recipe consisting of carburizing and diffusion pulses, and the program calculates the resulting carburizing profile. The simulation program is preferably used in the first mode. As shown in Fig. 26.6, the simulation is quite accurate. The simulated carbon profile is very similar to the measured carbon profile. Some simulation programs also include a quenching module for calculation of the case-hardening depth instead of the carburizing depth.

FIGURE 26.6 | Comparison of simulated and measured carbon potential for 20MnCr5 (~SAE 5120) material (courtesy of ALD Vacuum Technologies GmbH)

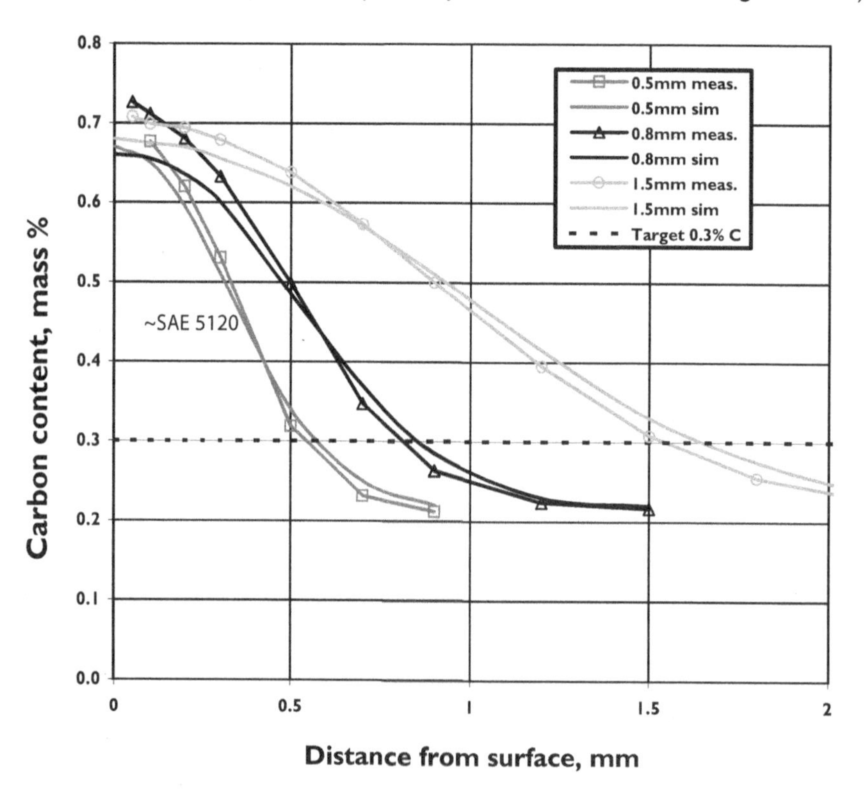

The use of a simulation program to create a carburizing recipe is well accepted for LPC applications in all manufacturing industries, including automotive and aerospace. Using these simulators, a test load is run and the parts are checked for correct case profile. The achieved case profile is normally within the specified range. If not, the parameters are adjusted slightly and a second cycle is run. Since the LPC process offers consistent case uniformity, it can be advantageous to perform additional simulation runs so as to adjust the case depth toward the lower end of the case specification. This is done to reduce the overall process time, thereby increasing the productivity of the system. As soon as the required profile is achieved, the recipe is fixed and not changed.

Furthermore, the simulation program is a powerful tool to achieve the desired microstructure after vacuum carburizing. The program shows the formation of the carbon profile as a function of time for different surface distances. Therefore, it is possible to create recipes that meet microstructural specification (i.e. the absence of large quantities of carbides or excessive amounts of retained austenite).

A further result of the simulation program is the calculation of the correct hydrocarbon flux, which depends on the actual load surface area and the carbon yield of the carburizing gas that is used. The carbon yield defines the amount of carbon transferred into the parts in relation to the amount of carbon supplied to the treatment chamber by injecting the carburizing gas. For acetylene, the carbon yield is in the range of 60-80%. This tool allows the user to reduce the amount of carburizing gas to a minimum, which is beneficial both economically and ecologically.

As the chemistry of the materials to be carburized has an influence on the carburizing profile to be achieved, it is possible for a user to enter the exact chemical composition of their own material into the program and store this data in the existing material data bank.

There have been numerous attempts to install sensors to further improve control of the LPC process. For this purpose, hydrogen or oxygen sensors that are placed in the hot zone or in the exhaust-gas pipe have been proposed. They measure the gas composition during the process in an attempt to control the carburizing process. To date, none of these sensors are used in production environments.

The following primary principles of atmosphere gas carburizing apply.

- Carburizing is performed within the austenitizing range for steels.
- The rate of carburizing increases with increasing temperature because the process is based upon diffusion of carbon within austenite.
- A suitable hydrocarbon is necessary to provide a source of carbon.

Vacuum processing minimizes or eliminates the formation of intergranular oxide (IGO), which, for example, decreases the bending fatigue life of atmosphere gas carburized gearing with unground tooth roots.

Surface preparation is very important for consistent and repeatable vacuum and plasma carburizing. Parts must be clean and free of oils, greases, dirt and rust. Cleaning may involve the use of aqueous, semi-aqueous or solvent-based chemistry, and it involves the proper application of time, temperature and energy. Surfaces that will be carburized may be abrasively blasted using garnet, aluminum oxide or other suitable blasting media. Glass beads are not recommended for cleaning prior to carburizing due to the difficulty of removing residue. Clean parts should be handled with clean, lint-free gloves to prevent unwanted surface contamination prior to carburizing.

PROCESS DESCRIPTION

LPC has been shown to improve the uniformity of carburizing, improve furnace cleanliness and reduce furnace maintenance. Vacuum carburizing of gears, for example, is normally performed at temperatures between 870-950°C (1660-1750°F). When approved by the customer, high-temperature vacuum carburizing above 950°C (1750°F) may be used.

Prior to the onset of carburization, the workload is heated to a preset temperature, usually at a controlled ramp rate in the neighborhood of 14-16°C/min (25-30°F/min). Thin, delicate parts will benefit from slower ramp rates and/or soaking at an intermediate temperature, typically below the lower critical temperature (on heating) of the material. Some processes utilize convection heating by introducing and circulating nitrogen at a slightly positive pressure of 0.5-2.0 bar (375-1,500 torr) to aid heat transfer and shorten heating times. Other processes introduce hydrogen during the heating stage to assist in surface conditioning or introduce ammonia to add nitrogen into the steel to limit grain growth prior to high-temperature carburizing.

After the workload reaches austenitizing temperature, it is held in the vacuum environment so as to achieve temperature uniformity throughout each individual part and the load. The carburizing cycle begins when a suitable hydrocarbon is introduced. The hydrocarbon dissociates, providing a source of carbon. The near surface becomes saturated with carbon to the limit of solubility of carbon in austenite. Planned interruptions in the addition of carburizing gas are necessary to allow carbon to diffuse into the steel. Therefore, vacuum carburizing is a boost/diffuse-type process. As with atmosphere gas carburizing, the rate of diffusion is temperature-dependent.

A typical vacuum carburizing cycle will balance the addition of carburizing gas with appropriate diffusion time to allow control of the microstructure (retained austenite and carbide formation). The required flow of the hydrocarbon source is dependent on the surface area of the load. In addition, while process uniformity is good, pulsing of the carburizing gas is often used to improve carburizing coverage (e.g., deep holes, splines, etc.). After completion of the carburizing-gas-addition portion of the process, a final diffusion step is normally performed to achieve the desired case depth and carbon profile prior to quenching in oil or high-pressure gas.

Case depth and hardness profiles can be accurately predicted by simulation software, but the final balance between carburizing-gas addition and boost/diffusion time is often determined by experience to assure that the desired microstructure is obtained. Numerous studies have been performed comparing vacuum and atmosphere processing,[3, 7, 12] and the results indicate vacuum carburizing is metallurgically superior (Fig. 26.7).

FIGURE 26.7 | Pitch line and root comparison of atmosphere and vacuum-carburized gears: a.) vacuum carburized and oil quenched – b.) atmosphere carburized and oil quenched

a.)

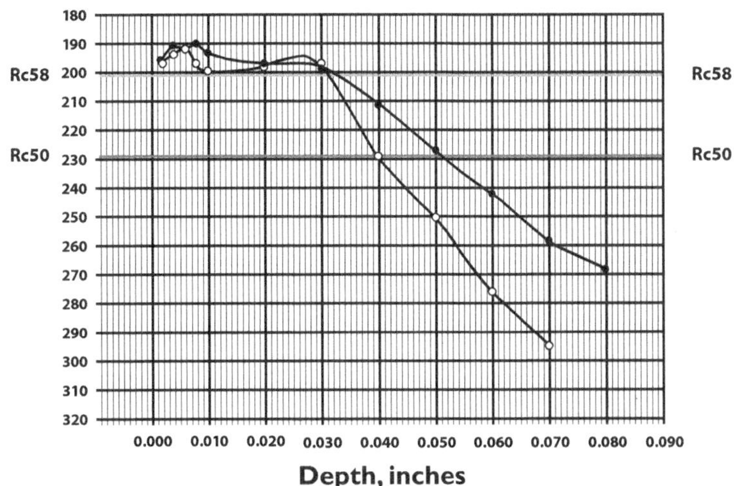

b.)

Effective case-depth comparison, oil quenched

180
190
Rc58 200
210
220
Rc50 230
240
250
260
270
280
290
300
310
320
Rc58
Rc50
0.000 0.010 0.020 0.030 0.040 0.050 0.060 0.070 0.080 0.090

Depth, inches

Atmosphere carburized at pitch line
Atmosphere carburized at root
Depth of Rc50
Depth of Rc58

LPC results show carburization to an effective case depth (50 HRC) of 0.76 mm (0.030 inches) in the root and 1.33 mm (0.052 inches) at the pitch diameter. Also of significance is the value for the depth of high hardness (≥58 HRC), namely 0.35 mm (0.014 inches) at both the gear-tooth pitch line and root. The hardness values quickly diverge from this depth. These results are typical of the vast majority of carburized gears currently in service.

PROCESS LIMITATIONS

Natural gas, since it is not 100% methane, should not be used in vacuum carburizing because its composition includes oxygen and saturated hydrocarbons at unacceptable levels. In addition, some older vacuum carburizing furnaces operating at pressures above 27 mbar (20 torr range) have a propensity to form soot and, in some cases, tar depositing on cold surfaces. These are by-products of the hydrocarbon gas choice and the vacuum carburizing process parameters. In these instances, rigorous maintenance is necessary to assure proper furnace operation.

EQUIPMENT AND CONTROLS

Instrumentation should be in place to record and control process variables, including time, temperature, gas pressure and gas volume. It is important to recognize that process parameters are affected by the surface area of the load. Furnaces should be capable of maintaining temperature uniformity of ±5.5°C (±10°F). The hydrocarbon source should be selected to match the furnace operating design, and the highest available grade of hydrocarbon should be used.

Examples of commonly carburized gear steels include AISI 1018, 3310, 4320, 5120, 8620 and 9310; and Pyrowear™ 53 and 675. Newer carburizing grades include XD15NW, X13VDW and Ferrium™ C69 to name a few (c.f. Chapter 28). European grades include 16MnCr5, 16MnCrB5, 18MnCrB5, 18NiCrMo5, 18MnCrMoB5, 20MnCr5, 20MoCr4, 20CrMo2, 20NiCrMo2, 27MnCr5, 27CrMo4 and Jomasco™ 23MnCrMo5 as well as specialty grades.

PROBLEM SOLVING

The problems associated with vacuum carburizing are well known. As such, corrective actions have been identified and implemented for most of the issues encountered (Table 26.3).

TABLE 26.3 | LPC problem-solving matrix

Problem \ Cause	Temperature too high	Temperature too low	Time (carburizing) too short	Time (carburizing) too long	Oxygen present in carburizing gas	Dwell time of carburizing gas too short	More carburize/diffuse steps required	Carburizing pressure too low	Load too dense	Insufficient pre-soak time	Non-uniform heat distribution	Carburizing gas flow too high	Carburizing gas flow too low	Unstable hydrocarbon gas	Carburizing gas pressure too high	Wrong gas used below temp threshold	Circulation inadequate	Parts have surface contamination	Surface carbon too high	Surface carbon too low	Alloy contains austenite formers	Cryogenic treatment recommended	Ammonia flow too high	Air leak in quench	Air leak in backfill gas stream	Contamination in quench
Case too shallow		x	x																							
Case too deep	x			x																						
Case uneven					x	x	x	x	x																	
Load non-uniform – case too deep						x	x	x	x	x	x	x														
Load non-uniform – case too shallow							x		x		x		x													
Load non-uniform (center)							x	x	x	x			x													
Soot present					x							x		x	x	x	x	x								
High % retained austenite	x																		x		x					
Soft spots (case not hardened)		x											x					x		x		x			x	x
Pitting - microscopic (carbonitrided parts)																						x				
Parts not bright		x																x					x	x	x	x

Plasma Carburizing Methods

Plasma (ion) carburizing processes are performed in specialized vacuum furnaces operating under a partial pressure (i.e. less than atmospheric pressure). Vacuum carburizing can be performed within a pressure range from slightly below atmospheric pressure, approximately 800 mbar (600 torr), to significantly below atmospheric pressure, typically less than 27 mbar (20 torr). By contrast, plasma carburizing is typically performed at lower pressures, approximately 0.133-13.33 mbar (0.1-10 torr), and the equipment employs a DC power supply to generate the plasma. The choice of plasma carburizing technology depends on a number of factors, including selective carburizing requirements, part geometry, load configuration, etc.

PROCESS DESCRIPTION

Plasma carburizing (Fig. 26.8) is a process that takes place in an ionized gaseous environment within a specialized vacuum furnace and, as previously stated, typically operates in the 0.133-13.33 mbar (0.1-10 torr) pressure range. The plasma is referred to as a glow discharge. In the process, the workload is heated by the plasma or by external means. Prior to carburizing, cleaning (i.e. sputtering) of the surface occurs due to the interaction of the part surface with the activated plasma. Hydrogen is commonly added during this step to aid the cleaning process. The length of time required for sputtering is a function of the cleanliness of the parts and fixtures being introduced. This cleaning step aids the removal of contaminants, such as oxides, that may inhibit carburizing.

FIGURE 26.8 | Plasma-carburized workload (courtesy of Surface Combustion Inc.)

Two plasma carburizing methods are in common use today. One uses methane as the source of carbon, while the other relies on propane (or propane diluted with hydrogen gas). The carbon-transfer characteristics of the two gases differ. Dilution gases may be argon, hydrogen or nitrogen (in some cases). The rate of attraction may be adjusted by varying process parameters such as time, temperature, gas pressure, current density or other plasma conditions. These parameters are adjustable to better carburize recessed areas and deep holes.

Plasma carburizing temperatures are normally in the range of 850-1090°C (1550-2000°F). Only surfaces that come in contact with the plasma glow discharge will be carburized. For this reason, mechanical (physical) masks may be used in plasma carburizing. Copper plating is not a suitable mask for the plasma carburizing process because the copper will be ionized.

PROCESS LIMITATIONS

The calculation of surface area is essential to proper process control in plasma carburizing. In addition, the boost/diffuse relationship must be developed for each of the carburizing-gas types. System cost, in relationship to vacuum carburizing, may also be a factor in decisions to use this technology.

Gases may be added by either continuous flow or pulsing to change the furnace pressure. Carburizing gas is introduced with the vacuum evacuation valve closed, causing an increase in pressure. When gas is introduced into the vacuum furnace, the gas rapidly expands within the vacuum chamber around the workload. After a pressure setpoint or flow time has been achieved, the gas valve is closed and the vacuum valve reopens to evacuate the furnace to either a pressure setpoint or for a set time period. The gas addition/gas evacuation cycle repeats until the desired case depth has been achieved. The continuous-flow method of carburizing introduces an inert gas such as nitrogen or argon into the chamber during the diffusion step.

The furnace is designed to create an anode/cathode relationship between the furnace and workload. A regulated power supply is used to establish an electrical potential between the anode (furnace) and the cathode (workload). When a carburizing gas is introduced into the furnace, a hydrocarbon-rich plasma is created, and that plasma is attracted to the workload. Carbon transfer with propane is so high that the carbon content of the workpiece quickly reaches the limit of saturation of carbon in austenite. A series of short boost/diffuse cycles are used to develop the required case depth and surface carbon level. In contrast, plasma carburizing with methane is typically done in a single boost/diffuse cycle, where the boost portion is about one-third of total cycle time.

EQUIPMENT AND CONTROLS

In plasma furnaces, a power supply independent of the power supply for heating the furnace is necessary for generating the glow discharge. Current density is used to control the amount of carbon available for carburizing. The area of all-metallic surfaces (baskets, screens, parts and masking materials) must be included in the calculation for required current density. It is advantageous to run standard loads or to provide an easy method of surface-area calculation.

Vacuum Carbonitriding

Carbonitriding is a modified carburizing process, not a form of nitriding. This modification consists of introducing ammonia into the carburizing atmosphere in order to add nitrogen to the carburized case as it is being produced. Carbonitriding is a surface-hardening treatment that introduces carbon and nitrogen into steel above the austenitizing temperature (Ac_3). A martensitic case is achieved upon quenching. The hardness of the case is dependent on the carbon and nitrogen concentration of the case.

Carbonitriding is done at a lower temperature than carburizing, typically between 790-900°C (1450-1650°F), and for a shorter time. Combine this with the fact that nitrogen inhibits the diffusion of carbon, and the result is a shallower case than is typical for carburized parts. A carbonitrided case is usually between 0.075-0.75 mm (0.003-0.030 inches) deep. The process creates a desirable epsilon-nitride phase that improves case hardness and provides excellent wear and anti-scuffing characteristics.

It is important to note that a common contributor to non-uniform case depths is to begin the ammonia addition before the load is stabilized at temperature. This mistake often occurs in furnaces that start gas additions as soon as the setpoint recovers. It is better to introduce a time delay for the entire load to reach temperature. Remember that when the ammonia addition ends, desorption of nitrogen begins.

The temperature range for carbonitriding is not arbitrary. The thermal decomposition of ammonia is too rapid at higher austenitizing temperatures, limiting nitrogen availability. At lower temperatures, a more brittle structure is formed.

Vacuum Nitriding

Two forms of nitriding are commonly performed in vacuum furnaces: low-temperature nitriding of a variety of materials and high-temperature nitriding (solution nitriding) of stainless steels. The nitriding process can be accomplished by two different methods: gas nitriding and ion/plasma nitriding.

GAS NITRIDING

Vacuum gas nitriding has been advanced with the development of new vacuum-furnace designs with better control systems for the nitriding atmosphere (Fig 26.9). Vacuum gas nitriding is an alternative to traditional retort-type gas nitriding systems, and claimed benefits include reduced cycle times (by as much as 50% when compared to the older systems).

FIGURE 26.9 | Gas nitriding system (courtesy of Solar Atmospheres Inc. and Winston Heat Treating Inc.)

In a typical vacuum gas nitriding system, the work is placed in the furnace, the furnace is evacuated to approximately 0.027 mbar (0.01 torr) and then backfilled with nitrogen to 1,067 mbar (800 torr). The furnace is heated to a nitriding temperature of 455-580°C (850-1075°F), a portion of the nitrogen is pumped out and ammonia is introduced to maintain pressure. The partial pressures of the nitrogen and ammonia are continuously controlled at a fixed ratio throughout the cycle to provide the final desired result. During the nitriding cycle, the gas is circulated throughout the hot zone by a circular fan.

The vacuum gas nitriding furnace can provide a precise and accurate programmed nitriding potential for repeatable results on critical case depth and white-layer control. White layer, a by-product of the nitriding process, can be minimized for the part's final use.

ION (PLASMA) NITRIDING

Ion nitriding introduces nascent (elemental) nitrogen to the surface of a metal part for subsequent diffusion into the material. In actual practice, the injected gas is a mixture consisting of 25% N_2 and 75% H_2. In a low vacuum, high-voltage electrical energy is used to form plasma, through which nitrogen ions are accelerated to impinge on the workpiece. This ion bombardment heats the workpiece, cleans the surface and provides active nitrogen.

Ordinary nitrogen is introduced, with the pressure held between 1 and 10 torr. A voltage potential of about 450 volts (DC) is applied between the inside of the positively charged vessel and the work support on which the parts are placed. This causes the parts to be negatively charged, and they become the cathode in the electrical circuit formed. Because of the high voltage, the nitrogen ions become positively charged, move to the negatively charged part surfaces, pick up an electron and diffuse into the part surfaces.

The ionization process is very efficient. Very little nitrogen is required, and the surface is sputter-cleaned during nitriding. The oxide that is always present on stainless steels is removed by sputtering, which allows the nitrogen to easily penetrate the surface. The oxide would need to be chemically removed or mechanically blasted in gas or salt nitriding. In addition, because the process is electrical in nature, the composition and thickness of the white layer can be easily controlled and even eliminated.

Ion nitriding, however, is not without its own challenges. Materials to be processed must be thoroughly cleaned prior to placement in the vacuum furnace. Unwanted or remaining particles left on the workpieces can cause arcing, with resulting damage to the work. Placement of parts within the furnace is critical, and any holes or cavities must be properly masked to prevent overheating.

Plasma (ion) nitriding (Fig. 26.10) using pulsed power generators is an alternative to the traditional gas nitriding process. Nitriding is used in many applications for increased wear resistance and improved sliding friction as well as in components where increased load-bearing capacity, fatigue strength and corrosion resistance are important. Corrosion resistance can be particularly enhanced by a plasma post-oxidation treatment. Dimensional changes are minimal, and the masking process for selective nitriding is simple and effective.

FIGURE 26.10 | Ion nitriding of automotive components (courtesy of Surface Combustion Inc.)

SEE FULL COLOR IMAGE ON PAGE E

Plasma nitriding uses nitrogen gas at low pressures in the range of 1-10 mbar (0.75-7.5 torr) as the source for nitrogen transfer. Above 1000°C (1832°F), nitrogen becomes reactive when an electric field in the range of 300-1,200 V is applied. The electrical field is established in such a way that the workload is at the negative potential (cathode) and the furnace wall is at ground potential (anode). The nitrogen transfer is caused by the attraction of the positively charged nitrogen ions to the cathode (workpieces), with the ionization and excitation processes taking place in the glow discharge near the cathode's surface. The rate of nitrogen transfer can be adjusted by diluting the nitrogen gas with hydrogen (above 75%). The higher the nitrogen concentration, the thicker the compound layer.

The compound layer consists of iron and alloy nitrides that develop in the outer region of the diffusion layer after saturation with nitrogen. According to the iron-nitrogen phase diagram, basically two iron nitrides are possible, the nitrogen-poor gamma prime (γ') phase (Fe_4N) and the nitrogen-rich epsilon (ε) phase ($Fe_{2\text{-}3}N$).

The temperature of the workpiece is another important control variable. The depth of the diffusion layer also depends strongly on nitriding temperature, part uniformity and time. For a given temperature, the case depth increases approximately as the square root of time. A third process variable is the plasma power or current density, a function of surface area that has an influence on the thickness of the compound layer.

SOLUTION NITRIDING

A typical solution nitriding (high-temperature nitriding) process takes place in a vacuum furnace in the temperature range of 1050-1100°C (1925-2015°F). Nitrogen is introduced at pressures in the range of 1.1-1.9 bar (Fig. 26.11) for a period of 10-12 hours. The workload is then rapidly vacuum quenched at pressures in the range of 10 bar.

FIGURE 26.11 | Nitrogen solubility as a function of temperature (420 SS)

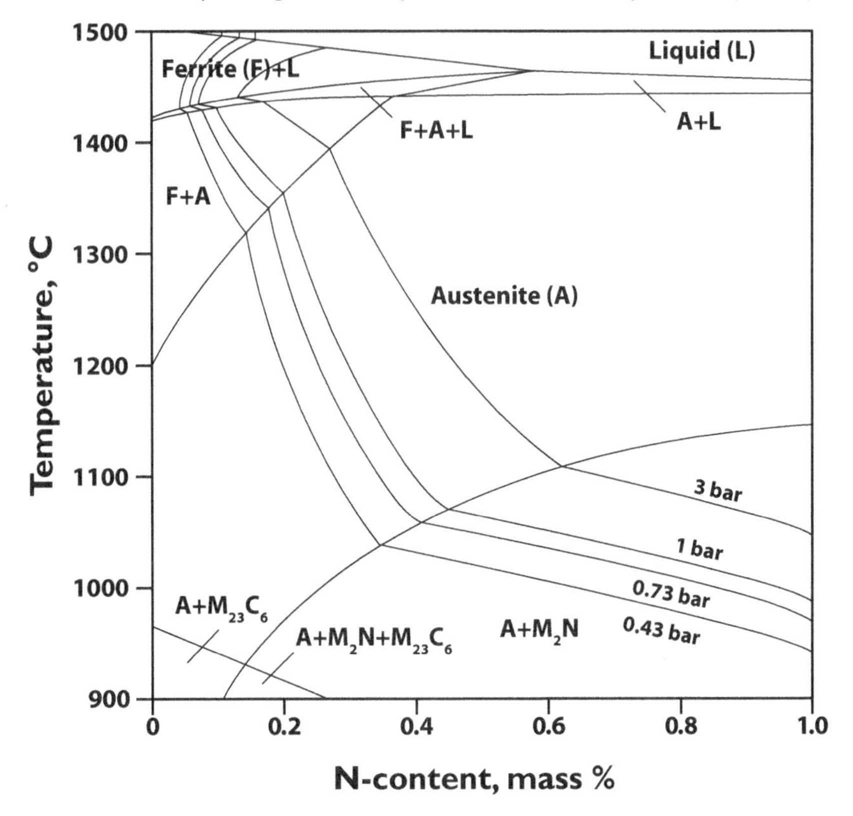

Solution nitriding (Fig. 26.12) has been found to extend the corrosion properties of 400 series stainless steels, a characteristic highly desirable to bearing manufacturers supplying product to the aerospace industry. The microstructure (Fig. 26.13) typically consists of martensite and (coarse) ferrite to a depth of 2.3

mm (0.090). The martensite development begins at the outer layer and extends into the core as the nitriding time is increased. Partial martensite formation has been reported to take place in the core, which is initiated at ferrite grain boundaries. The existence of the martensite structure contributes to the hardness improvement of the steel. However, nitrogen diffusion is also believed to increase the hardness of the martensite. Prolonging the nitriding time diffuses more nitrogen into the material.

Process variables include:

- Surface area – a function of geometry, mass and surface roughness
- Nitrogen concentration – a function of pressure, flow, time and concentration
- Material – a function of chemistry and form
- Cleaning method – a function of time, temperature, chemistry and energy
- Hydrogen cleanup – a function of time and surface contamination
- Temperature loss – a function of transfer time to quench, cooling rate, backfill speed and temperature uniformity in the heating chamber
- Vacuum hold – a function of load position, time and level
- Loading – a function of part size

In the solution nitriding process, it is known that:

- Higher temperatures promote greater nitride formation
- Nitrogen addition results in nitride formation and solid-solution diffusion
- Variation in nitriding temperature changes the thickness and composition of the nitrided layer
- Nitrogen present in the lattice is captured by chromium (Cr:N ratio 1:1 or 1:2)
- Chromium nitride (CrN) results from expanded austenite decomposition
- Nitrogen will absorb faster into martensite than into austenite
- Nitrogen absorption is dependent on material composition and crystallographic plane orientation to the surface
- Nitrogen absorption in austenite is up to 40 atomic %. (typically 16.7-37.8% nitrogen)

FIGURE 26.12 | Typical equipment for solution nitriding (courtesy of ALD Thermal Treatment)

FIGURE 26.13 | Typical solution nitriding microstructures (400-series stainless steel): a.) surface – low magnification; b.) surface – high magnification (courtesy of ALD Thermal Treatment)

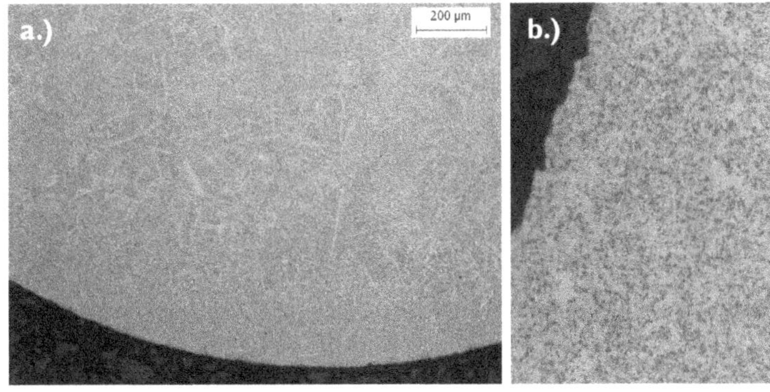

Vacuum Nitrocarburizing

Nitrocarburizing is a modification of nitriding, not a form of carburizing. Two types of nitrocarburizing are common: ferritic and austenitic. In this process, nitrogen and carbon are simultaneously introduced into the surface.

Nitrocarburizing is not a process traditionally associated with vacuum processing. It is carried out either by combining a vacuum purge (to eliminate oxygen) with conventional gas nitrocarburizing in a controlled-atmosphere furnace (typically, but not exclusively, hot-wall designs) or via plasma technology. Plasma nitrocarburizing is achieved by adding small amounts (1-3%) of methane or carbon dioxide gas to the nitrogen-hydrogen gas mixture to produce a carbon-containing epsilon (ε) compound layer ($Fe_{2\text{-}3}C_XN_Y$). It is commonly used only for unalloyed steels and cast irons.

FERRITIC NITROCARBURIZING

Ferritic nitrocarburizing involves the addition of nitrogen and carbon while the part is below the lower critical temperature (Ac_1) of the material (i.e. at a temperature below which austenite begins to form during heating). The process is typically performed at 550-600°C (1020-1110°F) in an atmosphere of 50% endothermic gas + 50% ammonia or 60% nitrogen + 35% ammonia + 5% carbon dioxide. Other atmospheres, such as 40% endothermic gas + 50% ammonia + 10% air, are also used. The presence of oxygen in the atmosphere activates the kinetics of nitrogen transfer.

This process creates an epsilon-nitride phase that improves wear and scuffing characteristics of gear steels. Since nitrocarburizing is performed at lower temperatures than carbonitriding and carburizing, it is possible to achieve better distortion control. This process can be applied to inexpensive steels to create high-hardness, shallow cases.

AUSTENITIC NITROCARBURIZING

Austenitic nitrocarburizing is the addition of carbon and nitrogen at temperatures in the 675-775°C (1250-1425°F) temperature range. The process can be controlled to produce a surface compound layer of epsilon carbonitride, and it produces a subsurface layer of bainite and/or martensite upon quenching. This microstructure is particularly useful in intermediate-stress or point-contact-resistance applications such as helical gears.

Part Hardness Measurement of Case-Hardened Gears

Hardness inspection should be performed after the last thermal operation that could affect hardness has been completed. An exception to this rule is hardness testing before silver plating and baking, because hardness testing will damage silver plating. Hardness testing should be performed after nital-etch inspection (if required) and prior to final magnetic-particle inspection, shot peening and chemical finishing (e.g., plating, black oxide).

Hardness inspection locations may include the gear-tooth root, the end-face of the gear teeth or the adjacent gear rim. When gear-teeth roots are inspected, testing of three azimuth positions (0-, 120- and 240-degree locations) and at each end of the face width of a gear will help verify uniformity of stock removal on parts with ground gear roots.

The applied load must be appropriate for case depth at the test location. Carburized or nitrided gears with 0.5 mm (0.020 inches) or less case depth require using the HR15N scale. For case depths over 0.5 mm (0.020 inches), the HR30N scale is suitable. Do not use superficial hardness scales for testing core hardness. Test locations on case-hardened surfaces should be agreed upon with the design organization.

If surface hardness inspection of the product is not required, not possible or not permitted, the customer or design organization may approve use of representative test coupons or destructive testing of sample product.

Applications

LPC or low-pressure carbonitriding (LPCN) combined with either high-pressure gas quenching (HPGQ) or oil quenching has become increasingly popular over the past several decades. Industries such as aerospace, automotive, industrial and commercial heat treating have spearheaded the use of this technology.

MILITARY/DEFENSE INDUSTRY

Battlefield conditions tend to abuse firearms, particularly the components making up the trigger assembly (Figure 26.14), which were wearing out or malfunctioning at an alarming rate in desert warfare and causing life-threatening conditions and undue stress in combat. LPC of the SAE 9310 triggers (Fig. 26.15) using acetylene improved wear resistance and case uniformity (Fig. 26.16) and achieved minimal part distortion. Carbide formation and high levels of retained austenite were avoided by proper selection of carburizing parameters and post processes including deep freeze and tempering.

FIGURE 26.14 | Squad automatic weapon RPK 74 trigger

FIGURE 26.15 | Typical production load of triggers (courtesy of Solar Atmospheres Inc.)

FIGURE 26.16 | Trigger case depth and hardness profiles: a.) profile; b.) microstructure

a.)

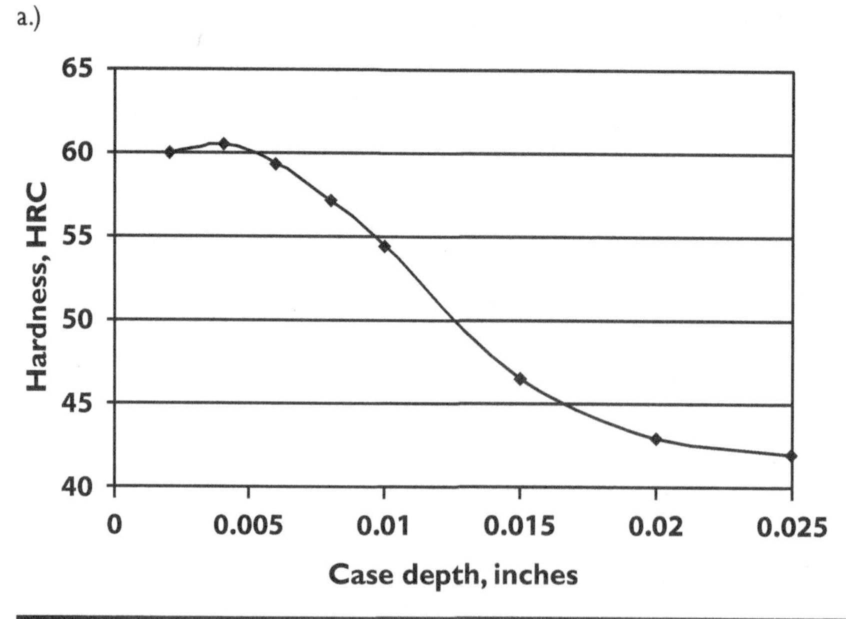

AUTOMOTIVE INDUSTRY

Automobiles rely on transmission gears (Fig. 26.17) to maximize horsepower and torque at a given speed. As the vehicle speeds up and slows down, the transmission allows the gear ratio between the engine and drive wheels to change (manually or automatically) so that the engine stays below its redline value and near the RPM band that maximizes its performance.

Examples include sun gears (Fig. 26.18) made of SAE 5120M material. They are acetylene vacuum carburized to an effective case depth of 0.3-0.5 mm (0.012-0.020 inches) followed by an 8-bar helium quench. The load weight is approximately 275 kg (610 pounds). Similarly, main drive gears (Fig. 26.19) of SAE 4121M material are acetylene vacuum carburized to a case depth of 0.7-1.1 mm (0.028-0.043 inches) followed by an 18-bar helium quench. Carbon/carbon composite fixtures are used to hold dimensional tolerances and minimize part distortion. The load weight is approximately 225 kg (500 pounds).

FIGURE 26.17 | Family of vacuum-carburized six-speed transmission gears (courtesy of ALD Thermal Treatment)

FIGURE 26.18 | Load of sun gears (courtesy of ALD Thermal Treatment)

FIGURE 26.19 | Load of main drive gears (courtesy of ALD Thermal Treatment)

Powder-metal bushings manufactured from FLN2-4405 material (Fig. 26.20) having a composition of 2% Ni, 0.85% Mo, 0.6% graphite and a density of 7.75 g/cc are typical industrial components that are vacuum carburized. These parts are processed at 930°C (1700°F) using cyclohexane at 10 torr to a case depth of 0.36 mm (0.014 inches). They are oil quenched from 860°C (1575°F) and subsequently tempered at 175°C (350°F).

FIGURE 26.20 | Automotive powder-metal bushings (courtesy of Surface Combustion Inc.)

FIGURE 26.21 | Bushing case depth and hardness profiles: a.) profile; b.) microstructure (courtesy of Surface Combustion Inc.)

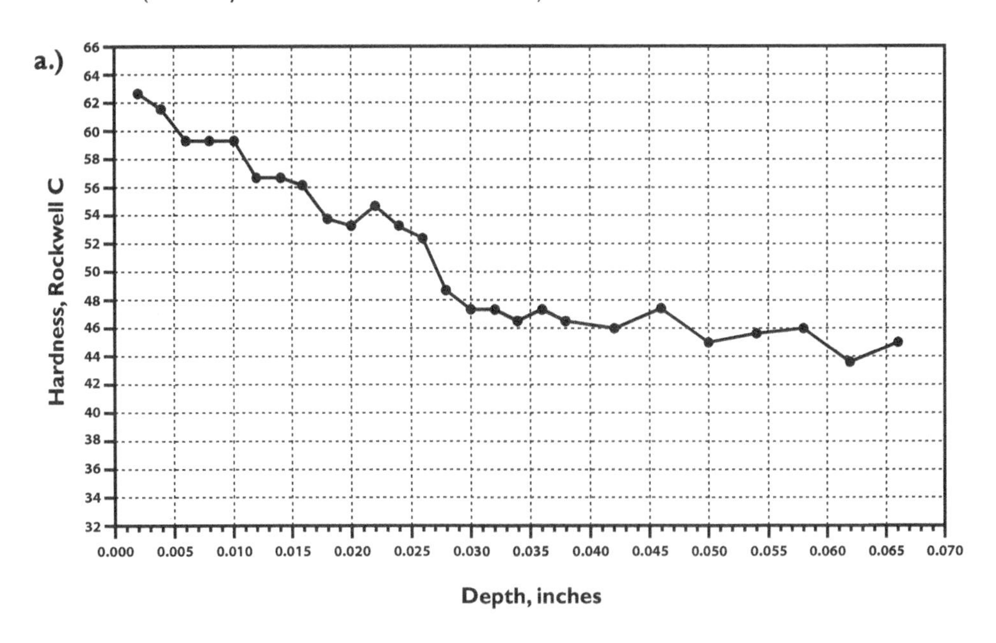

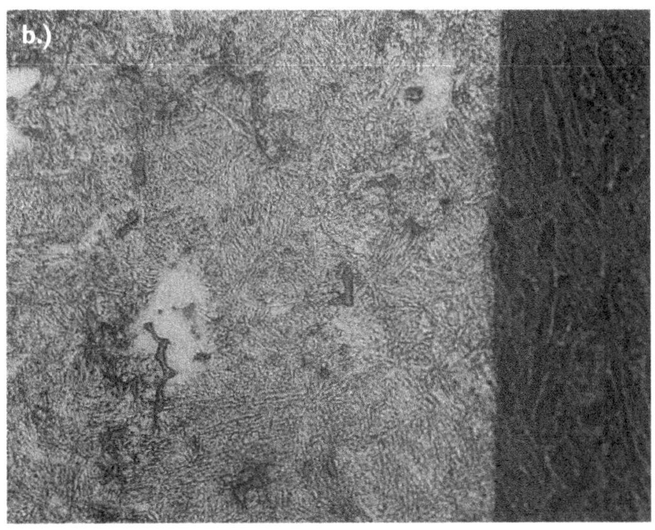

HEAVY-TRUCK INDUSTRY

Medium- and heavy-duty commercial vehicles, including off-highway vehicles, require heavy-duty and high-performance transmissions (Fig. 26.22a) in order to perform under both normal and extreme service conditions. Transmission gears (Fig. 26.22b) of 8822H material are vacuum carburized at 960°C (1750°F) to an effective case depth of 1.10-2.25 mm (0.040-0.090 inches) and pressure quenched at 20 bar using helium. Load weights up to 1,000 kg (2,200 pounds) are possible.

FIGURE 26.22 | Truck drivetrain/ring and pinion set: a.) drivetrain; b.) truck ring and pinion set (courtesy of Dana Holding Corp.)

a.)

b.)

Truck drive shafts (Fig. 26.23) of SAE 9310 are vacuum carburized at 960°C (1750°F) to an effective case depth of 1.15-1.65 mm (0.045-0.065 inches) and oil quenched into 65°C (150°F) oil. The load weight is 225 kg (500 pounds).

FIGURE 26.23 | Truck drive shafts (courtesy of ECM USA)

MOTORSPORTS INDUSTRY

Motorsports encompasses a wide variety of events, including Formula 1, sports-car racing, dragsters, off-road racing, trucks, top-fuel funny cars and power-boats. Race teams are using LPC technology almost exclusively. Highly loaded gear applications, such as those in 3,500-hp top-fuel dragsters, require crankshaft fracture toughness values two or three times normal values. If the bottom end fails in one of these races, the engine literally explodes.

For example, main shafts (Fig. 26.24) of SAE 8620 material are vacuum carburized at 960°C (1750°F) to an effective case depth of 1.00 mm (0.040 inches) and pressure quenched at 7-10 bar using nitrogen.

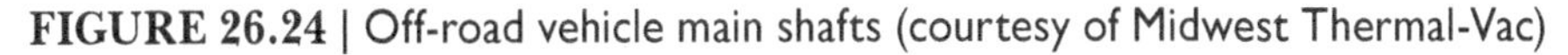

FIGURE 26.24 | Off-road vehicle main shafts (courtesy of Midwest Thermal-Vac)

AEROSPACE INDUSTRY

An aircraft landing gear consists of eight main components: struts and links, locks, retraction system, gearboxes, brakes, wheels and steering. Controls provide for a safe landing and make maneuvering on the ground possible. The gearbox is the primary drive mechanism and consists of planet gears, sun gears and shafts. It is used to transfer the power drive from either a hydraulic motor or the emergency electric motor to the right-angle drive assemblies. The gearbox typically operates at high speed (low torque) and low speed (high torque), and it allows the nose and main gear actuators to extend or retract an aircraft's landing gear.

Transmission gears (Fig. 26.25) of SAE 9310 are selectively vacuum carburized at 960°C (1750°F) to an effective case depth of 1.40-1.65 mm (0.055-0.065 inches) and pressure quenched at 6 bar using nitrogen. The load weight is 360 kg (790 pounds).

FIGURE 26.25 | Vacuum-carburized gearbox actuator gears (courtesy of Midwest Thermal Vac)

SEE FULL COLOR IMAGE ON PAGE E

In another example, nozzle blanks (Fig. 26.26) of Ovako 477Q material were vacuum carburized at 845°C (1550°F) with cyclohexane at 10 torr, gas cooled, then reheated to 925°C (1700°F) followed by a 2-bar nitrogen quench, deep freeze and temper. Effective case depth (50 HRC) requirements are 0.25-0.50 mm (0.010-0.020 inches) for the full length of the blind hole, hardness of 57-63 HRC at a depth of 0.05 mm (0.002 inches) and a core hardness of 40-48 HRC.

FIGURE 26.26 | Precision carburization of Ovako 477Q nozzle blanks (courtesy of Specialty Heat Treating Inc.)

INDUSTRIAL-PRODUCTS INDUSTRY

Hydraulic-pump gears (internal or external) rely on the principle of gears coming in and out of mesh to produce flow. As the gears come out of mesh, they create expanding volume on the inlet side of the pump. Liquid flows into the cavity and is trapped by the gear teeth as they rotate. Liquid travels around the interior of the casing in the pockets between the teeth and the casing (it does not pass between the gears). Finally, the meshing of the gears forces liquid through the output port under pressure.

Ring gears (Fig. 26.27) of SAE 9310 are vacuum carburized at 960°C (1750°F) to an effective case depth of 1.75-2.30 mm (0.070-0.090 inches) and pressure quenched at 11 bar using nitrogen. The load weight is 500 kg (1,100 pounds).

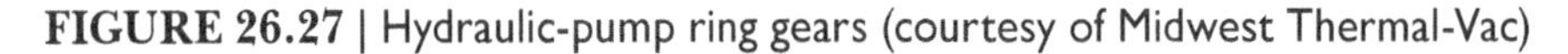

FIGURE 26.27 | Hydraulic-pump ring gears (courtesy of Midwest Thermal-Vac)

Die-cutting punches (Fig. 26.28a) are used in manufacturing industries to cut or shape material using a press. Similar to molds, dies are generally customized to the item they are used to create and can be used to cut shapes in materials such as paper, plastic, rubber, Kevlar® and the like.

SAE 12L14 punches are vacuum carbonitrided at 815°C (1500°F) to a case depth of 0.07 mm (0.003 inches) followed by oil quenching (Fig 26.28b) to achieve a surface hardness at the cut edge above 60 HRC.

FIGURE 26.28 | Family of die-cutting punches: a.) vacuum heat-treated die-cutting punches; b.) load of side-outlet die-cutting punches after hardening and tempering (courtesy of AmeriKen)

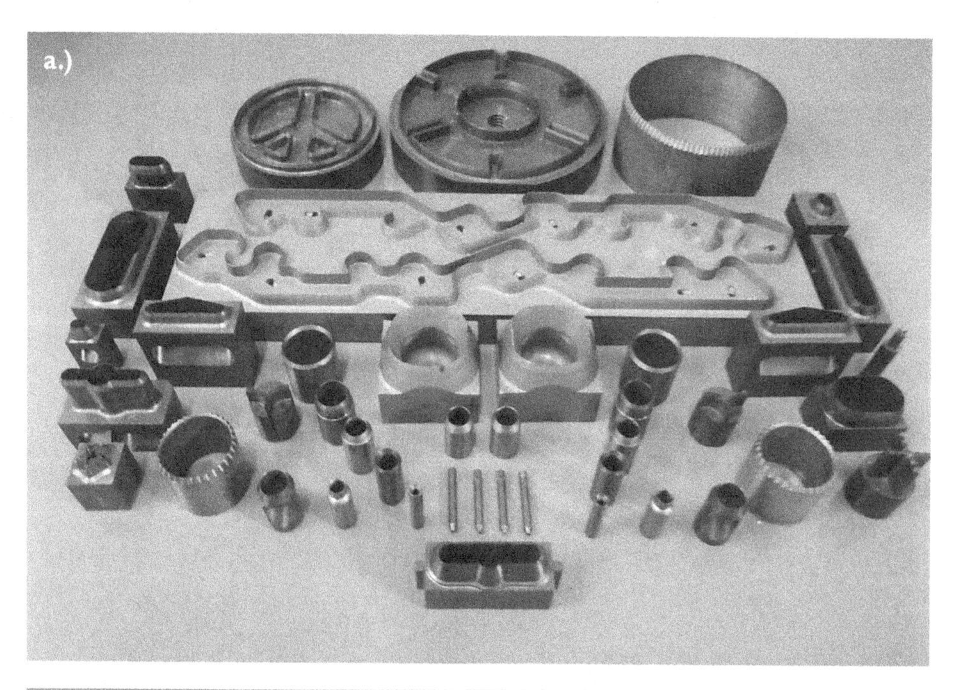

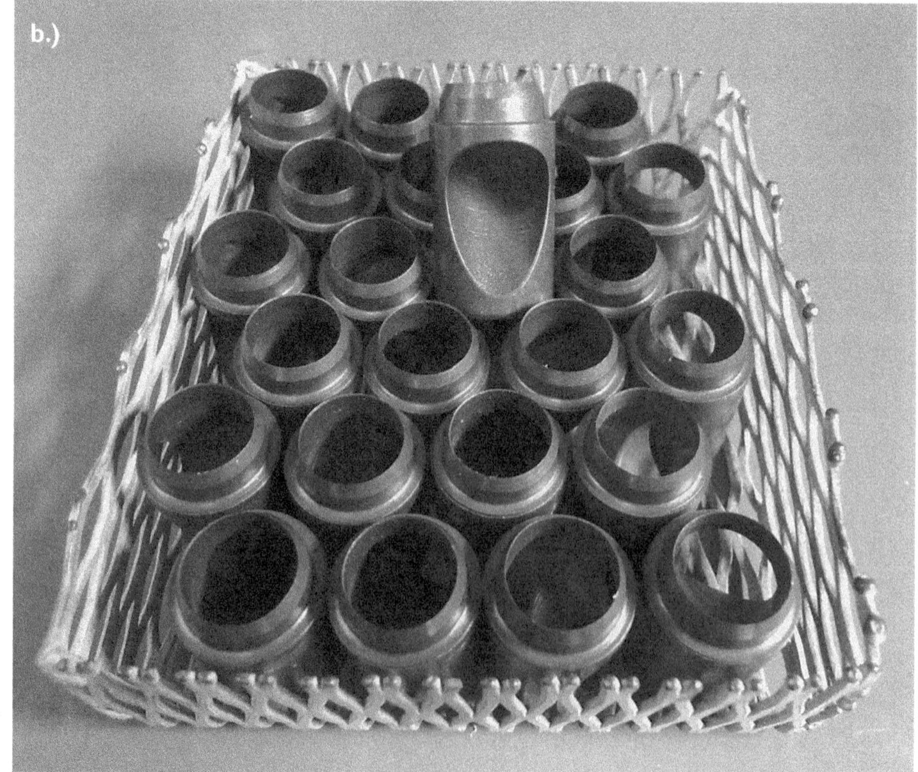

MEDICAL/DENTAL INDUSTRY

Medical implants are a growing segment of the medical-device industry. After hip surgery, for example, some patients experience continued hip dislocations, necessitating an additional surgery that involves screwing a titanium retention plate across the ball area. Screws (Fig. 26.29) weighing 2.8 grams (0.10 ounces) hold the plate in position. These are vacuum carburized due in part to the precise case-depth requirements. The addition of carbon produces a high-strength, high-hardness surface with excellent wear and abrasion resistance. Residual compressive stress at the root of the threads is an important contributor to fatigue strength. A typical screw material is Böhler-Uddeholm N360, and effective case depths are typically in the 0.0038 mm (0.0015 inches) range.

FIGURE 26.29 | Hip retention fastener

OFF-HIGHWAY INDUSTRY

Marine transmission gears take one-way rotational power from the engine crankshaft and allow it to be converted into the three modes necessary for propulsion of the craft through water: forward, reverse and neutral. Secondly, they cut the number of crankshaft rotations per minute by a fraction. Transmission cases (Fig. 26.30) are manufactured of SAE 8620 material and vacuum carburized at 980°C (1800°F) to an effective case depth of 1.15-1.65 mm (0.045-0.065 inches). Individual parts weigh 29 kg (64 pounds) each,and the total load weight is 475 kg (1,050 pounds). Parts can be either oil or high-pressure gas quenched (20 bar) using nitrogen.

FIGURE 26.30 | Marine transfer cases (courtesy of Midwest Thermal-Vac)

Advanced Techniques

Methods are under development to help control grain growth and allow for higher carburizing temperatures. One such technique involves pre-nitriding such as is done in the patented PreNitLPC® process (Patent Nos. 7,513,958 and 7,550,049) for extending the application range toward higher carburizing temperatures and a wider range of steel grades. The process involves the addition of ammonia (NH_3) gas to the vacuum-furnace chamber during heating of the workload in the temperature interval from 400°C (750°F) to the moment the workload reaches the carburizing temperature. The nitrogen diffuses into the surface layer, helping to pin the grain boundaries and limiting austenite grain growth before the carbon diffusion takes over this function. Nitrogen then diffuses deeper into the case, but it does not result in higher retained-austenite levels (Fig. 26.31). As a consequence, it has been reported that LPC temperature may be increased to as high as 1000°C (1850°F) without affecting the microstructure or mechanical properties (Fig. 26.32).

FIGURE 26.31 | Nitrogen profiles after various thermal treatments [25]

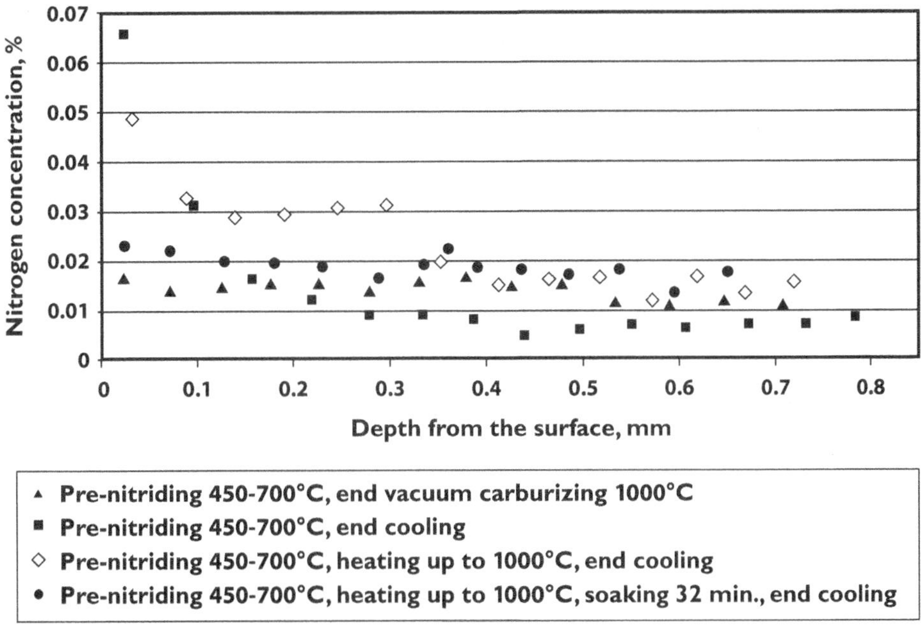

Nitrogen concentration, %

0.07
0.06
0.05
0.04
0.03
0.02
0.01
0

0 0.1 0.2 0.3 0.4 0.5 0.6 0.7 0.8

Depth from the surface, mm

- ▲ Pre-nitriding 450-700°C, end vacuum carburizing 1000°C
- ■ Pre-nitriding 450-700°C, end cooling
- ◇ Pre-nitriding 450-700°C, heating up to 1000°C, end cooling
- ● Pre-nitriding 450-700°C, heating up to 1000°C, soaking 32 min., end cooling

FIGURE 26.32 | Effect of austenite grain growth on a 17CrNiMo6 material vacuum carburized at 1000°C (1850°F): a.) without pre-nitriding; b.) with pre-nitriding [25]

a.) b.)

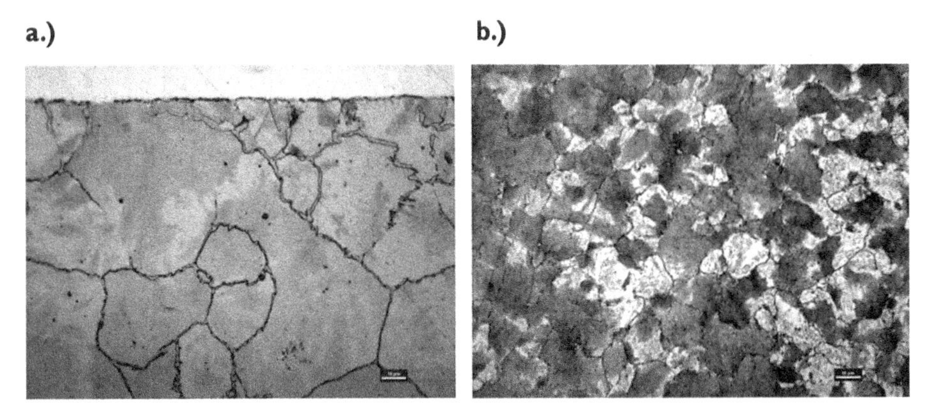

The resultant cycle reduction from high-temperature processing has an economic advantage even against LPC without pre-nitriding (Table 26.4).

TABLE 26.4 | Time savings – conventional LPC vs. pre-nitriding

Process	LPC @ 920°C (1690°F)	PreNitLPC® @ 950°C (1750°F)	PreNitLPC® @ 980°C (1800°F)	PreNitLPC® @ 1000°C (1832°F)
Effective case depth, mm (inches)	0.6 (0.024)	0.6 (0.024)	0.6 (0.024)	0.6 (0.024)
Total boost time, hours	0.383	0.283	0.211	0.190
Total diffusion time, hours	1.86	1.40	1.18	0.90
Total carburizing time, hours	2.24	1.68	1.39	1.09

New Technology Trends

Carburizing cycles for new materials, especially using carburizing temperatures above 980°C (1800°F), continue to be of great R&D interest. In some cases, cycle times have been reduced by 33-50% over atmosphere carburizing – processes requiring several hundred hours. Short-duration boost cycles (in the range of seconds) and long diffusion times (in the range of hours) provide just enough carbon to the surface of the part while avoiding the formation of retained austenite, carbide networks and necklaces.

Some of these materials include:

- Aubert & Duval – X13 VDW and XD15NW
- Carpenter Technology - Pyrowear® 53, 675 (Fig. 26.33), AF1410, HY180, HP-9-230, HP-9-430 and AerMet 100
- Questek Innovations – Ferrium® C61, CS63, C69 (Fig. 26.34), M60S and S53
- Böhler-Uddeholm – N360 Iso Extra, N695, R250 and R350
- The Timken Company - VacTec® 200, 250, 275L, 300, 325L, 350, 400, CSS-42L, M50NiL, CBS-600, BG42VIM/VAR, CBS 223 and 16NCD13 VAC-ARC
- Atlas Specialty Steels – BS970, EN30B
- VSG Essen – Cronidur 30
- Teledyne Corp. – VascoMax C-250, C-300 and C-350

A new generation of materials (Table 26.5) is emerging to meet service and performance needs. They are designed specifically for high-temperature service applications, retaining their hardness and mechanical properties well into a service range of 600-950°F (315-510°C) and higher. Many of these materials are similar in chemistry to that of stainless steels and tool steels to take advantage of better corrosion and wear resistance. These alloys (Table 26.6) produce impressive mechanical properties.

TABLE 26.5 | Advanced carburizing alloys [11]

Material	%C	Significant alloying additions (%)	Material	%C	Significant alloying additions (%)
XD15NW	0.37	Cr (15.5), Mo, V, Ni	CSS-42L	0.12	Cr (14.0), Co, Mo, Ni, V, Cb
X13VDW	0.12	Cr (11.5), Ni, Mo, V	CSB-50NIL	0.13	Mo (4.25), Cr, Ni, V
			AF1410	015	Co (14.0), Ni, Cr, Mo
Pyrowear 675	0.07	Cr (13.0). Ni, Mo, V	CBS 223	0.15	Cr (4.95), Co, Mo
Pyrowear 53	0.10	Mo (3.25), Ni, Cr, Si	CBS-600	0.19	Cr (1.45) Mo, Si
AerMet 100	0.23	Co(13.4), Ni, Cr, Mo	BG42	1.15	Cr (14.5) Mo, V
			HP 9-4-30	0.30	Ni (9.0), Co, Cr
N360 Iso Extra	0.33	Cr (15.0), Mo, Ni	HY-180	0.13	Ni (10.0), Cr, Mo
N695	1.05	Cr (17.0), Mo, Si			
R250	0.83	Mo (4.30), Cr, V	EN30B	0.32	Cr (1.30), Mo, Si
R350	0.14	Mo (4.30), Cr, Ni			
			Cronidur 30	0.31	Cr (17.0), Mo, Si, Ni
C61	0.16	Co (18.0), Ni, Cr, Mo			
CS62	0.08	Co (15.0), Ni, Cr, V	VascoMax C-250	0.03 max	Ni (18.5), Co, Ni, Cb
C69	0.10	Co (28.0), Cr, Ni, Mo, V	VascoMax C-350	0.03 max	Ni (18.5), Co, Ni, Cb

TABLE 26.6 | Achievable performance results with advanced materials

Material	Surface hardness (HRC)	UTS, MPa (ksi)	YS 0.2% offset, MPa (ksi)	Charpy V-Notch, Nm (ft-lbs)	Fracture toughness, MPa (ksi)	Temper temperature, °C (°F)
Hardening (oil quench)						
4340	53	1,979 (287)	1,862 (270)	20.3 (15)	331 (48)	205 (400)
4340	46	1,496 (217)	1,365 (198)	29.8 (22)	489 (68)	425 (800)
4340	40	1,241 (180)	1,158 (168)	47.5 (35)	689 (100)	540 (1000)
H11	56	2,006 (291)	1,675 (243)	20.3 (15)	331 (48)	540 (1000)
H11	48	1,641 (238)	1,413 (205)	27.1 (20)	427 (62)	580 (1075)
H11	44	1,427 (207)	1,276 (185)	31.2 (23)	483 (70)	595 (1100)
300M	56	2,344 (340)	1,241 (180)	17.5 (13)	310 (45)	95 (200)
300M	54	2,137 (310)	1,655 (240)	21.7 (16)	345 (50)	205 (400)
300M	45	1,793 (260)	1,482 (215)	13.6 (10)	235 (34)	425 (800)
300M	40	1,586 (230)	1,358 (197)	42.0 (31)	689 (100)	540 (1000)
EN30B	52	1,793 (260)	1,489 (216)	36.6 (27)	565 (82)	205 (400)
EN30B	40	1,400 (203)	1,289 (187)	44.7 (33)	689 (100)	540 (1000)
Carburizing (vacuum)						
300M	63/65	2,344 (340)	1,241 (180)	17.5 (13)	310 (45)	95 (200)
300M	61/63	2,137 (310)	1,655 (240)	21.7 (16)	345 (50)	205 (400)
AerMet 100	61/63	1,931 (280)	1,724 (250)	56.9 (42)	827 (120)	480 (900)
AerMet T	61/63	1,965 (285)	1,724 (250)	33.9 (25)	558 (81)	480 (900)
AerMet 310	61/63	2,172 (315)	1,896 (275)	27.1 (20)	448 (65)	480 (900)
AF1410	59/62	1,813 (263)	1,586 (230)	80.0 (59)	1,255 (182)	495 (925)
AF1410	58/61	1,710 (248)	1,551 (225)	93.6 (69)	1,469 (213)	510 (950)
EN30B	62/65	1,793 (260)	1,489 (216)	36.6 (27)	565 (82)	95 (200)
EN30B	58/62	1,400 (203)	1,289 (187)	44.7 (33)	689 (100)	205 (400)
Pyrowear 53	59/63	2,000 (290)	1,620 (235)	38.0 (28)	517 (75)	315 (600)
VascoMax C-250	61/63	1,792 (260)	1,758(255)	50.2 (37)	786 (114)	480 (900)
VascoMax C-300	61/63	2,027 (294)	1,999 (290)	28.0 (28)	517 (75)	480 (900)

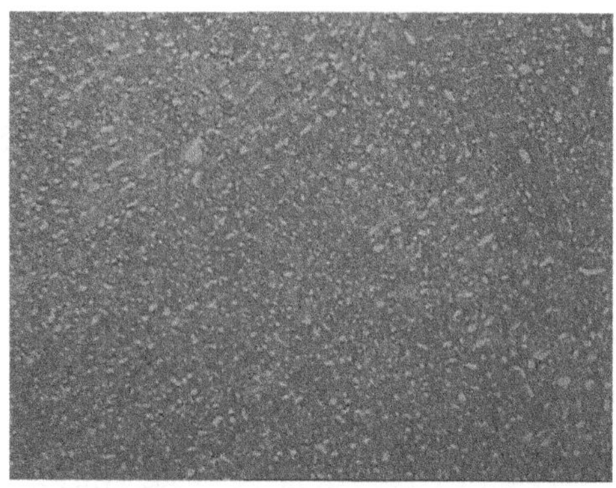

FIGURE 26.33 | Vacuum-carburized Pyrowear® 675 microstructure 1065°C (1950°F)

FIGURE 26.34 | Ferrium® C69 gear microstructure (0.040-inch ECD @ 53 HRC)

Effective case-depth ranges for vacuum-carburized parts routinely vary from 0.25-6.35mm (0.010-0.250 inches). A maximum case variation within a load can be held to within 0.125 mm (0.005 inches) and is routinely done for aerospace and motorsports applications. In one particular vacuum carbonitriding application, the specification called out an extremely shallow case, namely 0.0005-0.0025 inches, while the final part achieved 0.0018-0.0022 inches.

Surface carbon is controlled in the range of 0.60-0.80% for most conventional materials and between 0.45-0.75% for many of the advanced materials. Controlling retained austenite levels is critically important.

These new materials demand better control of hardness and carbon distribution (case/core hardness, surface/near surface), optimized microstructures and control of such items as retained austenite, carbide size (type and distribution) and non-martensitic phases. Property control (residual-stress patterns, surface finish and mechanical properties such as toughness, impact and wear resistance) is also critical.

The relationship between carburizing pressure and gas flow (carrier and hydrocarbon) as a function of load surface area is now understood so as to avoid carbide necklaces and to optimize carbide formation. Pressure ranges must be allowed to vary between 5-20 mbar (3.75-15 torr) and gas flow, a function of the load surface area, typically is controlled between 0.05-0.20 cfh (1,500-6,000 ml/h). Carbon flux is a function of the type of hydrocarbon gas used.

Final Thoughts

Finally, it is worth noting that aerospace and automotive specifications are in the process of being reviewed in light of rapid developments with new materials and new carburizing techniques. The industry appears to be headed toward shorter cycle times, which necessitates the use of an elevated carburizing temperature (at or above 980°C/1800°F) and direct quenching (oil or high gas pressure) to minimize cycle times and reduce carbon absorption without adversely affecting microstructure. Once this is accomplished, low-pressure (vacuum) carburizing technology offers almost limitless possibilities.

REFERENCES

1. Vacuum Carburizing Patents
   a. SECO/WARWICK (U.S. Patent No. 7,513,958 and U.S. Patent No. 7,550,049)
   b. Solar Atmospheres Inc. (U.S. Patent No. 7,541,035)
   c. Surface Combustion Inc. (U.S. Patent No. 7,267,793)
2. Dr. Klaus Loser, ALD-Vacuum Technologies GmbH, technical contribution and private correspondence
3. Herring, D.H, David J. Breuer and Gerald D. Lindell, "Selecting the Best Carburizing Method for the Heat Treatment of Gears," AGMA Fall Technical Meeting, 2002
4. ANSI/AGMA 2004-C08, *Gear Materials, Heat Treatment and Processing Manual,* American Gear Manufacturing Association, Fall Technical Conference, 2008
5. Herring, D.H., "Technology Trends in Vacuum Heat Treating, Part Two: Processes and Applications," *Industrial Heating,* November 2008
6. Ptashnik, W.J., "Influence of Partial Pressure Carburizing on Bending Fatigue Durability," Proceedings of Vacuum Metallurgy Conference, Science Press, 1977, p. 157-204
7. Herring, D.H., "Why Vacuum Carburizing is Effective for Today and Tomorrow, I-III," *Industrial Heating,* June, September, October, 1985
8. Herring, D.H. and G.P. Read, "Vacuum Carburizing Developments," *Metallurgia,* May 1986
9. Titus, Jack W., "Considerations When Choosing A Carburizing Process: A Comparison – Traditional Batch Atmosphere Carburizing Versus Vacuum and Ion Carburizing," *Industrial Heating,* September 1987
10. Edenhofer, B., "Overview of Advances in Atmosphere and Vacuum Heat Treatment," *Heat Treatment of Metals,* 1998.4 and 1999.1

11. Lohrmann, M. and D.H. Herring, "Heat Treating Challenges in the 21st Century," *Heat Treating Progress*, June/July 2001
12. Lindell, Gerald D., D.H. Herring, David J. Breuer and Beth Matlock, "An Analytical Comparison of Atmosphere and Vacuum Carburizing Using Residual Stress and Microhardness Measurements," *Heat Treating Progress*, November 2001
13. Kubato, Ken, U.S. Patent No. 5,702,540, December 1997
14. Gräfen, W. and B. Edenhofer, "Acetylene Low Pressure Carburizing – A Novel and Superior Carburizing Technology," *Heat Treatment of Metals*, 1999
15. Hebauf, Tim and Aymeric Goldsteinas, "Experience in Low-Pressure Vacuum Carburizing," *Industrial Heating*, September 2003
16. Yu, T., T. Wu and D.H. Herring, "Low Pressure Vacuum Carburizing + High Pressure Gas Quenching," *Heat Treating Progress*, August/September 2003
17. Poor, Ralph, "Impact of Delivery and Control in Vacuum Carburizing," *Industrial Heating*, September 2003
18. Kula, Piotr, Josef Olejnik and Janusz Kowalweski, "New Vacuum Carburizing Technology," *Heat Treating Progress*, February/March 2004
19. Osterman, Virginia, "Development Experience on Low-Torr Range Vacuum Carburizing," *Industrial Heating*, September 2005
20. Herring, D.H., "A Comparison of Atmosphere and Vacuum Carburizing Technology," *Industrial Heating*, January 2002
21. Otto, Frederick J. and D.H. Herring, "Vacuum Carburizing of Aerospace and Automotive Materials," *Heat Treating Progress*, January/February 2005
22. Herring, D.H., "A Perspective from the Americas: Low Pressure Vacuum Carburizing/Carbonitriding and High Pressure Gas Quenching," Nowoczesne Treny W Obróbre Cieplnej, September 2006
23. Otto, Frederick J. and D.H. Herring, "Advancements in Precision Carburizing of New Aerospace and Motorsports Materials," *Heat Treating Progress*, March/April 2007
24. Binoniemi, Bob, "Lessons Learned – Vacuum Carburizing & High Pressure Gas Quenching vs. Gas Carburizing and Oil Quenching," ALD-Holcroft Vacuum Carburizing Conference Proceedings, 2006
25. Kula, Piotr, Maciej Korecki, Robert Pietraski, Emila Stanczyk-Wolowic, Konrad Dybowski, Lukasz Kolodziejczyk, Radomir Atraskiewicz and Michal Krasowski, *FineCarb – The Flexible System for Low Pressure Carburizing*
26. Private correspondence with equipment manufacturers offering low-pressure carburizing technology

CHAPTER 27

VACUUM APPLICATIONS: MORE STANDARD, CUSTOM PROCESSES

In this chapter, we focus on examples of certain standard and special processes run every day in vacuum equipment. The types of materials, products and processes vary depending on the needs of the industry being serviced, but all take advantage of vacuum's unique ability with respect to quality and repeatability of cycles and results.

Annealing

Annealing treatments are undertaken primarily to soften a material, to relieve internal stresses and/or to modify the grain structure. These operations are carried out by heating to the required temperature followed by soaking at this temperature for sufficient time to allow the material to stabilize followed by slow cooling (except for solution annealing) at a predetermined rate. The choice of vacuum annealing is primarily influenced by the cleanliness and high quality of surface finish that can be obtained relatively easily compared to other controlled-atmosphere heat-treatment operations.

COPPER AND COPPER ALLOYS

Annealing of copper (Fig. 27.1) and copper alloys (Figs. 27.2, 27.3) is normally performed to soften the material after work (strain) hardening and to retain bright surface finishes.

FIGURE 27.1 | Bright annealing of copper blanks for diode heat sinks

SEE FULL COLOR IMAGE ON PAGE F

FIGURE 27.2 | Brass wire for bright annealing (courtesy of Solar Atmospheres Inc.)

FIGURE 27.3 | Beryllium copper wire for bright annealing (courtesy of Surface Combustion Inc.)

CARBON AND LOW-ALLOY STEELS

These materials are only economically processed in applications where cleanliness of the products or the prevention of carburization/decarburization of the part surfaces is absolutely critical.

STAINLESS STEELS

Annealing of stainless steel components in vacuum furnaces is often specified not only because of the cleanliness of the finished product, but also because of the fast gas-quench capability in vacuum. Stainless steel grades are commonly gas quenched in nitrogen for general commercial applications. However, austenitic steel grades stabilized with titanium or columbium (niobium) require argon or helium quenching to avoid sensitization in the form of nitrogen pickup, which degrades the corrosion resistance, particularly for nuclear, medical and aerospace applications.

A number of different annealing methods (full, isothermal, subcritical) are commonly used for stainless steel. Austenitic stainless steels cannot be hardened by heat treatment, but they do harden by cold working. Annealing not only allows recrystallization of the work-hardened grains but also places chromium carbides (precipitated at grain boundaries) back into solution. Annealing can also be used for homogenization of castings or welds and to relieve stresses from cold working.

Time at temperature is often kept short to minimize surface oxidation and to control grain growth, which may lead to a surface phenomenon called orange peel. Some chromium evaporation can take place during the annealing of stainless steels, but the amount lost is normally not significant because of the short time at heat and the slow diffusion rates of chromium in steel. Partial pressure can be used in cases where this is a concern.

Annealing temperatures range from 630-900°C (1150-1650°F) for ferritic and martensitic stainless steels to above 1040°C (1900°F) for austenitic (stabilized and unstabilized) alloys. When fine grain size is desired, the annealing temperature must be closely controlled. For example, a load of 304 stainless steel steam reactor boiler tubes (Fig. 27.4) are solution annealed at 1065°C (1950°F) +14°C/-0°C (+25°F/-0°F) and 2-bar gas quenched in nitrogen. The tubes are 57-mm (2.25-inch) outside diameter x 11.2 mm (0.440 inches) thick and 7 meters (23 feet) long. The load weight is approximately 12.5 tons (25,000 pounds), and the parts are stacked to minimize distortion.

FIGURE 27.4 | Stainless steel boiler tubes (courtesy of Solar Atmospheres Inc.)

ZIRCONIUM ALLOYS

Annealing of nuclear-industry materials such as Zircaloy tubing requires precise workload temperature-uniformity control and usually involves processing heavy loads. Loads as large as 1,155 kg (2,500 pounds) are vacuum processed vertically in furnaces up to 760 mm (30 inches) diameter x 7 meters (23 feet) in length (Fig. 27.5). These furnaces have all-metal hot zones with metallic (e.g., nichrome ribbon) heating elements.

A multi-heating-zone temperature-control approach is needed due to the size of the furnace along with complex temperature-control algorithms to ensure large, tightly packed loads are held to tolerances of ±5°C (±10°F). SCADA-based control systems that track the temperature of multiple workload thermocouples and automatically adjust the power input to discrete control zones of the heating elements are common.

In most cases, directional cooling is also a requirement due to part configuration. In the design shown in Figure 27.5, cooling gas enters the work zone from the bottom of the chamber, travels the length of the tubes picking up heat, exits near the top of the chamber and is then recirculated through a water-cooled heat exchanger. This allows the tubes to be cooled uniformly on their internal and external surfaces to minimize distortion.

FIGURE 27.5 | Vacuum furnace for annealing zirconium tubing (courtesy of VAC AERO International)

TITANIUM AND TITANIUM ALLOYS

Vacuum heat treating of titanium alloys is usually limited to aging and stress relief (Fig. 27.6). Stress relief can also serve as an intermediate heat treatment between welding and the various stages of cold forming. The temperatures employed lie below the annealing ranges for titanium alloys. They generally fall within 480-730°C (900-1350°F) with times ranging from 30-60 minutes, depending on the workpiece configuration and degree of stress relief desired.

Because of its propensity to react with contaminants, the heat treating of titanium alloys requires tight control of cleanliness, vacuum levels and leak-up rates (e.g., 5-10 microns/hour). Titanium alloys will react with residual water, oxygen, hydrogen and carbon dioxide to produce a brittle surface condition known as alpha-case, which must be removed before the part goes into service. Pre-cleaning of workpieces is important, and care should be taken to remove even fingerprints before heat treating. Precautions must be taken when processing titanium and titanium alloys when using nickel-alloy fixtures because of the low melting point of the titanium/nickel eutectic, which will occur at 943°C (1730°F). Exceeding this temperature can cause melting in both the titanium parts and the fixtures, resulting in a catastrophic meltdown.

FIGURE 27.6 | Stress relief of a titanium part (courtesy of Solar Atmospheres Inc.)

TOOL STEELS

Vacuum annealing is often used on tools (Fig. 27.7) that have been improperly hardened so that they can be reworked to meet required specifications on re-hardening. Vacuum processing avoids such issues as intergranular oxidation or carburization/decarburization on the working surfaces of the tools from a furnace atmosphere, thus maintaining dimensional precision.

FIGURE 27.7 | Typical M-series high-speed steel cutter hardness tested after vacuum annealing and re-hardening

SUPERALLOYS

Superalloys cover a wide classification of materials that are typically nickel-, cobalt- or iron-based and generally intended for high-temperature applications. Certain superalloys, however, require special treatments to develop required performance properties, including annealing and stress relief (Fig. 27.8).

FIGURE 27.8 | Typical load of Inconel F15/F16 jet-engine afterburner components for vacuum annealing and stress relief (courtesy of Solomon Engineering Inc.)

Chemical Conversion

One of the applications not commonly considered in vacuum processing is that of chemical conversion. Sample material is loaded in graphite trays (Fig. 27.9) and stacked inside large graphite retorts. The material is thermally processed in these retorts under controlled temperatures and pressures to chemically convert a mixture of elemental materials into a compound. A typical chemical conversion process is run at 1370°C (2500°F) for several days.

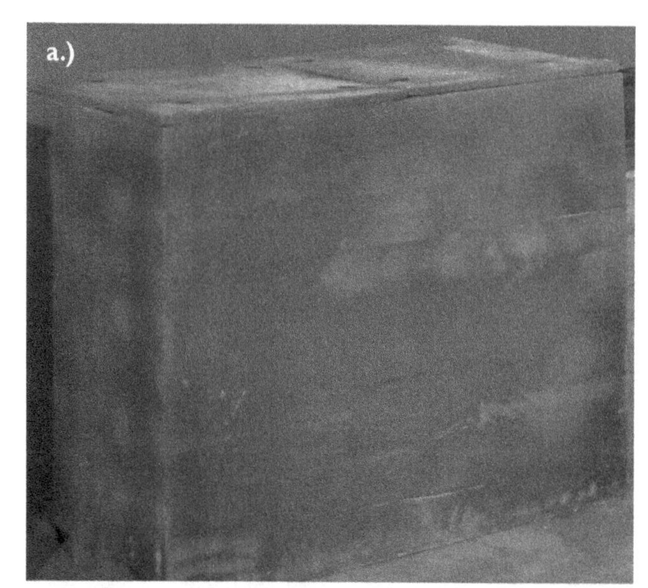

FIGURE 27.9 | Chemical conversion process: a.) graphite retort; b.) elemental materials for conversion (courtesy of Solar Atmospheres Inc.)

Creep Forming

Creep forming or hot sizing (Fig. 27.10) is often used to flatten or form to a near-net shape and for correcting spring-back and inaccuracies in shape and dimensions of pre-formed parts. The part is suitably fixtured such that controlled pressure is applied to certain areas of the part during heating. This fixtured unit is then placed in a furnace and heated at temperatures and times sufficient to cause the metal to creep until it conforms to the desired shape. Creep forming is used in a variety of ways with titanium, often in conjunction with compression forming.

Titanium and its alloys can be cold and/or hot formed. Titanium possesses certain unique characteristics that affect formability, and these must be considered when undertaking titanium forming operations. For example, the room-temperature ductility of titanium and its alloys is generally less than that of the common structural metals, including stainless steels. Titanium has a relatively low modulus of elasticity (about half that of stainless steel). This results in greater spring-back during forming and requires compensation either during bending or in post-bend treatment. Titanium in contact with itself or other metals exhibits a greater tendency to gall than does stainless steel. Thus, sliding contact with tooling surfaces during forming calls for the use of lubricants.

Heating titanium increases its formability, reduces spring-back and permits maximum deformation with minimum annealing between forming operations. Mild/warm forming of most grades of titanium is carried out at 205-315°C (400-600°F), while more severe forming is done at 480-790°C (900-1450°F).

Any hot forming and/or annealing of titanium products in air at temperatures above approximately 120°C (250°F) produces a visible surface-oxide scale and diffused-in oxygen layer (alpha-case) that will require removal on fatigue- and/or fracture-critical components. Oxide-scale removal can be achieved mechanically (e.g., grit-blasting or grinding) or by chemical descaling treatments (e.g., pickling). Vacuum creep forming avoids these issues.

FIGURE 27.10 | Creep-formed titanium components

Degassing

Vacuum degassing (Fig. 27.11) is a term often used to describe improved cleanliness in the steelmaking process. However, it is also used to reduce the hydrogen levels in many alloys of titanium, tantalum and niobium to avoid hydrogen embrittlement. Hydrogen is imparted into these alloys during ingot, rolling and forging operations. Hydrogen can also be diffused into these alloys during pickling or other chemical processes. Newer aerospace specifications demand that hydrogen levels be no greater than 70 ppm. Vacuum degassing, which is usually performed between 535-790°C (1000-1450°F) depending on the alloy, can achieve hydrogen levels of less than 20 ppm.

FIGURE 27.11 | Degassing of titanium logs (courtesy of Solar Atmospheres Inc.)

Homogenization

Vacuum homogenization (Fig. 27.12) is performed primarily to eliminate or decrease chemical segregation in castings and ingots by diffusion (prior to hot working). Homogenization typically has one or more purposes depending on the alloy, product and fabricating process involved. For example, one of the principal objectives is improved workability since the microstructure of most alloys in the as-cast condition is quite heterogeneous. In addition, a properly homogenized ingot will enhance the validity of ultrasonic testing conducted downstream in the manufacturing process for that material.

FIGURE 27.12 | Homogenization of titanium ingots (courtesy of Solar Atmospheres Inc.)

Hydriding and Dehydriding

The hydride/dehydride process is used during the manufacture of transition metal powders such as tantalum, niobium, vanadium and titanium. These particular metals have a unique characteristic: They undergo a reversible reaction (Equation 27.1) during hydriding. Vacuum purification and activation of the metal is necessary before hydrogen is readily absorbed to form the metal hydride. Absorbed hydrogen leads to an expanded metal lattice, which promotes cracking and embrittlement. Once embrittled, the material can be easily crushed into various mesh-size powder.

The reversible hydride/dehydride (HDH) process is one in which both an exothermic and endothermic reaction occur (Fig. 27.13). The metal hydride formation is exothermic, and heat is released during the reaction.

27.1) $M + X/2\ H_2 \leftrightarrows MH_x + \text{Heat}$

where:

M is the metal

MH_x is the metal hydride

FIGURE 27.13 | a.) Hydrogen molecules adsorb onto the metal lattice; b.) absorption and chemisorption lead to an expanded metal lattice

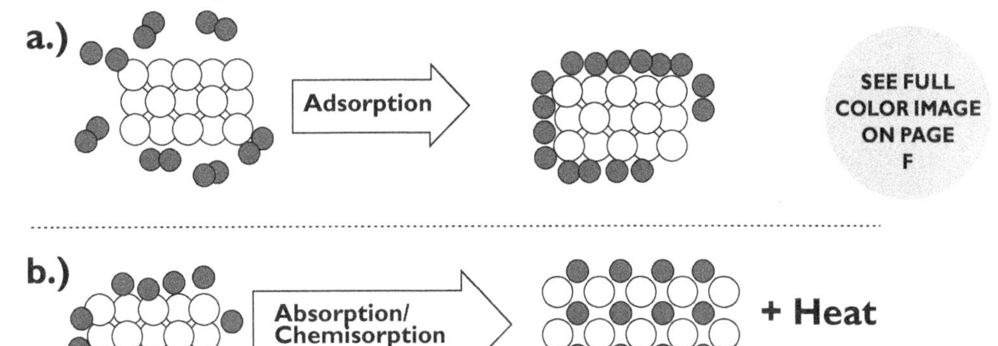

SEE FULL COLOR IMAGE ON PAGE F

The dehydriding step of the HDH process is endothermic. Therefore, heat must be provided to the reaction, which is run in vacuum to convert the powdered-metal hydride to the pure-metal powder. Heating in vacuum helps remove absorbed hydrogen quickly and efficiently without contamination from oxygen and nitrogen. Careful control of heating rates and final dehydriding temperature will result in a powder with minimal sintering.

Feedstock (Fig. 27.14) such as solid scrap, billet or machined turnings are processed to remove contaminants, hydrogenated to produce brittle material then ground under argon in a vibratory ball mill, typically at 400°C (750°F) for four hours at a pressure of 1 psi for Ti Grade 5. The resulting particles are angular and measure between 50 and 300 μm. Cold compaction after dehydrogenation of the powder followed by either vacuum hot pressing (in this case, the dehydrogenation process can be bypassed as hydrogen is removed under vacuum) or HIP, and a final vacuum anneal produces powders with hydrogen below 125 ppm (Fig. 27.15) and with special procedures to 10 ppm.

FIGURE 27.14 | Hydrided metal

FIGURE 27.15 | Load ready for dehydriding

Sintering and Sinter Compaction Techniques

Vacuum sintering, compaction (CIP, HIP) and secondary heat-treatment operations are performed on both conventional powder-metal (PM) as well as particulate (CIM, PIM, MIM) materials (Fig. 27.16). In an industry dominated by atmosphere processing, increased interest in controlled vacuum sintering arises from factors such as:

- The purity of the vacuum environment and its effect on part microstructure
- The use of sub-atmospheric (partial) pressure to improve the efficiency of the sintering reactions, especially with highly alloyed materials that require elevated sintering temperatures
- The use of reducing gases at temperatures less than 500°C (930°F) to minimize retained oxides and assist in clean binder removal and to reduce A-type porosity
- The ability of the vacuum process to reduce pore size and improve pore-size distribution
- The higher furnace temperature capabilities that permit faster sintering reactions carried out much closer to the melting point and with alloys of higher melting-point interstitial elements or liquid-phase metal sintering itself

- ❖ The ability to modify, in-situ, carbon balance and affect carbon additions or reductions
- ❖ The use of over-press sintering (HIP) to densify the materials and close A- and B-type porosity without formation of cobalt lakes
- ❖ Single-cycle debind, sinter and HIP with rapid cooling for rapid floor-to-floor times

The limitation on the application of sintering in vacuum furnaces is the vapor pressure of the metals being processed at the chosen sintering temperature. If the vapor pressure is comparable with the working pressure in the vacuum furnace, there will be considerable loss of metal by vaporization unless a sufficiently high partial pressure of inert gas is used. In certain instances, the partial-pressure gas can react with the surface of the part, creating a surface layer that may need to be removed.

FIGURE 27.16 | Typical sinter HIP furnace: a.) hot zone; b.) workload (courtesy of AVS Inc. and HB Carbide)

STAINLESS STEEL

Vacuum sintering of stainless steel powder-metal parts is a common process, employed for 300 series (304, 316) and 400 series (410, 420) as well as precipitation-hardening grades (e.g., 17-4 PH, 17-7 PH, 13-8 Mo). These products are very often superior to those sintered in hydrogen or dissociated ammonia atmospheres with respect to their corrosion resistance and mechanical properties.

HIGH-SPEED TOOL STEEL

Powder-metallurgy manufacturing methods have been developed for producing finished and full-density cutting tools of high-speed tool steel. Applications include such items as complex-geometry hobs, pipe taps and reamers. Special isostatic compacting techniques have been developed that use neither lubricants nor binders for these types of components. The pressed compacts are sintered in vacuum furnaces under precise control of heating rate, sintering time, temperature and vacuum pressure in order to eliminate porosity. The result is predictable densification of the pressed compact with final size tolerances of ±0.5-1.0%. Full-density, sintered high-speed steel tools have been shown to be at least equivalent to conventional wrought material in cutting performance. Grindability is dramatically improved, particularly for high-alloy grades such as M4 and T15. This is attributed to a finer and more uniform carbide distribution.

Solution Treating and Aging

Inconel® 718 is another example of a material that is usually used in the solution treated and aged condition. The exact temperatures, times and cooling rates depend on the application and desired mechanical properties. Many aerospace applications requiring high tensile strength, high fatigue strength and good stress-rupture properties use a solution treatment and a two-step aging treatment that involves the following steps.

1. Solution heat treating at 925-1010°C (1700-1850°F) for one to two hours or faster, followed by air cooling
2. Age at 720°C (1325°F) for eight hours, followed by furnace cooling to 620°C (1150°F)
3. Hold at 620°C (1150°F) for a total aging time of 18 hours (air cooling)

Tempering

Where surface finish is critical and clean parts are desired to avoid any post-heat-treat processing, many heat treaters, especially commercial shops, now employ vacuum furnaces for tempering and stress relief. These units typically operate in the temperature range of 130-675°C (275-1250°F), below which radiant energy is an efficient method for heating. As such, heating by convection is utilized. The furnace is normally evacuated to below 0.10 mbar (0.075 torr) then backfilled with an inert gas such as nitrogen, argon or even 97% nitrogen/3% hydrogen mixtures to a pressure slightly above atmospheric, typically in the range of 0.5-2 bar. A fan in the furnace recirculates this atmosphere, and parts are heated by both convection and conduction. Temperature uniformity in the range of ±5°C (±10°F) is common, with tighter uniformities possible.

Certain horizontal single- and multiple-chamber furnaces have been designed to perform single or multiple tempering treatments after hardening or case hardening without having to remove the workload from the equipment. This process is in common use for tempering high-speed steel components and a variety of other materials.

Stress Relief

Vacuum stress-relief operations can be performed on a variety of materials, especially those where zero oxidation is allowed. An example is the stress relief of nickel aluminum-bronze impellers (Fig. 27.17) used to propel the nuclear-powered Virginia Class Attack Submarines (SSN 774 Class). These impellers are located on the side of the vessel and provide tremendous maneuverability (side-to-side thrust). The impellers are 1,370 mm (54 inches) in diameter x 610 mm (24 inches) wide and weigh 1.5 tons (3,000 pounds). The stress-relief cycle involves holding at 315°C (600°F) ±9°C (15°F) for 60 hours after a long, controlled ramp heat-up, not to exceed temperatures on the part as determined by actual thermocouples located on the part. (A total of 10 work thermocouples are involved.) After stress relief, a slow rate of cooling is used to prevent reintroduction of stresses into the parts.

FIGURE 27.17 | Nickel aluminum-bronze impellers (courtesy of Solar Atmospheres Inc.)

Summing Up

The use of vacuum equipment and vacuum-processing techniques is commonplace for processing materials and component parts that require precise control of variability and optimum metallurgical properties. Vacuum processing's high level of repeatability ensures that once a process recipe has been determined it can be reproduced time after time.

REFERENCES

1. Robert Hill, Solar Atmospheres of Western Pennsylvania, technical review and private correspondence
2. Herring, D.H., "Technology Trends in Vacuum Heat Treating, Part One: Markets, Processes and Applications," *Industrial Heating*, October 2008
3. Hill, Bob, "Titanium Use in Aerospace Applications," *Industrial Heating*, November 2007
4. *Vacuum Brazing & Heat Treating Training Manual*, VAC AERO International Inc.
5. Osterman, Virginia M., "Utilization of Vacuum Technology in the Processing of Refractory Metals, Titanium and Their Alloys for Powder Applications," *Industrial Heating*, May 2008
6. Professor Philip Nash, Illinois Institute of Technology/Thermal Processing Technology Center, private correspondence

CHAPTER 28

VACUUM MARKETS, TRENDS AND FUTURE DIRECTION

Vacuum Markets

The early 21st century has seen double-digit growth in the use of vacuum heat treating and an increased vacuum market share throughout the Americas (Figs. 28.1-28.5). Vacuum processing is growing more rapidly than any other technology due in large part to the demand for high-quality, precision processes and repeatability of part performance in more sophisticated and demanding service applications.

FIGURES 28.1-28.5 | Vacuum market share in the Americas

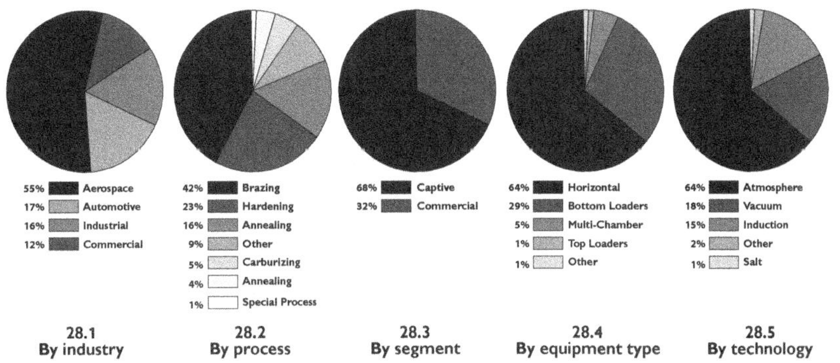

SEE FULL COLOR IMAGE ON PAGE G

Industries Serviced

Traditional industries include: aerospace, automotive, commercial heat treating (i.e. outsourcing or tolling) and industrial products. The strategy necessary for these industries involves the employment of vacuum processes and equipment by:

❖ Process substitution – Older process technologies and the equipment associated with it are being replaced by vacuum technology. The justification is reduced unit cost achieved by lowering the overall manufacturing cost and/or by material savings.

❖ Process replacement – Older products are being replaced by new engineering designs, allowing improvements in product performance at the same or reduced unit cost.

Furthermore, companies performing vacuum heat treating have identified those equipment types, features and processes required by each industry. In the automotive industry, for example, high-volume throughput requires automated systems integrated into the manufacturing flow. This is essential to keep up with the manufacturing philosophy and production demands (Figs. 28.6, 28.7).

FIGURE 28.6 | Heat-treat equipment servicing the automotive industry (courtesy of ALD Thermal Treatment)

FIGURE 28.7 | Typical production loads ready for thermal processing (courtesy of ALD Thermal Treatment)

An analysis of industry use in North America is shown based on a survey of 85 heat-treatment companies. This study was conducted to understand the industry uses of vacuum furnaces and the types of processes performed in them. Most end-users service several industries and perform more than one type of heat-treating operation within their facilities (Figs. 28.8, 28.9).

Typical processes performed, by industry, include:

- ❖ Aerospace (aircraft, rotorcraft and space)
  - Brazing
  - Casting/melting
  - Coating
  - Solution treating
- ❖ Automotive (heavy truck and off-highway, including racing)
  - Carburizing/carbonitriding
  - Hardening and tempering
- ❖ Commercial heat treating (outsourcing or tolling)
  - Annealing
  - Brazing
  - Carburizing/carbonitriding
  - Hardening and tempering
  - Sintering
  - Solution treating
- ❖ Industrial products (appliance and tools)
  - Annealing
  - Brazing
  - Carburizing/carbonitriding
  - Hardening and tempering
  - Sintering

FIGURE 28.8 | Industrial use of vacuum furnaces by industry (North America)

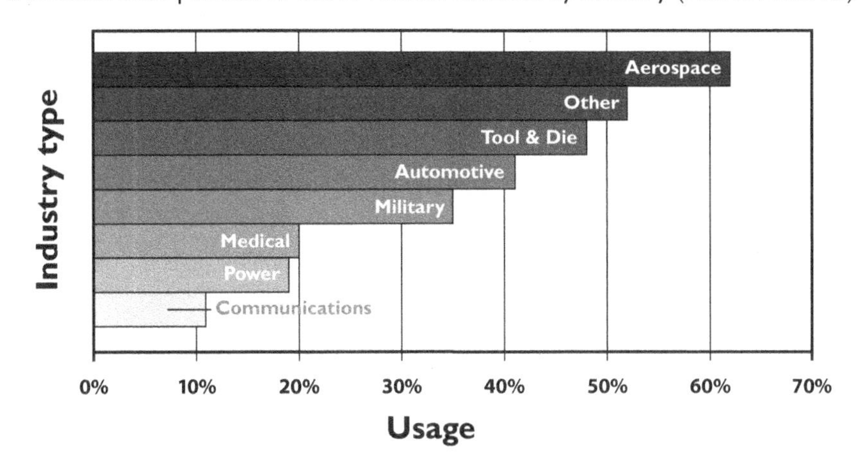

FIGURE 28.9 | Industrial use of vacuum furnaces by process (North America)

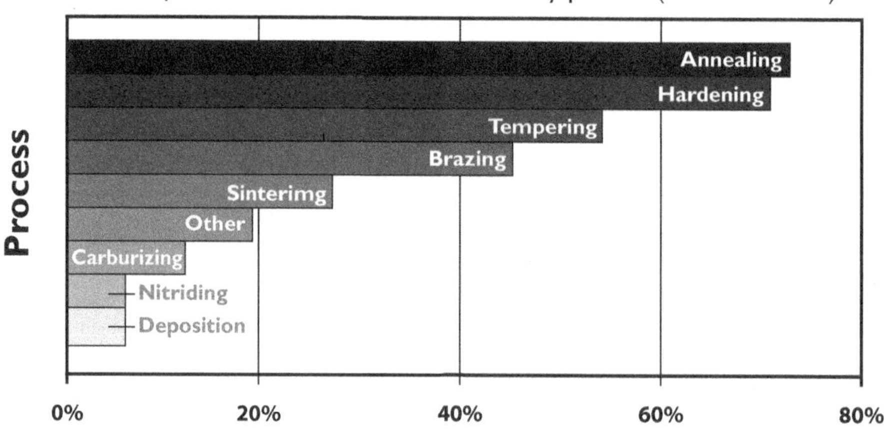

Heat-Treatment Technologies

Examples of processing industrial parts used in critical applications – ranging from highly stressed structural members to close-tolerance, high-performance products and high-performance components – cover a wide range of process technologies, including:

- Annealing (Fig. 28.10)
- Brazing (Fig. 28.11)
- Case hardening (Fig. 28.12)
- Hardening (Fig. 28.13)
- Sintering (Fig. 28.14)
- Stress relief (Fig. 28.15)
- Special processes (Fig. 28.16)

FIGURE 28.10 | Annealing of nickel-based superalloy F15/F16 afterburner housings (courtesy of Solomon Engineering)

FIGURE 28.11 | Copper brazing of 304 SS housings (courtesy of VAC AERO International Inc.)

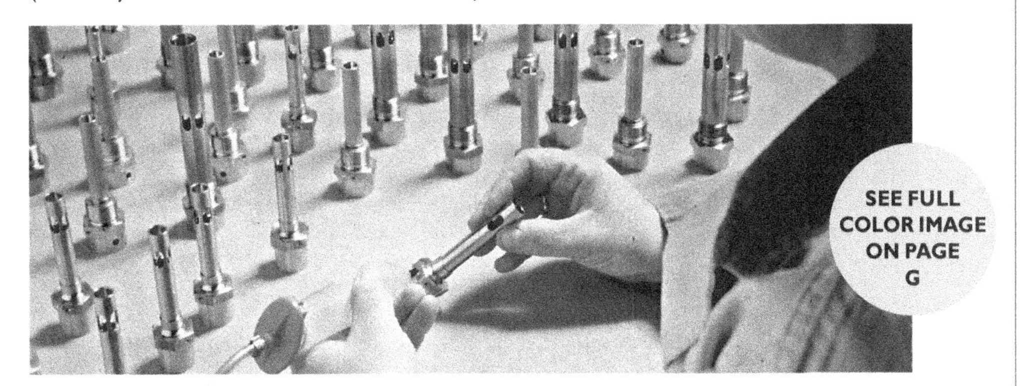
SEE FULL COLOR IMAGE ON PAGE G

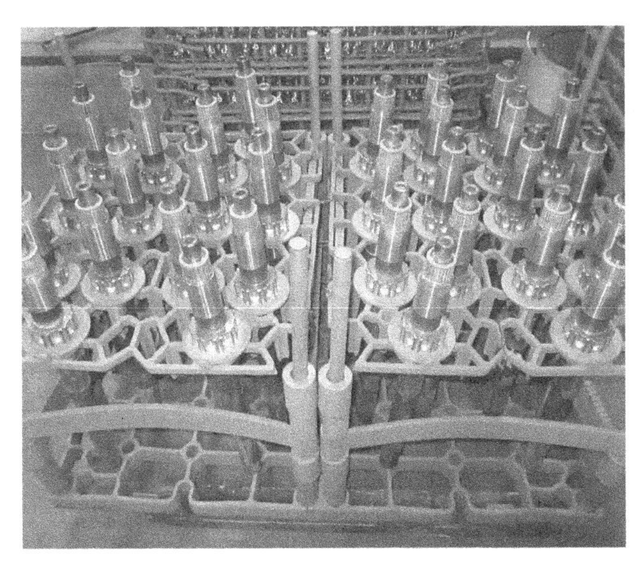

FIGURE 28.12 | Main drive shafts ready for LPC (courtesy of Mid-west Thermal-Vac)

FIGURE 28.13 | Vacuum hardening of H11 tool steel die (courtesy of Ipsen Inc.)

FIGURE 28.14 | Sintering of tungsten-carbide rock-bit inserts

FIGURE 28.15 | Stress relief of nickel aluminum-bronze impeller (courtesy of Solar Atmospheres Inc.)

FIGURE 28.16 | Hydriding/dehydriding of titanium powder (courtesy of Solar Atmospheres Inc.)

NEW TECHNOLOGY TRENDS

New technology applications are constantly being developed in vacuum heat treating. Several of these technologies are highlighted here.

BASCA Process

Titanium and titanium alloys are annealed (Fig. 28.17) to produce an optimum combination of ductility, machinability, dimensional stability and structure stability. Like recrystallization annealing, beta annealing improves the fracture toughness of titanium alloys. Beta annealing is performed at temperatures slightly above the beta transus of the alloy being annealed to prevent excessive grain growth. Annealing times are dependent on the section thickness and are sufficient for complete transformation. Time at temperature after transformation is held to a minimum to control beta grain growth.

Russian and U.S. aerospace manufacturers have collaborated to develop a unique beta annealing process involving slow cooling and age hardening. The BASCA 160 process (Fig. 28.18) for alloys such as Ti-5553 (5%Al-5%V-5%Mo-3%Cr) is used to achieve tensile values in the 1,100 MPa (160,000 psi) range. The key to the process is a very precise and controlled slow cool through a defined critical range then age hardening to avoid both distortion and alpha (α) case formation.

FIGURE 28.17 | Annealing of titanium sheet (courtesy of Solar Atmospheres Inc.)

FIGURE 28.18 | Ti castings (courtesy of Solar Atmospheres Inc.)

Combination Sintering and Hardening

High-temperature sintering of automotive transmission clutch and synchronizer hubs (Fig. 28.19) coupled with vacuum hardening and high-pressure gas quenching provides an example of the synergy between material selection and the heat-treating process. Product specifications mandate finish machining and heat treatment to 35 HRC. Conventional process solutions, including hard-machining operations, involve:

- FLN2-4405 (0.6%C, 1.85%Ni, 0.85%Mo) material hardened, oil quenched and tempered
- FLNC-4408 (0.75%C, 0.85%Mo, 2.0%Cu, 2.0%Ni) material sinter hardened

Finding a cheaper technology alternative led to the investigation of SL-5506 (0.6%C, 0.5%Mn, 0.5%Cr, 0.4% Ni) material. High-temperature sintering was required at 1280°C (2340°F) followed by vacuum hardening, 5-bar helium quenching and tempering. Benefits included a reduced number of manufacturing steps (from nine to seven) and the avoidance of the hard-turning operation.

FIGURE 28.19 | Clutch and synchronizer hubs after LPC + HPGQ SL-5506 material [1]

A comparison of the technologies (Fig. 28.20) revealed significant quality improvement (less statistical scatter). No qualification of the spline form was necessary between sinter and heat treat. The profile was held within 10 µm (0.0004 inches) and the lead within 15 µm (0.0006 inches) over 14 mm (0.55 inches).

FIGURE 28.20 | Statistical process-analysis measurement over balls 0.036578 mm (0.0015 inches); two locations [1]

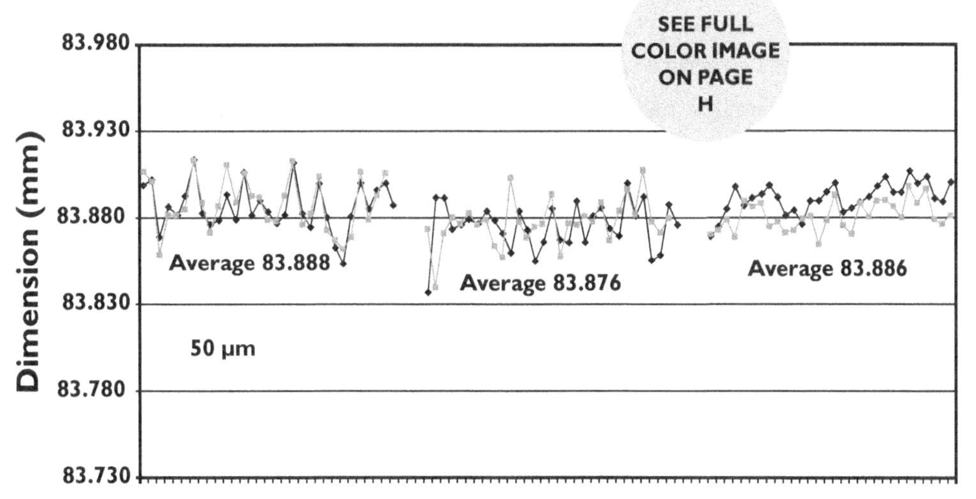

Other Technologies

VACUUM THERMAL RECYCLING

A commercialized recycling process called vacuum thermal recycling (VTR) is available to industry. This technology is relatively new and can be used on certain industrial and special wastes (e.g., mercury-containing waste, PCB-contaminated materials and oil-laden grinding sludge) for the separation of products with the simultaneous recovery of certain base materials. The technology was awarded the 1996 Philip Morris Research Prize.

VACUUM FREEZE DRYING

Equipment for freeze drying is another technology available to industry. The principle of freeze/sublimation drying is based on the fact that below its triple point, water only exists in the solid and the gaseous states. The freeze-drying process involves two distinct process steps. Step one is the freezing of the material below the triple-point temperature, and step two involves removing the ice crystals. Since the freezing process has a great influence on the quality of the finished product, this technology ensures that the product will be kept in the freshest possible condition.

VACUUM METALLURGY

Assorted technologies fall under the umbrella of vacuum metallurgy. These processes are used for the production of substances of higher purity and for materials and alloys requiring extremely high performance characteristics. Applications include aircraft turbine blades, ingots of tool steels for mining and milling cutters, superalloy ingots for aerospace and power turbines, alloys for the chemical industry, and high-purity metals for electronics and telecommunications. All of these processes use various methods for melting and solidifying materials. The various methods and equipment for this industry can be summarized as:

- Vacuum induction melting (VIM)
- Vacuum induction degassing and pouring (VIDP)
- Electroslag remelting (ESR) and various proprietary process variations
  - Increased pressure (PESR)
  - Inert gas atmosphere (IESR)
  - Reduced pressure (VAC-ESR)
- Vacuum arc remelting (VAR)
- Vacuum precision casting (VPC)
- Vacuum inert-gas metal-powder technology (VIGMPT)
- Vacuum isothermal forging (VIF)
- Vacuum turbine-blade coating using EB/PVD coating applications

The vacuum-metallurgy equipment industry deals with a client base having diverse purchasing philosophies (Fig. 28.21). North American firms buy new equipment sporadically, and their orders during peak periods typically involve large (20-30%) overcapacity, while Europe and Asia purchase smaller systems on a more consistent basis in response to increasing capacity issues. The retrofit market is quite large, especially in North America.

FIGURE 28.21 | Vacuum-metallurgy equipment market (by process)

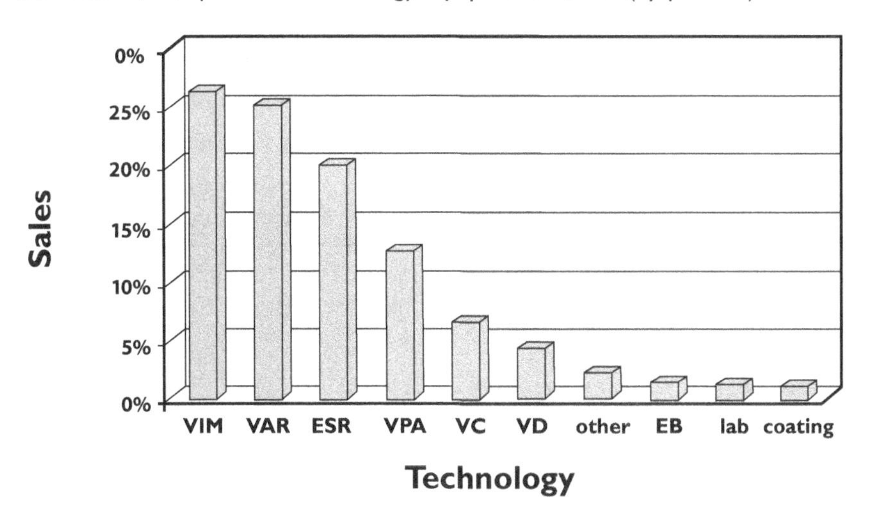

Vacuum metallurgy is an exciting growth technology for the future. New developments in materials, manufacturing methods and product performance will fuel growth in this area. The range of applications is extremely diverse, but all share the commonality of a high-performance expectation. Several examples designed to highlight the diversity of materials and processes are shown in Table 28.1.

TABLE 28.1 | Examples of vacuum-metallurgy applications and industries

Application	Typical industry
Engine and land-based turbine blades, vanes and discs	Aerospace, industrial
Biomedical implants and devices (orthopedic)	Industrial (medical)
Pump castings and valves	Industrial (nuclear)
Rare-earth magnets	Automotive (sensors, control devices)
Semi-conductor and fiber-optic devices	Industrial (telecommunications, electronics)
Structural components	Aerospace, industrial (military, nuclear)
Sputter targets	Industrial (electronics)
Golf-club heads	Industrial (recreational)

The melting of metals and metal alloys (Fig. 28.22) is a broad and diverse field. The materials that are melted range from aluminum to zirconium and almost all metals in between. Melting is the application of heat energy to change the state of a material, or combination of materials, from solid to liquid. Depending on the mix and the resultant properties desired, melting could be performed in a wide variety of ways in a vast array of furnace equipment.

FIGURE 28.22 | Metal melting

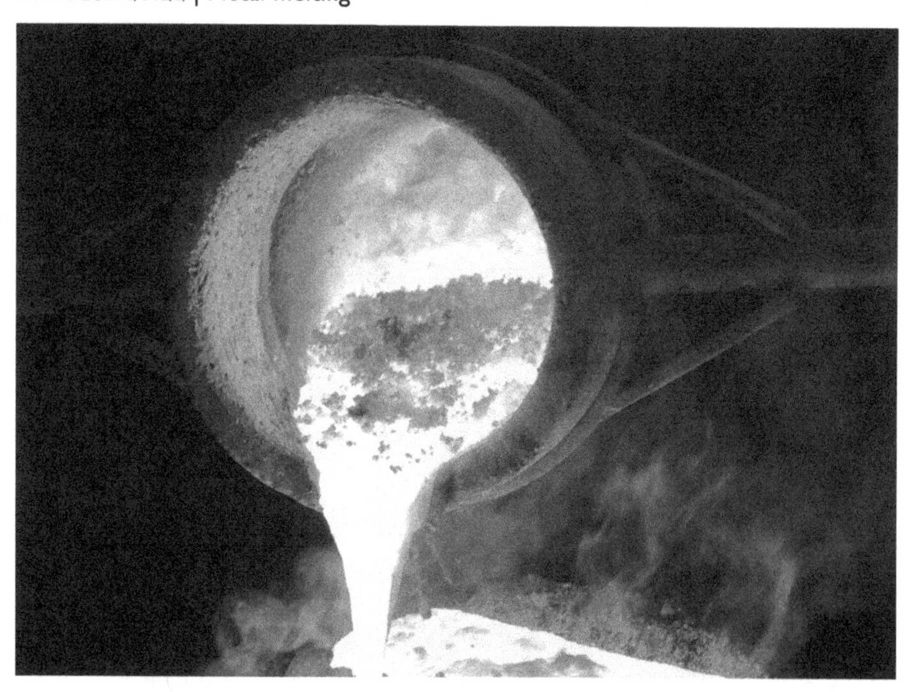

Metal-melting techniques can be used to make products as diverse as single crystals and automobile engine blocks, and they can produce components weighing a few grams or thousands of tons. Techniques involve the production of both wrought and powder-metal materials. Virtually any shape that can be cast can be produced. Metal-melting materials are typically divided into ferrous (iron and steel) and nonferrous (aluminum, copper, nickel and cobalt alloy) categories.

A wide variety of furnaces are used for melting metals. They are divided into two classes: conventional (non-vacuum) and vacuum. For ferrous materials, these vary from electrically heated induction units processing specialty grades to cupola furnaces fueled by coke. Nonferrous operations are performed in fuel-fired reverb furnaces, blast furnaces and electric units. Typically, the molten metal is poured from the furnace by tilting or tapping, and the metal is then continuously cast or introduced into a mold for cooling.

Vacuum Melting

Vacuum melting (Fig. 28.23) is the melting of material in a vacuum environment. Although more costly than conventional melting technology and therefore vastly limited in terms of overall tonnage produced, the advantages of these processes are in the production of substances of great purity for industries such as telecommunications, optics, biomedical and electronics. There are also process advantages for materials and alloys having extremely high performance characteristics for industries such as aerospace, nuclear, military, chemical, transportation and power distribution.

FIGURE 28.23 | Vacuum melting

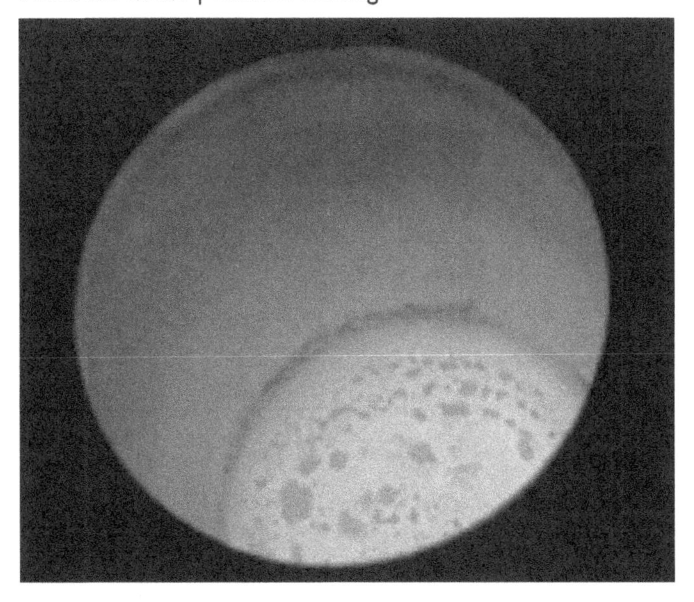

These various applications require different methods for melting and solidifying materials. The general categories are:

- Melting
  - Induction (ingot production, casting, degassing/pouring)
  - Plasma (arc, centrifugal)
  - Electron beam
- Remelting
  - Vacuum arc
  - Electroslag
- Casting
  - Consumable and non-consumable electrode
- Metal-powder production
  - Powder atomization

- ❖ Forging
- ❖ Coating
- ❖ Laboratory and light industrial
  - Directional solidification and single crystal

FIGURE 28.24 | Typical vacuum melting installation – 16-ton vacuum induction degassing and pouring unit shown in the pouring position (courtesy of ALD Vacuum Technologies GmbH)

Vacuum Induction Melting (VIM)

Vacuum induction melting (VIM) is one of the most commonly used processes in secondary metallurgy. It is applied for refining, treatment in the liquid state and adjustment of chemical composition of alloys (Fig. 28.25). VIM can be used for the refining of metals that contain significant quantities of reactive elements such as titanium and aluminum. Impurities are removed by chemical reaction, dissociation, flotation and volatilization. The result is a product that is both homogeneous and low in inclusions.

FIGURE 28.25 | Schematic illustration of a vacuum induction melter (courtesy of PV/T Inc.)

The process involves melting of metal via electromagnetic induction under vacuum in a refractory-lined crucible surrounded by an induction coil connected to a power source. The metal is melted, refined and then poured into molds under vacuum or inert gas.

VIM applications include:

- ❖ Directional solidification
  - Single-crystal casting
- ❖ Alloy or electrode casting
- ❖ Semi-finished products
  - Wire, strip, rod
  - Ingots and electrodes
  - Targets and flakes
  - Powders
- ❖ Distillation
- ❖ Degassing and pouring

Vacuum Arc Remelting (VAR)

Vacuum arc remelting (VAR) is a technology typically utilized for the production of titanium, nickel and steel ingots. The process is widely used to improve the cleanliness and refine the structure of standard air-melted or vacuum induction-melted ingots, which become the consumable electrodes. In addition, homogeneity and mechanical properties such as fatigue and fracture toughness of the final product are enhanced.

The primary benefits of remelting a consumable electrode under vacuum include the removal of dissolved gases (such as hydrogen, nitrogen and CO), reduction of undesired trace elements with high vapor pressure, improved oxide cleanliness, avoidance of macrosegregation and reduction of microsegregation. The technology is particularly well suited for the melting of reactive metals (titanium, zirconium and their alloys) used in the aerospace, chemical and nuclear industries.

VAR is in direct competition with electroslag remelting (ESR) for production of specialty-steel ingots. VAR and ESR both use an electric current to gradually remelt a solid electrode into a water-cooled crucible. VAR is the dominant process. Capital and production costs for the two processes are roughly equal, so there is no clear-cut cost incentive for a rapid shift to ESR. The higher metallurgical qualities of ESR may provide the rationale for such a shift in the future.

VAR applications include:

- Superalloys for aerospace
- High-strength steels for rocket booster rings and high-pressure tubes
- Ball-bearing steels
- Tool steels (cold- and hot-work steels) for milling cutters, drill bits, etc.
- Die steels

Electroslag Remelting (ESR)

Electroslag remelting (ESR) techniques are used by the ferrous and nonferrous metals industry to produce high-purity, structurally homogeneous metal shapes with smooth surfaces.

In ESR, the metal to be processed is formed into an electrode and dipped into a chemical slag pool within a water-cooled mold. An electric current is run from the electrode through the slag pool to a base plate, which acts as a second electrode. Heat produced by this current melts the tip of the metal electrode, forming molten droplets. These droplets fall into and through the slag pool, where impurities are chemically removed. An underlying mass of molten metal is then formed in a mold at the bottom of the slag pool. This material solidifies by controlled conduction to the cool walls of the mold, producing a homogeneous grain structure. As the mold is withdrawn, a thin slag layer solidifies on the outer surface of the metal, producing a smooth surface finish.

ESR ingot production is widely used to refine nickel and cobalt-based alloys for aerospace applications. The process is less widely used for stainless, carbon and low-alloy steels. It provides some advantages for production of large ingots for gears, rolls, rotors and dies. ESR is also used as a secondary remelting process for high-alloy and specialty steels (i.e. tool steels). ESR may be employed to produce ingots of certain hard-to-machine refractory, reactive and heavy metals such as titanium, zirconium and molybdenum. The advantage in this latter application is that the high surface qualities achieved can significantly reduce machining requirements.

ESR applications include:

- Tool steels for milling cutters, mining, etc.
- Die steels for the glass, plastics and automotive industries
- Ball-bearing steels
- Steels for turbine and generator shafts
- Superalloys for aerospace and power turbines
- Nickel-based alloys for the chemical industry

Several process variations have been introduced into ESR technology, which include remelting under inert-gas atmosphere, increased pressure and reduced pressure. The use of inert-gas atmosphere techniques appears to have significant growth potential in the ESR market.

Electron-Beam Melting

Electron-beam (EB) melting is distinguished by its superior refining capacity and offers a high degree of flexibility of the heat source. Therefore, it is ideal for remelting and refining of metals and alloys under high vacuum in water-cooled copper molds. The process is mainly employed for the production of refractory and reactive metals (tantalum, niobium, molybdenum, tungsten, vanadium, hafnium, zirconium, titanium) and their alloys. It plays an important role in manufacturing ultra-pure sputtering target materials and electronic alloys and in recycling titanium scrap.

EB applications include:

- Remelting of high-purity materials such as Nb and Ta
- Production of titanium for the chemical and aerospace industries
- Production of zirconium for the chemical industry
- Production of high-purity metals for electronic applications (e.g., sputtering targets)

Vacuum Precision Casting

The type of product to be produced has a great influence on the type of vacuum precision melting equipment required (Fig. 28.26).

FIGURE 28.26 | Schematic of vacuum precision casting unit (courtesy of PV/T Inc.)

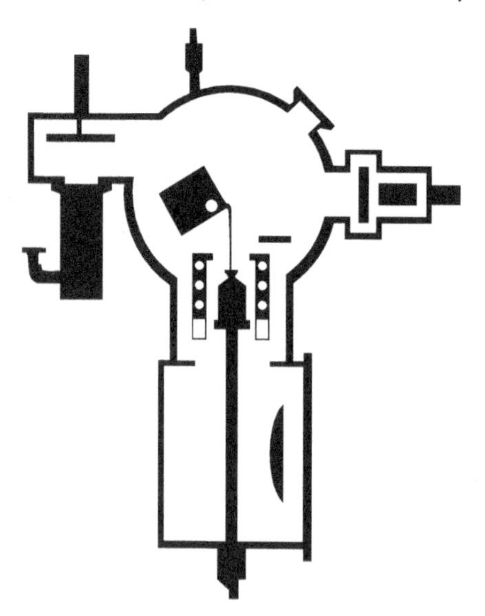

Investment Castings

The majority of vacuum investment castings (e.g., turbine and vanes for the aircraft and industrial gas-turbine industries made from Ni-base superalloys) are produced in vacuum induction melting and precision-casting furnaces using tilt pour as well as ceramic crucible options with bottom-pouring techniques (Fig. 28.27). Static- and centrifugal-casting techniques can be applied.

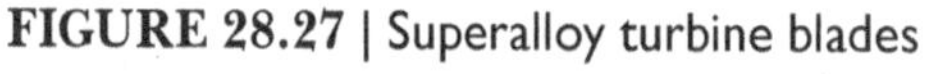

FIGURE 28.27 | Superalloy turbine blades

High melt volumes are typically required for investment and permanent-mold casting technology using ceramic-free, water-cooled copper crucibles. Tilt-pour and bottom-pouring systems are used. Static and centrifugal investment molds or permanent molds made of special alloys are widely used.

Precision Casting Molds

Titanium, with its high affinity for oxygen, is often poured into precision-casting molds. A consumable titanium electrode is melted with an electric arc into a water-cooled crucible. When the desired fill level in the crucible is obtained, the electrode is automatically and quickly retracted, and the molten titanium is poured into the mold, usually by tilting.

The applications for vacuum precision casting include:

- Structural and engine parts (titanium castings) for the aerospace industry
- Golf-club heads (titanium)
- Titanium-aluminide automobile valves
- Biomedical implants
- Zirconium pump casings and valves (chemical industry and offshore drilling)

Electroslag Casting (ESC)

Electroslag casting (ESC) is used for the manufacture of finish-shaped or near-net-shape components by remelting consumable electrodes in water-cooled molds of special designs. Because molds in electroslag furnaces may be widely varied in shape, ESC can produce complex geometries that minimize or eliminate subsequent processing steps such as rolling, forming and forging. ESC can produce very large and complex castings, such as nuclear valves and tube fittings, crankshafts for marine diesel engines, gears and rolls, tools, dies, and die blocks. ESC has several additional advantages over the more conventional processes of EAF melting, ladle refining and vacuum degassing followed by rolling, forging and casting.

Vacuum Isothermal Forging

The methods used to achieve full density and full strength for the final consolidation step in powder metallurgy include: sintering, hot isostatic pressing (HIP), hot extrusion and forging.

Of these methods, isothermal forging uses super-plastic deformation, which is done by exerting extremely low deformation strain rates over a tightly controlled temperature band. If the forging is done under these conditions, the resulting stresses that occur in the component are quite small, and the grain size remains nearly unchanged. The technology offers near-net-shape potential with resultant savings in materials and subsequent machining operations. High-frequency induction systems are used to accomplish this process.

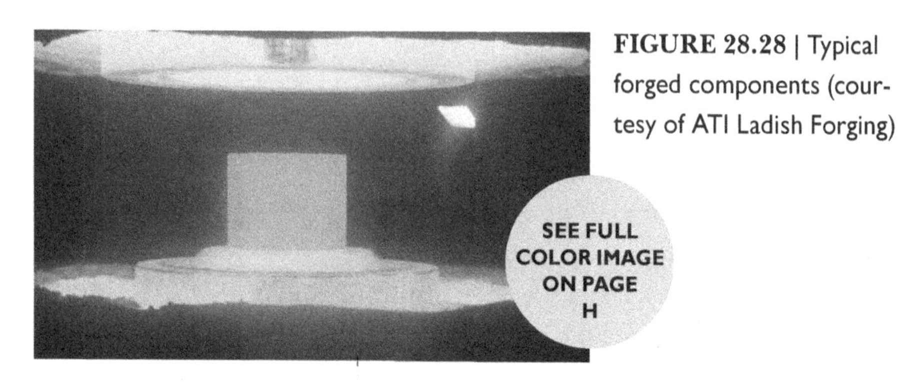

FIGURE 28.28 | Typical forged components (courtesy of ATI Ladish Forging)

Vacuum Powder Production

The process steps involved in the production of metal powders are melting, atomizing and solidifying of the respective metals and alloys. Vacuum metal-powder technology under inert gas is an established production method. Other metal-powder production methods (such as ore reduction and water atomization) are limited with respect to special powder blends, achievable quality criteria (such as particle geometry, particle morphology) and chemical purity.

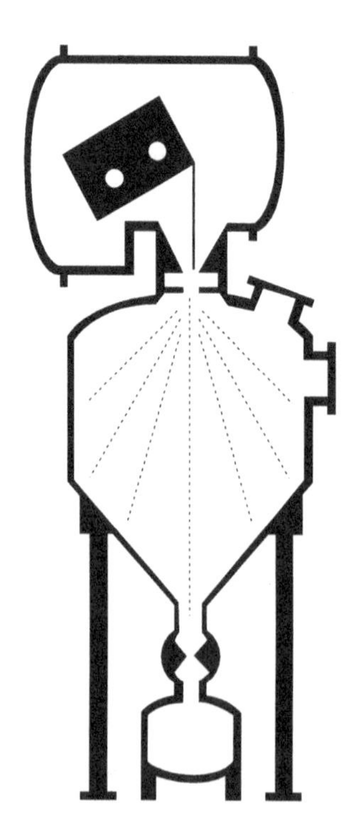

FIGURE 28.29 | Schematic illustration of vacuum powder production unit (courtesy of PV/T Inc.)

Melting techniques are used in conjunction with three powder production technologies: gas atomization, spin-cup atomization and spin-wheel casting. The spin-wheel casters produce a very fine flake, which is usually further reduced in size. Metal can be melted by plasma, consumable arc, and hot- or cold-wall induction.

Metal-powder applications include:

- Nickel-based superalloys for the aviation industry and power engineering
- Solders and brazing metals
- Hydrogen-storage alloys
- Wear-protection coatings
- MIM powders for components
- Magnetic alloys
- Sputter target production for electronics

Laboratory and Light Industrial

Laboratory and light industrial furnaces (Fig. 28.30) are suited for many common heat-treatment, melting and distillation processes in which reactivity or the special properties of the material, metal-based alloy or mixture require a complete or partial treatment under vacuum or protective gas. A major use of this equipment is for directionally solidified (DS) and single crystals (SX).

In order to produce these components, high thermal gradients and constant-temperature solidification fronts are required.

Laboratory and light production applications include:

- Rare-earth metal production (yttrium, samarium, uranium aluminide)
- Optical coating materials
- Battery recycling
- Sintering of ceramics
- Distillation of metallic scrap and metals
- Ultra-pure materials for semiconductor devices and fiber optics
- Heat treatment of ceramics and glass

FIGURE 28.30 | Typical schematic of laboratory and light-industrial vacuum system (courtesy of PV/T Inc.)

Vacuum Coating

Among various vacuum-coating methods, electron beam/physical vapor deposition is characterized by the use of a focused high-power electron beam that melts and evaporates metals as well as ceramics (Fig. 28.31). The process results in high deposition speed in many cost-effective applications.

Coating applications include:

- Optical coatings for lenses and filters
- Coating of packaging webs in semiconductor manufacture
- Coating of blades and vanes

FIGURE 28.31 | Single-crystal coated turbine blades

Turbine blades and vanes are precision cast superalloy, single-crystal components with cooling air passages, a bond coating, a diffusion barrier and a thermal barrier coating.

Future Trends

Vacuum heat treating has been identified as a core technology of the future. It is positioned such that it will be able to take advantage of new developments in materials, manufacturing methods and product development. The range of applications is almost limitless. Emerging industries (Table 28.2) include:

- Energy and environmental (emissions reduction, materials recycling)
- Communications, biomedical, electronics
- Materials development (ceramics, composites, powder metallurgy)

These technologies are consistent with identified high-tech growth markets.

TABLE 28.2 | Examples of vacuum heat-treating applications and industries

Application	Typical industry
Batteries, fuel cells, solar-power devices	Automotive, industrial products (power generation)
Biomedical implants and devices (orthopedic, orthodontic, vascular)	Industrial products (medical, dental)
Engine and land-based turbine blades	Aerospace, industrial products
Fuel systems (diesel injectors)	Automotive
Nuclear fuel elements	Industrial products (including public utilities and research & development)
Sensors and permanent magnets including rare-earth materials	Aerospace, automotive
Transmission components (powertrain)	Aerospace, automotive
Tools and accessories (machine and poser tool devices)	Industrial products
Light guides (fiber optics)	Industrial products

The heat-treatment industry has also recognized the following industry technology development areas as key in the advancement of the industry as a whole. Areas where vacuum technology is applied as a solution are shown with an asterisk (*). They include:

- Energy-efficient equipment (goal: reduced energy consumption)
  - Accelerated-heating (high heat transfer) techniques and equipment*
  - High-efficiency insulation
  - Improved heat transfer during quench*
  - Advanced burner technology
  - Alternative heat sources
- Processing with minimal part distortion (goal: eliminate secondary manufacturing operations)
  - Gas pressure quenching*
  - Emulation of salt and oil quenching
  - Predicted quenching*
  - Probes/controls to measure heat-transfer characteristics
- Diffusion-related process optimization (goal: predict behavior)
  - Carburizing and carbonitriding*
  - Higher temperatures to shorten cycles*
  - Boost atmosphere potentials*
  - Gas/solid reaction kinetics
  - Nitriding and ferritic nitrocarburizing
  - Aging and tempering*
- Environmentally friendly by-products/emissions (goal: zero emissions)
  - Cleaning
  - Vacuum vapor-degreasing technology*

- Improve aqueous systems
- Water quality
- Air quality (NOx and CO emissions)
- Non-polluting quenchants

❖ Adaptability/flexibility for advanced materials (goal: improved product performance using advanced technology solutions)
  - Development of new carburizing alloys
  - Universal high-temperature materials
  - Grain-growth control technology
  - Steelmaking practices
  - Influence of chemistry/trace elements
  - Advancement of recent developments (cold-chamber quenching, low-pressure carburizing)*

❖ Process control anticipating intelligent sensors (goal: eliminate need for post-mortem failure analysis)
  - Automated systems
  - Anticipatory control
  - Artificial intelligence
  - Smart sensors
  - Predictable control behavior (heating rates, cooling rates)

❖ Use of heat-treatment modeling (goal: predict outcome before heat treatment)
  - Process prediction (composition, hardness, stress state, microstructure)
  - Planned preventive maintenance and prediction
  - Heat-treating behavior

❖ Integration into manufacturing (goal: lower manufacturing cost)
  - Batches of one
  - Continuous-flow equipment*
  - Lighter-weight fixtures (carbon/carbon composite, ceramics)*
  - Improved up-time productivity (advanced furnace materials, maintenance improvements)

MARKET-DRIVEN TECHNOLOGIES

Consumer demand has resulted in extremely high growth rates in a number of key industries where products experience rapid mortality rates. This has created a need for the continued development of new structural materials and manufacturing methods to play key roles in areas such as:

❖ Communication devices

- Biomedical products
- Microelectronic devices
- Power transmission equipment
- Power distribution equipment
- Powder-metallurgy components
- Coating technology
- Sensors
- Optical devices

Vacuum equipment best meets the needs of these rapid-growth industries.

Future Industry Needs

The future needs of the heat-treatment and vacuum-metallurgy segments are best viewed by the type of industry to be served.

ENERGY AND RAW MATERIALS

Energy conservation, the development of new energy-efficient equipment and reduced energy usage in the manufacture of raw materials are essential to the growth of the world economy as well as to contributing to the solution of ecological problems. The application of vacuum technology best meets these needs and continues the advancement of new, environmentally friendly technologies in a wide application market. The general economic trend for the use of vacuum technology in these areas has and will continue to exhibit remarkable growth.

AEROSPACE

Within the aerospace industry, high growth rates have been, and are expected to continue to be, found in the following technology areas:

- Development and manufacture of new types of high-temperature, high-performance alloys
- Manufacture of large aircraft turbine-engine components
- Creation of development and manufacturing partnerships

In addition, the need for and use of superalloys in aircraft engines has led to major investments in facilities for making these products in the North America, Asia and Europe.

AUTOMOTIVE

Technology trends in the automotive industry concentrate on reducing manufacturing costs by introducing new manufacturing methods, procedures and processes aimed at improving product quality. The integration of new heat-treatment technology into production lines is an essential element of this strategy.

New technology developments are resulting in improved engine performance (drive systems and power transmissions), lower fuel consumption (fuel systems and devices) and reduced exhaust emissions (exhaust systems).

Low-pressure carburizing combined with high-pressure gas quenching in a continuous flow through an automated manufacturing system is at the heart of the needs of the automotive industry at this time.

ENVIRONMENTAL (THERMAL RECYCLING)

Worldwide interest in recycling technology remains high, but it is limited by the large capital investments necessary to implement the technology as well as legal and legislative issues in the heavily industrialized countries where these facilities are necessary. The struggle between the balance of environmental concerns and manufacturing needs is a major issue. Recognized areas of need include:

- Development of environmentally friendly energy-recovery technologies
- Waste disposal and water purification
- Legislation as a prerequisite for the advancement of environmental-protection policies

SYNCHRONIZED MANUFACTURING

Many industries are interested in integrating heat-treatment operations into their manufacturing operations (Fig. 28.32). In the gear industry, for example, components are collected in large batches after soft machining and sent to a centralized heat-treatment shop (captive or commercial) for processing. Upon their return, operations such as hard machining take place. This approach is being replaced by in-line systems (Fig. 28.33) reportedly offering fast turn-around times and more predictable results. Both gas and oil quenching and even press quenching is possible in these units. Vacuum systems (Fig. 28.34) are a natural choice since they can be integrated into the manufacturing flow, offer high throughput capacity and are environmentally friendly.

FIGURE 28.32 | Synchronized manufacturing line [2]

FIGURE 28.33 | Work-flow comparison – centralized vs. integrated heat-treatment shop [2]

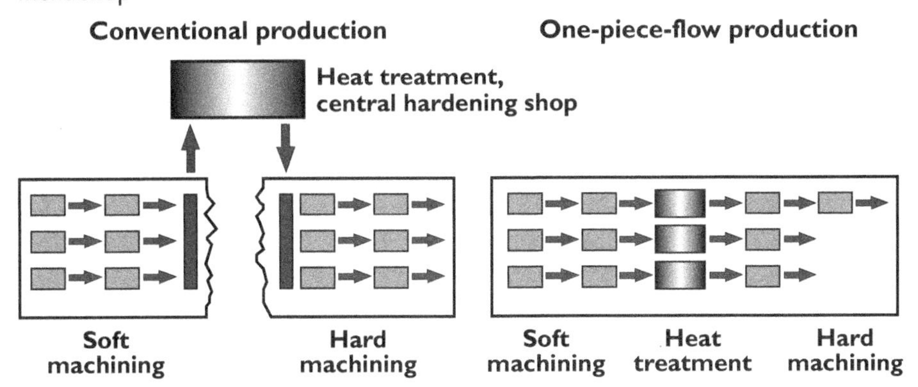

FIGURE 28.34 | Vacuum carburizing system for synchronized manufacturing (courtesy of ALD-Holcroft)

In Conclusion

Vacuum technology has a promising future in industries served by heat treatment and vacuum metallurgy due in large part to its inherent flexibility, repeatability and absolute process control. The type of vacuum equipment, processes and system configurations will be dictated by product performance demands and quality-assurance requirements.

REFERENCES

1. Shivanath, Rohith, "Vacuum Heat Treatment of Cr-Mn P/M Steels," Vacuum Carburizing Conference, Port Huron Mich., July 2007
2. Heuer, Volker, Klaus Loeser, Gunter Schmitt and Karl Ritter, "Integrated Heat Treating," *Gear Solutions*, June 2011

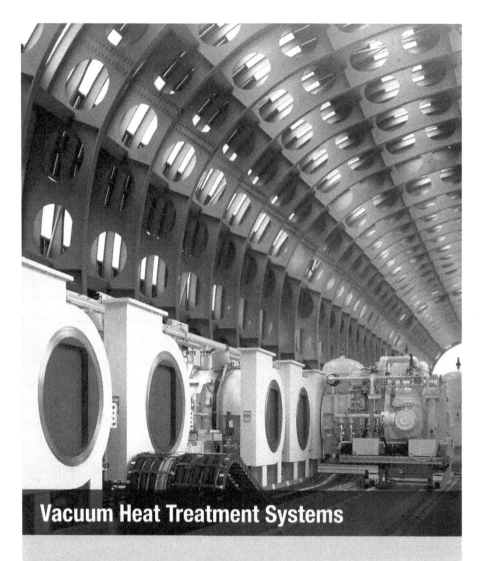

CHAPTER 29

VACUUM TERMINOLOGY AND USEFUL REFERENCE MATERIALS

Definitions: Vacuum Terms, Related Terminology

Absolute pressure – Pressure measured above the zero value of a perfect vacuum. It is designated psia (pounds per square inch absolute).

Absolute zero – The point at which atomic motion stops. The zero point on the absolute temperature scale (0 K).

Absorption – The binding of gas in the interior of a solid (or liquid).

Adsorbate – A material that takes in gas by adsorption.

Adsorption – The condensing of gas on the surface of a cooled solid.

Air-inlet valve – A valve used for letting atmospheric air into a vacuum system or chamber. It is also called a vacuum break or vent valve.

Alloy – The addition of other elements to a pure metal. Examples include steel (iron plus carbon and possibly other elements), brass (copper plus zinc), bronze (copper plus tin), solders (tin plus lead and possibly other elements) and nickel-silver (copper plus nickel and zinc).

Annealing – Heating above the critical (Ac_1) temperature and holding at a suitable temperature followed by cooling at a suitable (slow) rate for purposes such as reducing hardness, improving machinability, facilitating cold working, producing a desired microstructure (for subsequent operations), reducing internal stresses, or for obtaining desired mechanical, physical or other properties. Cooling is often performed in the furnace. Other common names and related subjects include black annealing, intermediate annealing, isothermal annealing, malleabilizing, process annealing, quench annealing, recrystallization annealing and spheroidizing.

Anti-backstreaming valve – A valve or other device that prevents the migration of oil and air from a mechanical vacuum pump into the system when the pump stops and the system is under vacuum.

Atmospheric pressure – The pressure exerted by a mercury column 760 mm high at 0°C under a standard acceleration of gravity (980.665 cm/sec^2); 14.7 psi at sea level.

Backstreaming – The backward flow of the working fluid of a pump upstream toward the vessel being evacuated.

Belt-drive pump – A pump with the motor drive and pulley system attached to the pump shaft. The ratios of the diameters of the pump and motor pulleys determine the actual rotational speed of the pump.

Boyle's Law – One of the gas laws. Boyle's law states that pressure and volume in a gas are inversely proportional (assuming constant temperature and mass).

Braze alloy – An alternative term for the brazing filler metal (BFM).

Braze joint – The junction of members or the edges of members that are to be bonded or have been bonded by brazing.

Brazing – Joining of materials by flowing a nonferrous filler metal into the space between them via capillary action. Bonding results from the intimate contact produced by dissolution of a small amount of base metal into the molten filler metal, without melting of the base metal. The term brazing is used when the filler metal has a liquidus temperature of 450°C (840°F) or higher.

Carbonitriding – The introduction of carbon and nitrogen into a solid ferrous alloy by holding above the temperature at which austenite begins to form – the critical (Ac_1) temperature of the material in contact with a suitable source of carbonaceous material and ammonia. The carbonitrided alloy is usually quenched to harden the case. Other common names and related subjects include case hardening, dry cyaniding, gas cyaniding, nicarbing and nitrocarburizing (obsolete).

Carburizing – The introduction of carbon into a solid ferrous alloy by holding above the temperature at which austenite begins to form – the critical temperature (Ac_1) of the material in contact with a suitable source of carbonaceous material, which may be solid, liquid or gas. The carburized alloy is usually quench hardened. Other common names and related subjects include case hardening, gas carburizing and cyaniding.

Carburization – A process that increases the carbon content of a steel surface.

Case hardening – Hardening a ferrous alloy so that the outer portion, or case, is made substantially harder than the inner portion, or core. This is accomplished by increasing the surface carbon content and quenching or by selectively hardening the surface using applied energy techniques (flame, induction, laser).

Cold trap (condenser) – A trap designed to hold a refrigerant, or cooled by coils in which a refrigerant circulates, inserted into a vacuum system for vapors present in the system to condense on.

Cold wall – A style of vacuum furnace in which the exterior vacuum-vessel wall is cooled by use of a liquid (e.g., water or a water-glycol mixture).

Compression ratio – The ratio between the outlet pressure and the inlet pressure of a pump for a specific gas.

Compound pump – A mechanical pump having two or more stages of compression.

Condensation rate – The number of molecules that condense on a surface per cm^2/sec ($feet^2/sec$).

Conductance – The actual capacity of a piping system in cubic meters (feet) per minute. Conductance in a vacuum system can be limited by line size and configuration.

Continuous Vacuum – A style of vacuum furnace in which the workload is conveyed through the furnace. Multiple loads are typically inside these types of furnaces at one time, and different processing conditions are possible.

Crystalline – A material consisting of particles arranged in a repetitive pattern is regularly known as crystalline. The atoms are arranged in fairly simple geometric patterns in all metals and metal alloys.

Degassing – The removal of gas and vapors from a substance under vacuum.

Diffusion bonding – A process by which two materials in contact with one another are joined by diffusion of atoms between the two materials.

Diffusion pump – A vacuum pump that uses an oil vaporizing/condensing action at low pressure to entrap gas molecules from the vacuum chamber and exhaust them from the system.

Direct-drive pump – A pump with the motor drive provided by a direct coupling to the pump rotor shaft. The rotational speed of the motor is the rotational speed of the pump.

Displacement – The geometric volume swept out per unit time by the working mechanical pumps at normal rotational speeds. It is also called free-air displacement. This value, being theoretical, is not usable by end-users. It is mainly a standard used by vacuum-pump manufacturers.

Double pump-down – A technique used to improve overall pump-down rate by first evacuating the vacuum vessel to a preset vacuum level, backfilling the vessel to near atmospheric pressure with an inert gas and again evacuating to remove unwanted gaseous species from the chamber.

Filler metal – The metal or alloy to be added in order to produce the braze joint.

Foreline – Vacuum line connecting the vacuum system to the inlet of a vacuum pump.

Gauge pressure – Pressure measured at atmospheric pressure as a reference point. Gauge pressure is designated psig (pounds per square inch gauge).

Gas – Gas is defined as the state of matter in which the molecules are practically unrestricted by intermolecular forces so that the molecules are free to occupy any space within an enclosure. In vacuum technology, the word gas has been loosely applied to both the permanent (non-condensable) gases and vapors, or condensable gases.

Gas ballast – The venting of the compression chamber of a mechanical pump to the atmosphere to prevent condensation of condensable vapors within the pump.

Gas quenching – A quenching process that rapidly cools the workload using either a gas (e.g., argon, helium, hydrogen, nitrogen) or gas mixture at pressures ranging from sub-atmospheric to 25 bar or higher.

Gate valve – A type of vacuum valve commonly located between the diffusion pump and the main chamber, between the mechanical pump and the vacuum chamber, and/or between the mechanical pump and a diffusion pump.

Hardening – A process producing an increase in the hardness of a material by suitable treatment, usually involving heating and cooling. Other common names and related subjects include neutral hardening, direct hardening, quench hardening and surface hardening.

Hearth – A support platform or mechanism upon which a workload is placed for processing.

High vacuum – 10^{-3} to 10^{-6} torr

Hot zone – The internal part of the vacuum furnace into which the workload is placed for heating (and perhaps subsequent cooling). The hot zone is typically an insulated chamber containing heating elements and some type of workload support structure.

Hot wall – A style of vacuum furnace in which the exterior vacuum-vessel wall is cooled without the use of a liquid medium. Alternately, a vacuum vessel that is placed inside an insulated furnace and heated by an external source.

Ideal gas – A gas that obeys Boyle's Law and has zero heat of free expansion (Charles' Law). It is also known as a perfect gas.

Ideal Gas Law – A mathematical expression for the relationship exhibited by any gas showing ideal gas behavior. This relationship is expressed by the formula PV = nRT; where P is pressure, V is volume, n is the number of moles of the gaseous species, R is the ideal gas constant and T is the temperature.

Impedance – The reciprocal of the conductance. It is also called resistance.

Implosion – The rapid inward collapsing of the walls of a vacuum system or device as the result of failure of the walls to sustain the external atmospheric pressure.

Inlet pressure – The total pressure measured at the inlet of a vacuum pump.

Ionization gauge – A vacuum measurement gauge containing a hot tungsten filament that is used to measure vacuum pressure below about 10^{-3} torr. Available in hot- or cold-cathode configurations.

Isolation valve – A valve that seals off a vacuum system from the vacuum pump when the pump is off.

Leak – An opening or porosity in the wall of a vacuum system.

Leak detection – A process by which a leak can be detected in a vacuum system by a variety of methods. The most common method in heat treating is by use of a helium mass spectrometer.

Leakage rate – The rate of gas flow through a leak or outgassing; pressure rise per unit time.

Liquidus – The lowest temperature at which a metal or alloy is completely liquid.

Low (rough) vacuum – Vacuum range from 1 torr to 10^{-3} torr.

Materials science – The science and technology of materials. Metals and nonmetals are included in this field of study.

McLeod gauge – A mercury-level vacuum gauge that measures the pressure in a vacuum system.

Mean free path – The average distance a gas molecule can travel before colliding with another gas molecule. The mean free path is dependent on the density of the gas and the diameter of the molecule.

Mechanical (wet) pump – A pump that moves the gas by the cyclic motion of a system of mechanical parts such as pistons, eccentric rotors, vanes, screws, valves, etc.

Metallurgy – A domain of materials science that studies the physical and chemical behavior of metallic elements. This is an all-inclusive definition that includes the study of processes involving the operation of various furnaces, the forging and rolling of steel, foundry operations, electrolytic refining of metals, sintering of metal powders, welding and heat treating.

Micron (of mercury) – Unit of pressure equal to 0.001 torr or 1 micron Hg. It is also known as a millitorr.

Millimeter of mercury – A unit of pressure (mm Hg) defined as that pressure that will support a column of mercury 1 millimeter high.

Multi-chamber – A style of vacuum furnace that consists of more than one chamber in which the processing cycle is conducted. Load movement between chambers is involved. A typical example would be a furnace in which the heating and cooling aspects of the cycle are conducted in separate chambers.

Nitriding – Introducing nitrogen into a solid ferrous alloy by holding below the critical (Ac_1) temperature in contact with a suitable nitrogenous material, which may be solid, liquid or gas. Quench hardening is not required to produce a hard case. Other common names and related subjects include case hardening and ion/plasma nitriding.

Nitrocarburizing – A case-hardening process similar to nitriding that involves the introduction of nitrogen and carbon into a solid ferrous alloy by holding below the critical temperature (Ac_1) in contact with a suitable nitrogenous and carbonaceous material, which may be solid, liquid or gas. A thin nitrogen- and carbon-enriched layer, possibly with accompanying carbonitrides and nitrides, is produced. The white, or compound, layer with an underlying diffusion zone contains dissolved nitrogen and iron (alloy) nitrides. Quench hardening is not required to produce a hard case. Other common names and related subjects include case hardening, ferritic nitrocarburizing (FNC) and austenitic nitrocarburizing (ANC).

Non-condensable gas – Permanent gases inside a vacuum system; not to be confused with vapors.

Normalizing – Heating a ferrous alloy to a suitable temperature above the transformation range (and typically above the suitable hardening temperature) and then cooling in air to a temperature substantially below the transformation range.

Oil quenching – A quenching process that rapidly cools the workload using a low vapor-pressure oil.

Oil separator – An oil reservoir with baffles to reduce the loss of oil by condensation in the exhaust of a mechanical vacuum pump.

Outgassing – The escape of gas from materials within a vacuum system. It can be a limiting factor in the ultimate pressure obtained.

Partial pressure – The pressure due to a gas or vapor component of a gaseous mixture when introduced into a vacuum vessel.

Pressure – The force per unit area a gas exerts. Common units are torr, millibar, microns, psia or millimeters of mercury.

Pump-down curve – A graph representing the relationship between pressure and time. It is used to determine the time required to achieve the desired operating pressure in a system with a given pump.

Pumping speed – The volume of gas per unit of time that the vacuum pump is able to remove from the system. Pumping speed is often expressed in CFM (cubic feet per minute), L/M (liters per minute) or L/S (liters per second).

Pumping system – The combination of vacuum pumps used to reduce the pressure inside the vacuum vessel from atmospheric pressure to processing pressure. A vacuum pumping system usually consists of one or more of the following pumps alone or in combination, namely a mechanical pump(s), booster (or blower) pump(s) and a diffusion pump(s) along with appropriate valving and controls.

Quenching – The process of rapidly cooling a heat-treated component from processing temperature in a suitable medium, including oil, high-pressure gas or water.

Rate of rise – The timed rate of pressure increase during a given interval in a vacuum system that is isolated from the pump by a valve.

Roots blower – A rotary blower pump that has a pair of rotating impellers. This type of pump is usually used in conjunction with a mechanical or backing pump to increase the pumping speed of a vacuum system over a certain pressure range.

Roughing time – The time required to pump a given system from atmospheric pressure to the crossover (cut-in) pressure for a vapor pump or other medium- or high-vacuum pump.

Rough vacuum – 760 torr to 1 torr

SCFM – Standard cubic feet per minute. The volume flow referenced at standard conditions, sea-level pressure (1013 mbar/760 torr) and 20°C (68°F).

Setting the braze – A term used to describe an intermediate cooling step whereby the filler metal is allowed to cool to a semi-stable (liquid/solid) phase prior to the onset of rapid cooling.

Single chamber – A style of vacuum furnace that consists of only one chamber in which the processing cycle is conducted. The workload remains stationary during the processing cycle.

Sintering – The bonding of adjacent powder particle surfaces in a mass of metal powders or a compact by heating. Also, a shaped body composed of metal powders and produced by sintering with or without prior compaction. Other common names and related subjects include cold/hot isostatic pressing, liquid-phase sintering and metal/powder injection molding.

Solidus – The highest temperature at which a metal or alloy is completely solid.

Stage – Operating unit of a vacuum pump. By combining stages in series, a higher compression ratio can be achieved.

Tempering – A heat-treating process primarily used to bring the metal to slightly lower hardness while improving ductility. Reheating a quench-hardened or normalized ferrous alloy to a temperature below the transformation temperature and then cooling at any rate desired. A reduction in strength and increase in ductility of the material generally results. Other common names and related subjects include draw, drawing and temper.

Thermocouple (T/C) – A device that measures temperature by the proportional change in electromotive force (EMF) generated by the junction of two dissimilar metal wires. Examples of thermocouple types include S, R, K, N and J.

Throughput – The quantity of gas in pressure-volume units at a specified temperature flowing per unit of time across an open cross section of a pump or pipe line.

Time of evacuation – The time required to pump a given system from atmospheric pressure to a specified base pressure. It is also known as pump-down time or exhaust time.

Total pressure – Sum of all the partial pressures in a gas mixture.

Trap – An accessory used to condense vapors present in the vacuum.

Torr – A unit of pressure defined as 1/760th of an atmosphere and equal to 1 mm of mercury.

Ultrahigh vacuum – 10^{-6} (Europe) or 10^{-7} to 10^{-10} torr (U.S.)

Ultimate pressure – The lowest attainable pressure in a vacuum system. In a vacuum pump, the lowest pressure that can be attained with that pump. Ultimate pressure is limited by the pumping speed of the vacuum pump and the vapor pressure of the sealing fluid, among other factors.

Vacuum – A space filled with gas at a pressure less than atmospheric pressure. Vacuum is classified as rough, medium, high, ultrahigh and extreme ultrahigh.

Vacuum cooling – A process for lowering the temperature of a material by subjecting it to vacuum conditions to cause vaporization of a liquid.

Vacuum drying – The removal of liquid by evaporation from a substance in a vacuum. When the liquid is water, the process is sometimes called vacuum dehydration.

Vacuum gauge – Any instrument used to measure pressure in a vacuum system. Examples include manometers, diaphragms, thermocouples and Pirani and McLeod gauges.

Vacuum manifold – Part of the vacuum-system piping; an enclosure with ports so that a number of vacuum processes can be operated simultaneously.

Vacuum system – A system designed for the manufacture of a product or the operation of a process. Consists of a vacuum pump or pumps, vacuum chamber, interconnecting piping and a variety of other components.

Vacuum vessel – The main chamber that houses the hot zone, workload and support components. It is evacuated prior to running a metallurgical process.

Vapor – A substance in gas phase that is condensable at ambient temperatures.

Vapor pressure – The partial pressure exerted by a vapor.

Wetting – The phenomenon in which a liquid filler metal or flux spreads or flows and adheres in a thin, continuous layer on a solid base metal.

Reference Material

FIGURE 29.1 | Vapor pressure curves (c.f. Chapter 7) [3]

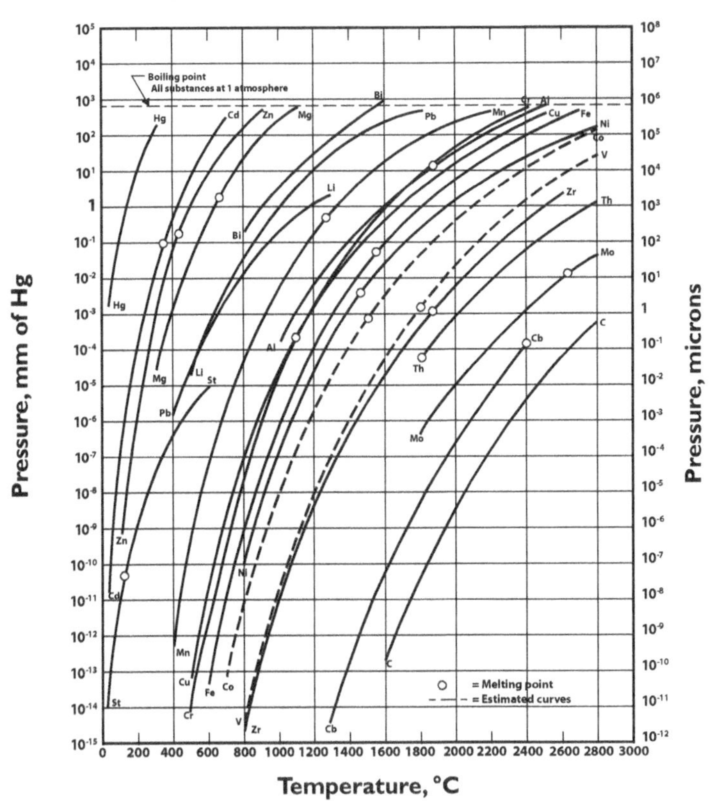

FIGURE 29.2 | Vapor pressure curves (c.f. Chapter 7) [3]

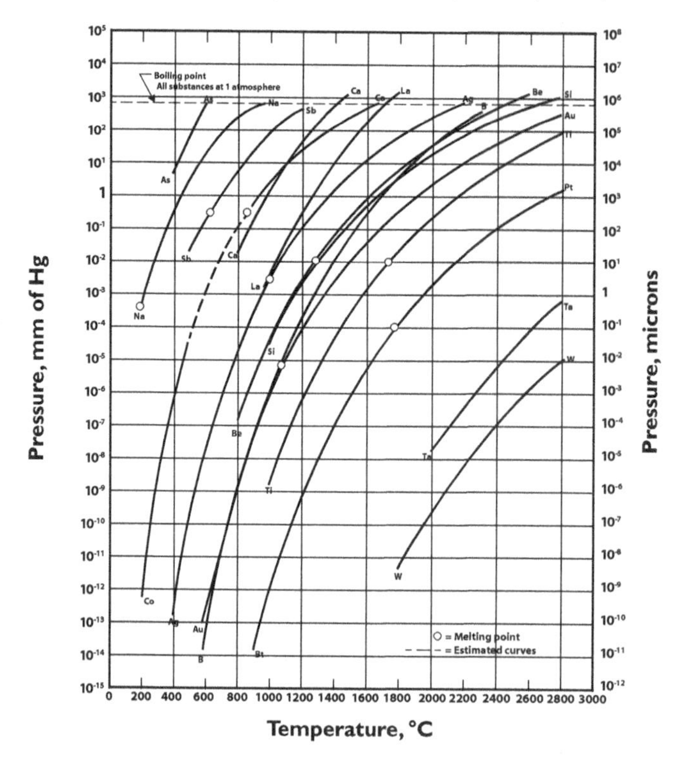

FIGURE 29.3 | Ellingham Diagram No. 1 (c.f. Chapter 7) [4]

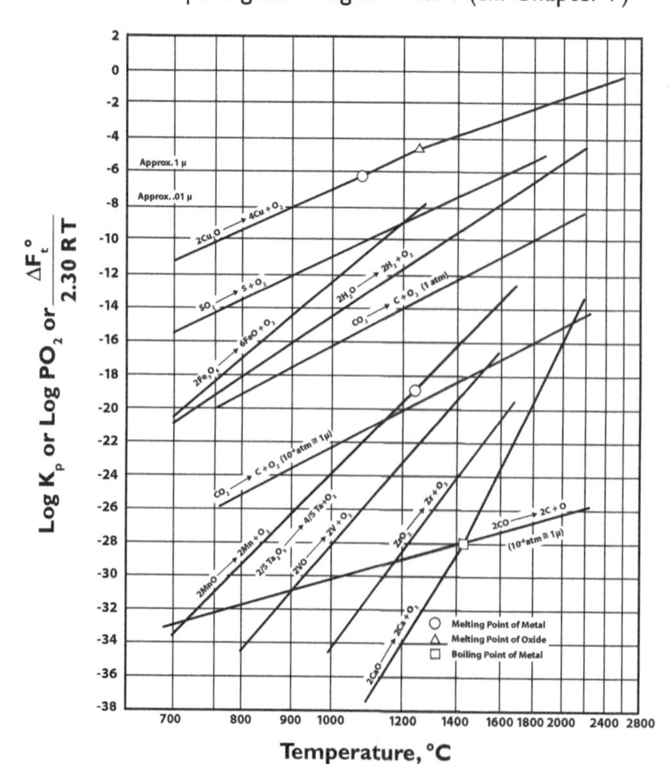

FIGURE 29.4 | Ellingham Diagram No. 2 (c.f. Chapter 7) [3]

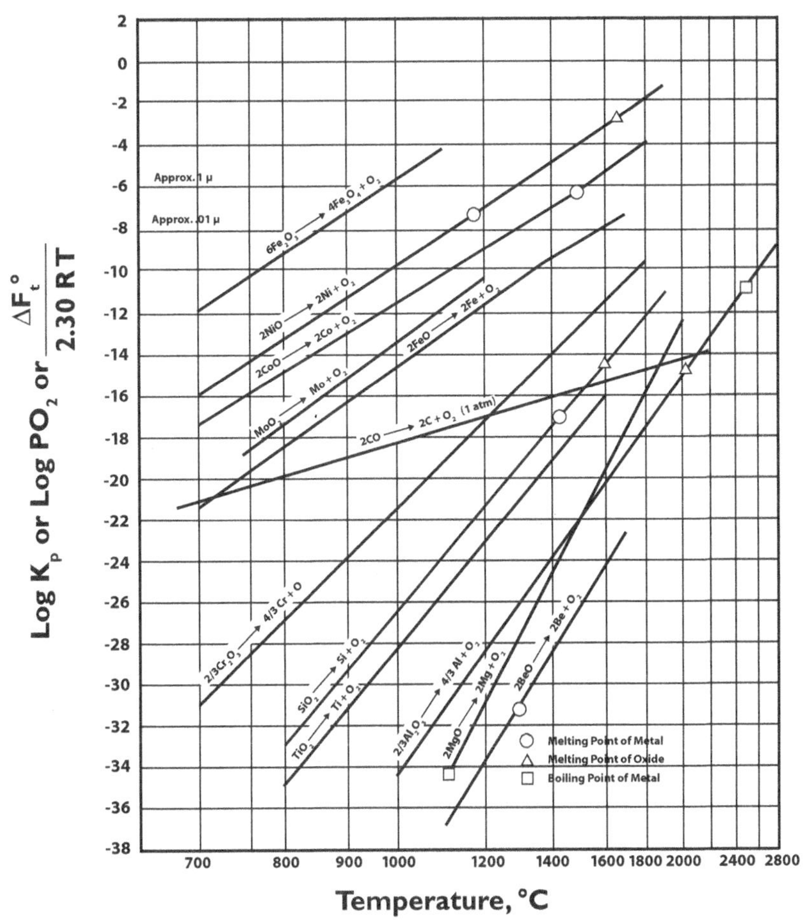

FIGURE 29.5 | Absolute pressure vs. relative pressure

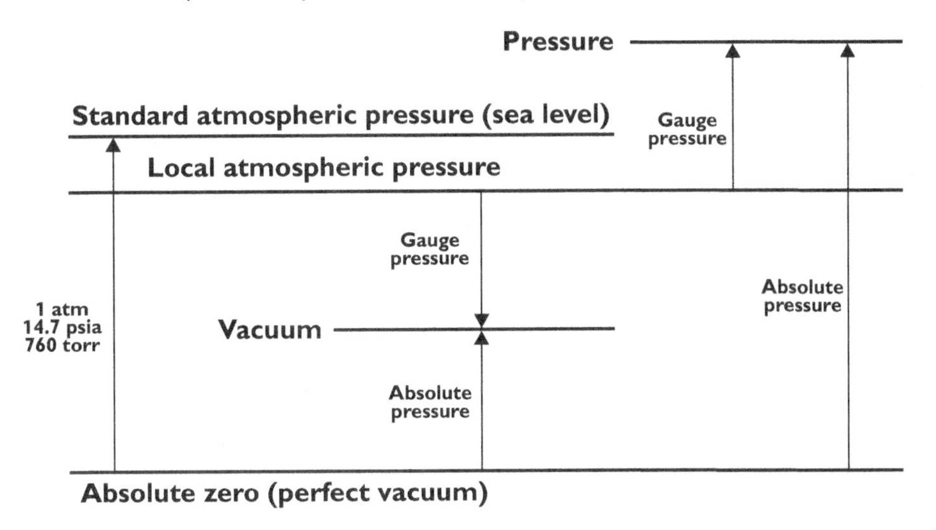

FIGURE 29.6 | Vapor pressure of selected oxides

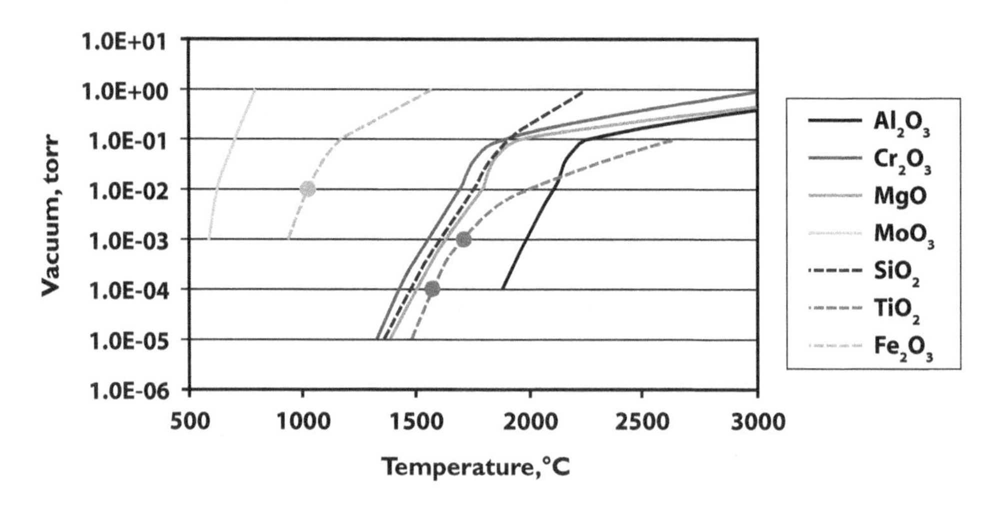

FIGURE 29.7 | Relationship between vacuum/pressure, water vapor and dew point [6]

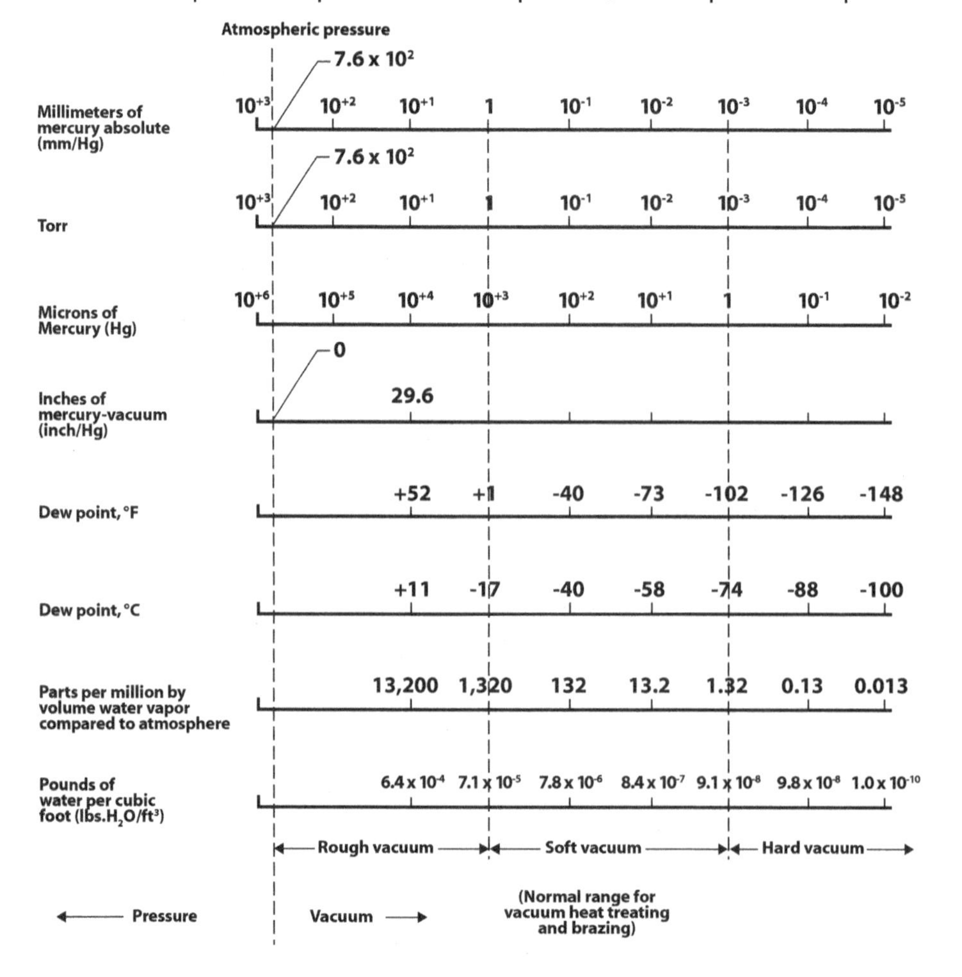

TABLE 29.1 | Vapor pressure of selected oxides [6]

Pressure (torr)	Temperature (°C) at which vapor pressures (torr) are:						
	Al_2O_3	Cr_2O_3	MgO	MoO_3	SiO_2	TiO_2	Fe_2O_3
0.00001		1320	1390		1365	1480	
0.0001	1880	1425	1510		1480	1575	
0.001	1980	1555	1630	595	1605	1705	935
0.01	2105	1695	1790	625	1750	1975	1025
0.1	2265	1875	1955	700	1910	2625	1175
1	3545	3030	3630	795	2235		1575

TABLE 29.2 | Vacuum furnace troubleshooting guide [7]

Symptom/ Problem	Possible cause	Possible solutions
Cannot hold proper vacuum level	Extensive outgassing	- Check for foreign matter within the vacuum chamber.
	Chamber leakage	- Check vacuum vessel with leak detector to locate leak. Seal small leak with liquid vacuum sealant; seal large leaks by welding, brazing or soldering.
	Bad pressure gauge(s)	- Replace pressure gauges as necessary.
Cannot hold proper temperature or over-temperature condition, causing component meltdown	Bad control thermocouple	- Check thermocouple for conductivity and/or excessive oxidation. - Replace thermocouple.
	Bad electrical heating element or radiant tube	- Broken element: replace or apply electrical patch to element. - Burner not firing; check pilot, gas or air source.
Over-carburization/ sooting (vacuum carburizing)	Carbon potential too high	- Adjust partial pressure of carbon-containing species.
Inside chamber coated with metallic film or possible "short-circuit" of electrical connections	Evaporation of high vapor pressure alloying elements in workpiece or other material in vacuum chamber	- Check all areas for foreign matter in vacuum chamber. Remove foreign materials if present. - If foreign matter not present, an adjustment of the processing parameters may be required.
Oily substance present in vacuum-chamber hot zone	Backstreaming	- Use of baffles or traps may be required.
	Oil vapor created by quenching	- Use of baffles or traps may be required. - Use of proper vapor-pressure oil.
Problems with improper joining operation	Parts contaminated; insufficient fluxing	- Check for contamination on parts before entering furnace. - Check that flux is properly applied and that the joint areas are clean.

Symptom/ Problem	Possible cause	Possible solutions
Insufficient hardness in finished parts	Improper quenching technique; quenching rate too slow Delay in getting to quench tank; gas injection rate too slow or too low a volume	- Check transfer mechanism for interference between heating chamber and quenching chamber. - Check all plenums to make sure they are clear. - Check for proper rotational speed of recirculating fan.
Surfaces of parts are oxidized upon removal from furnace	Vacuum chamber opened at too high a temperature or exposed to oxidizing environment before completion of processing cycle	- Allow vacuum chamber to cool to about 95°C (200°F) before opening chamber . - Keep vacuum level constant throughout entire process.

TABLE 29.3 | Emissivity values [8]

Emissivity of common materials			
Material	**Temperature, °C**	**Temperature, °F**	**Emissivity number**
Blackbody	24	75	1.0
Black paint, CuO	24	75	0.96
Carbon, lamp black	24	75	0.96
Soot, acetylene	24	75	0.97
Cement	0-200	32-392	0.96
White ceramic, Al2O3	24	75	0.90
Carbon, unoxidized	24	75	0.81
Water	38	100	0.67

Emissivity of select metals			
	Temperature, °C	**Temperature, °F**	**Emissivity number**
Aluminum, unoxidized	24	75	0.02
Aluminum, oxidized	199	390	0.11
Aluminum, heavily oxidized	93	200	0.20
Brass, unoxidized	25	77	0.035
Brass, oxidized	200	392	0.61
Bronze, polished	50	122	0.10
Chromium, unoxidized	100	212	0.08
Chromium, oxidized	316	600	0.08
Copper, cuprous oxide	38	100	0.87
Copper, heavily oxidized black	38	100	0.78
Copper, polished	38	100	0.03
Copper, Dow metal	38	100	0.15

Emissivity of select metals			
Gold, polished or unoxidized	100	212	0.02
Inconel X	24	75	0.19
Inconel B	24	75	0.21
Iron, cast unoxidized	200	392	0.21
Iron, cast oxidized	200	392	0.64
Iron, cast heavily oxidized	38	100	0.95
Iron, unoxidized	100	212	0.05
Iron, oxidized	100	212	0.74
Iron, heavily oxidized (rust)	25	77	0.65
Iron, wrought	25	77	0.94
Lead, oxidized	38	100	0.43
Magnesium	38	100	0.07
Mercury	25	77	0.10
Molybdenum	38	100	0.06
Molybdenum	1093	2000	0.18
Monel Ni-Cu	200	392	0.41
Monel Ni-Cu, oxidized	25	77	0.43
Nickel, unoxidized	25	77	0.05
Nickel, oxidized	38	100	0.31
Nickel, unoxidized	25-500	77-932	0.045-0.12
Nickel, oxidized	200	392	0.37
Niobium, unoxidized	816-1093	1500-2000	0.19-0.24
Niobium, oxidized	816	1500	0.73
Nichrome wire, unoxidized	50	122	0.65
Nichrome wire, oxidized	50	122	0.95
Platinum	38	100	0.05
Platinum, black	38	100	0.93
Silver, polished	38	100	0.01
Steel, unoxidized	100	212	0.08
Steel, oxidized	25	77	0.80
Steel, type 301, polished	24	75	0.27
Steel, type 303, oxidized	316	600	0.74
Steel, type 316	24	75	0.28
Tantalum, unoxidized	1500	2732	0.21
Tin, unoxidized	25	77	0.04
Titanium, C110M	149	300	0.08
Tungsten, unoxidized	25	77	0.02
Zinc, bright galvanized	38	100	0.23
Zinc, unoxidized	260	500	0.05
Zinc, oxidized	260	500	0.11

FIGURE 29.8 | Black-body radiation [8]

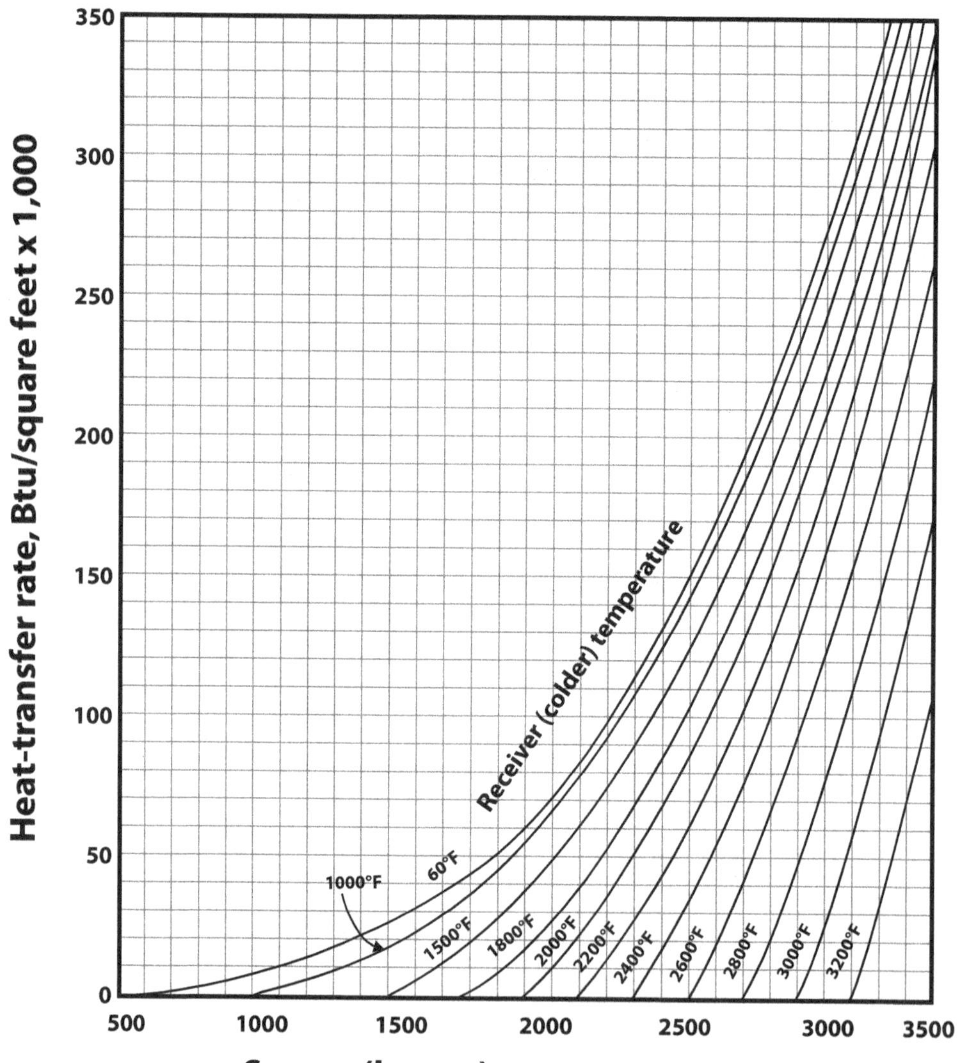

FIGURE 29.9 | Volume changes required to reach a given purity level [8]

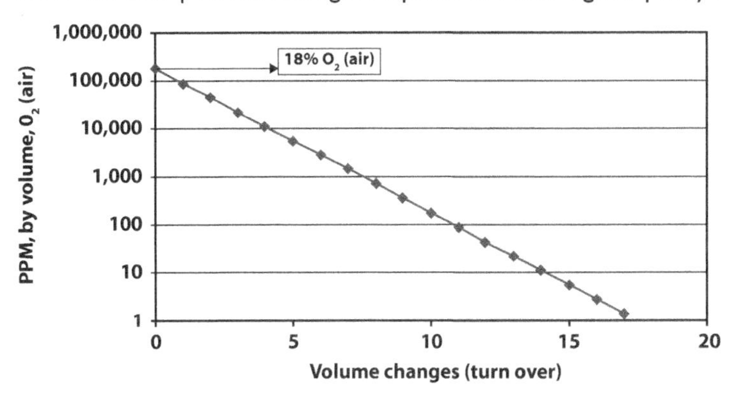

FIGURE 29.10 | Typical vacuum chamber blank-off pressure vs. leak rate [8]

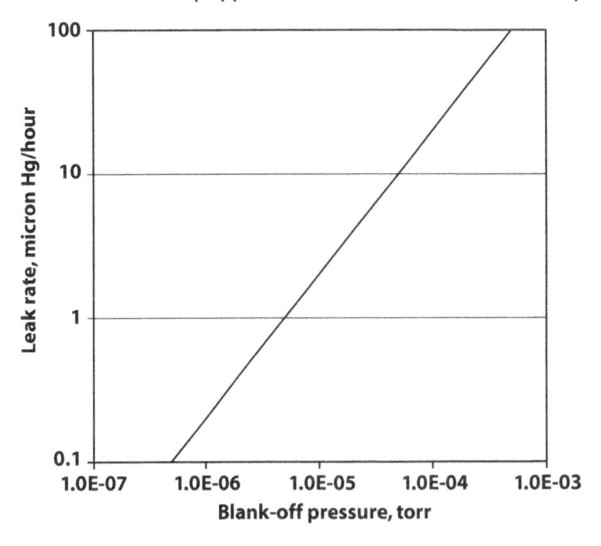

TABLE 29.4 | Common English to metric conversions [8]

Conversion factors for common units of length
1 inch = 2.54 centimeters (cm) = 25.4 millimeters (mm)
1 foot = 0.305 meters
1 centimeter = 0.394 inches
1 meter = 39.37 inches = 3.28 feet = 1.094 yards
Conversion factors for common units of area
1 square inch = 6.45 square centimeters
1 square foot = 0.093 square meter = 930 square centimeters
1 square centimeter = 0.155 square inches
1 square meter = 10.764 square feet= 1,550 square inches
Common conversion factors for units of volume (solid and liquid)
1 cubic inch = 16.4 cubic centimeters

1 cubic inches = 16.4 milliliters
1 cubic foot = 0.0283 cubic meters
1 cubic foot = 7.48 U.S. gallons
1 U.S. gallon = 3.78 liters
1 cubic foot= 28.3 liters
1 cubic centimeter = 0.061 cubic inches
1 millileter = 0.061 cubic inches
1 cubic meter = 35.314 cubic feet
1 liter = 35.314 cubic feet
Common conversion factors for units of mass (weight)
1 pound = 453.6 gram
1 ounce = 0.0625 pounds = 28.35 grams
1 gram = 0.0022 pounds = 0.0353 ounces

TABLE 29.5 | Adverse reaction table (graphite/oxides) and furnace components [8]

Onset temperature for adverse reactions between graphite or oxides and furnace components		
Surface-to-surface contact	Temperature, °C	Temperature, °F
Al_2O_3-C	<1600	<2912
Al_2O_3-Mo	1900	3452
Al_2O_3-Ta	1900	3452
Al_2O_3-W	1900	3452
BeO-C	2300	4172
BeO-Mo	1900	3452
BeO-MgO	1800	3272
BeO-W	2000	3632
C-Mo	1200*	2192
C-Ta	1000*	1832
C-W	1499*	2730
BeO-ZrO_2	1900	3452
MgO-C	1800	3272
MgO-Mo	1600	2912
MgO-Ta	1800	3272
MgO-W	2000**	3632
MgO-ZrO_2	2000	3632
ZrO_2-C	1600	2912
ZrO_2-Mo	2200	3992
ZrO_2-W	1600	2912

Notes:

*Strong carbide formation at higher temperatures

** Strong magnesia evaporation

REFERENCES

1. *Metals Handbook, Volume 1, Properties and Selection of Metals,* ASM International, 1961
2. www.airandvac.com
3. Honig, R.E., "Vapor Pressure Data for the More Common Elements," RCA Review 18 (1957): 195-204 (Curves shown are an adaptation from this work)
4. Gaskell, David R., *Introduction to the Thermodynamics of Materials,* 4th edition, Taylor & Francis, 2003
5. Trevor Jones, Solar Atmospheres Inc., private correspondence
6. *Vacuum Training Manual,* Ipsen "U", Ipsen Inc., 2000
7. *The Air Products Guide to Metal Processing, Volume Two: Vacuum Processing,* Air Products and Chemicals Company Inc., 2000
8. Critical Melting Points for Metals and Alloys with Other Important Metallurgical and Vacuum Heat Treating Reference Data, Osterman, Virginia and Harry Antes Jr., Eds., 2010, Solar Atmospheres Inc.

EQUATION INDEX

FIGURE INDEX

** Figures/captions in* **bold** *can also be found in the Color Image Section in the middle of the book.*

TABLE INDEX

TERM INDEX

TERM	PAGE NUMBER

B

C

D

E

F

G

H

I

M

N

O

P

Q

R

S

T

U

V

W

Y

Z